OPPOSITE

Color fringes in oblique illumination. No polarizers; stop introduced from upper left. *Top and middle:* 6.3×, N.A.=0.16 objective. *Bottom:* 2.5×, N.A.=0.08 objective (image of shadow reversed). Grains of apatite (hexagonal) immersed in an oil of $n = 1.630$. For explanation see Fig. 3-12, p. 63.

OVERLEAF

Top

Thin section of pyroxenite between crossed polars showing three generations of exsolution lamellae. The black background is a single crystal of hypersthene at extinction, containing fine "100" lamellae of augite (bright, vertical stripes). This crystal formed by the solid-state recrystallization of several variously oriented grains of pigeonite. In these, coarser "001" lamellae of exsolved augite survive to indicate the original size and orientation of the pigeonite grains before inversion to hypersthene. Bright, irregular patches of augite (also finely exsolved), and rounded black grains of magnetite also occur. Height of field about ½ cm. Pyroxenite body, Kiglapait Mountains, Labrador. See Fig. 14-16, p. 264.

Bottom

Interference color scale. Photo of quartz wedge between crossed polars, thin edge on left. Note scalloped edge where wedge is ground to zero thickness. Here the retardation equals zero, and darkness results. See Figs. 5-10 (p. 96) and 6-3 (p. 114). Numbers refer to orders 1-4 of interference colors.

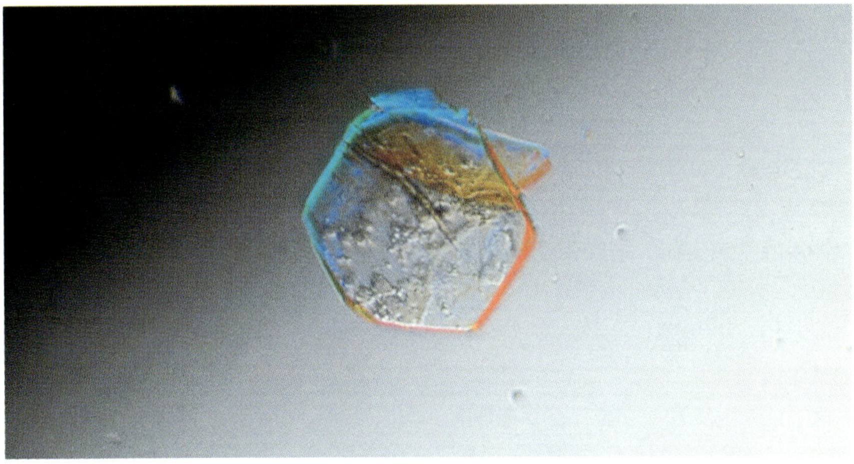

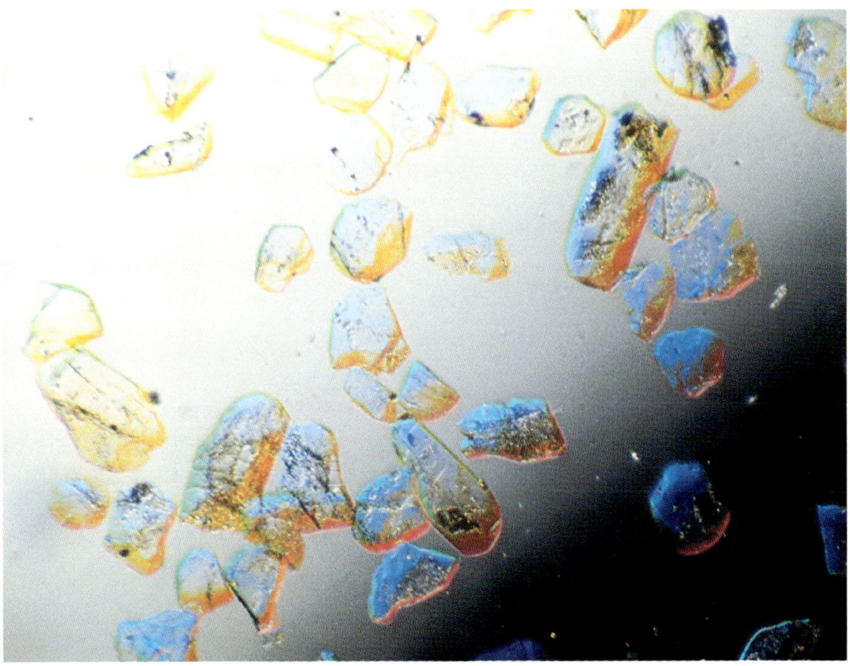

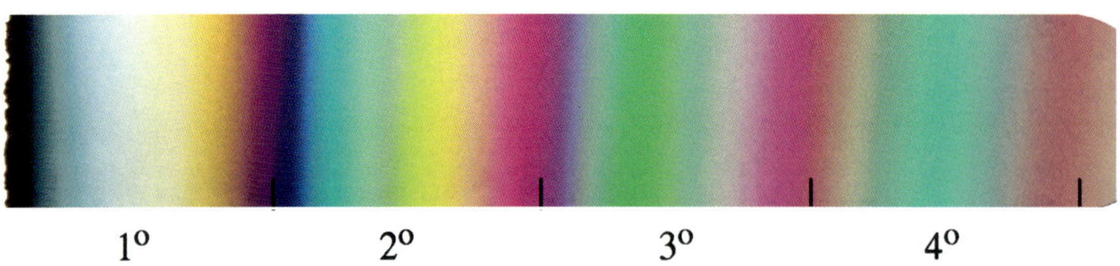
1°
2°
3°
4°

Crystal Identification with the Polarizing Microscope

Crystal Identification with the Polarizing Microscope

Richard E. Stoiber
F. Hall Professor Emeritus of
Mineralogy, Dartmouth College

Stearns A. Morse
Professor of Geology,
University of Massachusetts
at Amherst

CHAPMAN & HALL
New York • London

Published by
Chapman & Hall
One Penn Plaza
New York, NY 10119

Published in Great Britain by
Chapman & Hall
2-6 Boundary Row
London SE1 8HN

Printed in the United States of America

Library of Congress Cataloging-in-Publication Data

Stoiber, Richard E.
Crystal identification with the polarizing microscope / Richard E. Stoiber, Stearns A. Morse.
p. cm.
Includes bibliographical references and index.
ISBN 0-412-04821-3 (acid-free).
(pbk. : acid-free)
1. Crystal optics. I. Morse, Stearns Anthony, 1931– .
II. Title.
QD941.S713 1994
548′.9—dc20 93-27087
CIP

British Library of Cataloguing in Publication Data available

Please send your order for this or any Chapman & Hall book to **Chapman & Hall, 29 West 35th Street, New York, NY 10001, Attn: Customer Service Department.** You may also call our Order Department at 1-212-244-3336 or fax your purchase order to 1-800-248-4724.

For a complete listing of Chapman & Hall's titles, send your requests to **Chapman & Hall, Dept. BC, One Penn Plaza, New York, NY 10119.**

Contents

	Preface	vii
	Efficient Use of This Book	xi
	List of Symbols	xiii
1	Crystal Identification and Optical Principles	1
2	Equipment and Preparation of Materials	31
3	The Immersion Method	49
4	Isotropic Crystal Identification	76
5	Phase Relationships and Interference	87
6	Anisotropic Crystals	102
7	Uniaxial Crystal Optics	116
8	Uniaxial Interference Figures	123
9	Identification of Uniaxial Crystals	136
10	Biaxial Crystal Optics	159
11	Biaxial Interference Figures	172
12	Identification of Biaxial Crystals	199
13	Crystallographic Relations of Biaxial Crystals	227
14	Optic Orientation in Stereo	247
15	Special Orientation Methods	282
16	Use of the Dispersion Method	289
17	Crystal Identification Recipes	309
	Appendix A. Optical Properties of Common Rock-Forming Minerals	325
	Appendix B. Identification of Fibrous Asbestos	333
	References	337
	Index	343

Preface

Some of the simpler measurements of optical mineralogy are so precise and powerful that they give satisfaction to beginning students. Not long after mastering the strike and dip of rock surfaces with the Brunton compass, many geology students are able to determine precisely the identity of quartz, or the anorthite content of plagioclase, or the magnesium ratio of pyroxene with the polarizing or petrographic microscope, by means of measuring refractive index to better than one part in a thousand. Very little training and almost no theory are needed to achieve these skills. But there inevitably comes a time when theory is needed, either to get on with the art, or simply to reconstruct from first principles what is going on, when rote memory fails. In this book we hope to provide both the rote methods and the theoretical background for practitioners at all levels of experience. We draw from several careers—ours, our colleagues', and our students'—in teaching the subject at various levels of sophistication. Our book is intended to serve the needs of industrial and forensic scientists as well as petrographers who deal with rocks.

Much of our treatment is based on new research, both in matters of presentation and in the optical determination of minerals and other materials. In a new diagram we quantify the use of color fringes in oblique illumination, to aid the user in estimating not only in which direction, but how far away, a match between crystal and liquid is to be sought. We have incorporated original data on the errors of refractive index determination by the immersion method, as practiced by beginning and experienced students. We provide worked examples of refractive index measurement and crystal identification at several levels of experience. Included also are analytical (rather than graphical) solutions for the refractive index by the dispersion method, based on the Hartmann equation and the estimated dispersion of minerals. This procedure leads to precise results, as shown by examples. When a laboratory is provided with a microscope dedicated to use with monochromatic light, observations made in white light at another microscope can then be refined by the dispersion method using the same grains previously studied. This scheme can serve the needs of many investigators at once.

Our classification of interference figures emphasizes the recognition and use of those that are most useful in refractive index determination. A scheme is offered for the simultaneous determination of cleavage and crystal system in immersion mounts. Routine determinations of the standard compositions of plagioclase, olivine, orthopyroxene, and augite are described using algebraic solutions with the dispersion method. An elegant treatment of alkali feldspars is drawn from the recent literature and analytic equations are given for determination of their composition. Optic orientation in the five anisotropic crystal systems is presented with stereographic treatment of the crystallographic elements and interference figures, referred to perspective drawings and prominent crystal sections. Several mineral examples are chosen from each crystal system to illustrate these relationships.

In a final chapter, we provide a step-by-step recipe for crystal identification that will help the beginning practitioner to learn the most efficient path to a solution, and serve as a guide to those who have lost their way. Practical guidance is offered for the identification of separated mineral grains, crushed rocks, and minerals in thin section.

Our emphasis is practical, but we include enough theory to make rigorous understanding possible at several different levels of experience. Our goal is to provide a working knowledge of crystal identification suitable for the advanced graduate student, but also to guide the beginning and intermediate student who may not go on to specialize in petrography. Our explanations are based on the indicatrix and the directly observable vibration directions rather than velocity surfaces. Mineral examples are largely drawn from a standard reference work, the tables of Tröger (1979). We provide our own short determinative table of the common rock forming minerals, based on a compilation by our friend John Lyons, that has proved useful in beginning petrology laboratory studies. The identification of asbestos minerals is also discussed in a separate appendix.

Our main reference work, aside from the ever-useful Tröger, is the most welcome USGS Bulletin 1627 by Fleischer et al. (1984), which for the first time updates the seminal work of Larsen and Berman (1934). For advanced procedures and illustrations especially related to thin sections, we call attention to the excellent lab manuals of Jones and Bloss (1980), and of Philpotts (1989). Detailed mineral descriptions may be found in the growing treatise of Deer et al. (1962ff). Our discussion mainly concerns the flat stage, on which measurements of numerous grains may give a good sampling of mineral variability, but we draw attention to the masterful development of the spindle stage by

Bloss (1981) for the elegant pursuit of new mineral data of high precision, integrated with other methods of crystallographic study. In the hands of his research group, this instrument has also become a useful tool for the routine identification of crystals, particularly those available in small quantity.

The nucleus of this textbook has seen several incarnations, from an early mimeographed version half a century old, through various multilith editions known to some as "The Orange Pumpkin", to published editions of our *Microscopic Identification of Crystals* (1972). We are most grateful to Robert Krieger for maintaining a stock of the book at a modest price for many years while this edition was in preparation. We have now reverted to a version of the original (i.e., mimeograph and multilith) title, despite its length, because it tells everybody what we are trying to do. Much of the historical detail of the old edition has fallen by the wayside, but the core has been kept intact along with the new approaches and material mentioned above. We offer this new book in the firm belief that crystal identification with the polarizing microscope is not only still a fundamental skill in the earth sciences, but is also a tool of growing power and simplicity.

A word of warning. We have kicked over the traces and reverted to the use of *isometric* instead of *cubic* for the name of the crystal system with the highest symmetry. We did this because many crystals belonging to this system are not at all cubic in appearance (for example a dodecahedron), but all are isometric, and we seek not to confuse the reader with unnecessary euphemisms.

We are grateful to Richard P. Nickelsen for suggesting the calcite experiment in Chapter 2, and to the late Edward C. Taylor, Sr., for devising the presentation of 1-symmetry plane interference figures in Figs. 11-18 to 11-20. Our work has been aided and encouraged by many colleagues, especially Richard Birnie at Dartmouth and Howard Jaffe and Peter Robinson at U. Mass. We are especially grateful to our critical readers, Ray Wilcox for the old text and Shu-Chun Su for the new version. Both made many helpful suggestions. Dr. Su in particular hastened to furnish us with important new results used in Chapter 16. For encouragement we thank Don Bloss and Walter McCrone. S. A. M. wishes to acknowledge the gracious hospitality of the Department of Earth Sciences and Clare Hall at the University of Cambridge during the preparation of the manuscript.

Permissions to use published data or figures are gratefully acknowledged from M. J. Buerger, R. D. Butler, C. A. Chapman, H. W. Fairbairn, Barclay Kamb, J. S. Dickey, Jr., E. Schweizerbart'sche Verlagsbuchhandlung and the Tröger (1979) editors for parts of our Figs. 14-7, 14-8, 14-10, 14-11, 14-12, 14-13, 14-14, 14-15, 14-17, 14-18,

14-19, 14-20, and 14-28; *American Mineralogist* for Figs. 14-16, 14-22, 14-24, 14-25, 14-26, and 14-27; Shu-Chun Su, Paul Ribbe, F. D. Bloss, J. R. Goldsmith, D. B. Stewart, H. W. Jaffe, Peter Robinson, Malcolm Ross, and R. J. Tracy; *Reviews in Mineralogy* and Paul Ribbe for our Fig. 14-21; and *Microscope* and Shu-Chun Su for our Fig. 16-3.

RICHARD E. STOIBER
STEARNS A. MORSE

Hanover, New Hampshire
Amherst, Massachusetts
June 1993

Efficient Use of this Book

Crystal Identification with the Polarizing Microscope is aimed at readers ranging from the beginner to the professional. The art of crystal identification is practiced at many levels, and we firmly believe in repeated exposure to the procedures at successive levels of experience. Beginners can easily learn the practice of the polarizing microscope in a few lab sessions, and we advocate getting worthwhile results before theory. The key to rapid progress is the efficient use of Fig. 3–15, the calibration of color fringes. The strategy of crystal identification using the immersion method and thin sections is given in Chapter 17.

Students in mineralogy or petrology courses can learn to identify an isotropic mineral by rote use of the immersion method in the first microscope lab session, a uniaxial mineral such as quartz in the second session, a single refractive index and magnesium ratio of olivine in a third session, and a pyroxene or plagioclase composition in another session. For guidance to instructors or students, the worked examples at the ends of Chapters 4, 9, and 13 should be helpful. The brief table of common rock-forming minerals in Appendix A can be used as a source of data for crystal identification or as an entry point to more complete tabulations of data.

At some institutions, immersion methods are alternated with thin section study in dual two-hour lab sessions each week. A list of minerals commonly identified by immersion is sodalite, quartz, olivine, hypersthene, augite, plagioclase, staurolite, and gedrite. For all but the first and last two, only a single refractive index need be determined in order to find the composition of the mineral in terms of its two major end members. The results so achieved are comparable in quality (although not in scope) to any determinative procedure such as electron microprobe analysis, and we feel it imperative to equip students with the rudiments of this powerful and fundamental technique. A laboratory class can work effectively as a group, sharing the labor and the results. A video station can be provided with a monochromator and the dispersion method introduced as a demonstration. No theory need be taught at the beginning level except in response to questions.

For upper level courses or research in optical mineralogy, our book can be used in a linear fashion to acquire theory, and simultaneously (as for beginners, above) to review practical methods and gain skill in the laboratory. A single microscope fitted with a monochromator can be furnished at a dedicated work station in the lab, and refinement of a refractive index determination can be completed by the dispersion method. For those wishing to review practical matters relating either to the immersion method or to thin sections, Chapters 16 and 17 should be helpful. For identification of the less common minerals, reference should be made to the determinative tables of Tröger (1979) or Fleischer et al. (1984; "USGS 1627"), and to descriptive treatises such as that of Deer et al. (1962–92).

Although excellent results can be obtained without theory, in the long run an adequate theoretical background will be essential to efficient and reliable crystal identification. This background can be obtained from the body of the chapters. A good test of competence might be the ability to determine optic orientation and crystal composition, as illustrated by the examples given in Chapter 14. Perhaps the ultimate goal of efficiency in crystal identification is achieved with the identification of major and minor minerals in a crushed rock, according to methods described in Chapter 17. This skill powerfully complements the study of textures in thin sections, and for unfamiliar rocks it can save a great deal of time.

R. E. S. & S. A. M.

22 September 1993

List of Symbols

A	Amplitude.
Å	Ångstrom unit, 10^{-10} m.
a,b,c	The crystal axes; the axial lengths.
B	Birefringence.
Bxa	Acute bisectrix.
Bxo	Obtuse bisectrix.
C	Fraunhofer line in spectrum.
D	Fraunhofer line in spectrum; also distance between melatopes in a biaxial interference figure.
D	Any vibration direction. D_h,D_l are high and low index vibration directions. D_1 and D_2 imply distinction when the high-low designation is immaterial.
d	Distance; also minimum deviation angle.
E	Vibration direction of the extraordinary ray. Also, the half optic angle in air.
F	Fraunhofer line in spectrum.
f	Frequency.
H	The half optic angle in oil.
I	Intensity.
i	Angle of incidence.
K	Mallard's constant.
K	Specific refractivity of a substance.
k	Specific refractivity of a component.
k_D	Fractional Hartmann wavelength difference.
L	Hartmann wavelength function.
mμ	Millimicron; see nm.
N.A.	Numerical aperture.
n	Refractive index or an integer.
nm	Nanometer = mμ = 10^{-9} m = 10 Å.
O	Vibration direction of the ordinary ray.
oa	Optic axis.
ON	Optic normal.
OP	Optic plane.
r	Angle of refraction; also, the optic angle for red light.
SP	Symmetry plane.
t	Time; temperature; thickness.

V	The half optic angle in the crystal; also, reciprocal dispersion $(n_D - 1) / (n_F - n_C)$.
v	Wave velocity; also, the optic angle for violet or blue light.
WN	Wave normal; same as WPD.
WPD	Wave propagation direction.
X, Y, Z	Specific vibration directions in crystals; the axial directions of the indicatrix for biaxial crystals.
x, y, z	Rectangular coordinates.
α, β, γ	Least, intermediate, and greatest refractive indices in biaxial crystals; also, the interaxial angles.
Δ	Retardation (path difference); dispersion, $n_F - n_C$.
ϵ	Refractive index associated with the extraordinary ray in uniaxial crystals.
η	Arc between the median line of an isogyre and a crosshair, at the perimeter of the field of a centered optic axis interference figure at the 45° position of rotation.
θ	Glancing angle (in Bragg's Law).
λ	Wavelength; λ_0, observed wavelength of match.
μ	Micron, 10^{-6} m; also, half-angle of the cone of aperture of a lens; also, rotation angle of stage needed to bring the median line of an isogyre tangent to the edge of the field.
ρ	Density.
ω	Refractive index associated with the ordinary ray in uniaxial crystals.

1

Crystal Identification and Optical Principles

INTRODUCTION

This book is concerned with the use of the polarizing or polarized light microscope and the elements of crystal optics that make its use effective. The treatment is practical rather than theoretical. Crystal optics continues to be a developing branch of physics, and has many ramifications beyond determinative work. The nature of light and its behavior have formed a central part of physics since its earliest days, from Snell in 1621 through such celebrated figures as Fermat, Huygens, Newton, Biot, Fresnel, Cauchy, Maxwell, and Fraunhofer to Planck, Dirac, and Einstein. The contributions of these people to optics and to physics are concisely summarized by Born and Wolf (1964) in the historical introduction to their comprehensive treatise on optics. We shall later mention many of these and others in the context of their individual contributions.

The petrographic polarizing microscope is the fundamental instrument used by geologists for the identification of minerals and the description of rocks. Crystal identification accounts for most uses of the petrographic microscope. The study of rock textures, often the primary goal of an investigation, also includes identification of the minerals composing the rock. Included under "identification" are measurements of many optical and physical properties that shed some light on the composition or growth conditions (such as cooling history) of a mineral even though its species may already have been identified. The petrographic microscope is also used to estimate systematically the abundance of each mineral species comprising a rock, a process known as modal analysis. From such analysis and a knowledge of mineral composition, the chemical composition of the rock itself may be estimated.

Crystal identification with the polarizing microscope is also practiced by archaeologists, art conservationists, forensic scientists, environmental scientists, chemists, ceramists, gemologists, biologists, physicists, polymer scientists, and engineers. Among the foremost industrial uses of this instrument today is the identification of asbestos minerals, discussed briefly in Appendix B. The technique has application in chemistry as, for example, in distinguishing between crystallographically similar substances (isomorphs) and between chemically similar substances (polymorphs). Confirmation of the identity of a chemical compound or pigment may require only two or three minutes if the choice of possibilities is limited to compounds with sufficiently different optical properties. Applications to the conservation of works of art are described by Butler (1971), and to archaeology by Kemp and Harvey (1983).

In biology, the polarizing microscope finds a variety of applications (Hartshorne and Stuart, 1970). Among these are the study of urinary calculi (Lonsdale, 1968; Stevenson and Stevenson, 1973), the study of certain parts of mosses (Taylor, 1959), and the study of cells during mitosis.

NATURE OF DETERMINATIVE PROCEDURES

The polarizing or petrographic microscope is a compound microscope with which materials are studied in polarized light. These materials include not only crystals but also other materials whose visibility and properties are enhanced by polarized light, such as coal. The current treatment is concerned with translucent materials and transmitted-light optics, although some of the principles are pertinent to reflected-light optics used for studying opaque materials.

The materials to be identified with the polarizing microscope usually consist of small single crystals, or fragments broken from larger crystals, or thin sections of crystalline material. These are mounted on glass slides, with cover glasses, either in special liquids or in a transparent cement. A number of measurements are then made, consciously or subconsciously, that serve to identify the material. Among these are refractive indices and related quantities such as birefringence (the difference between the greatest and the least refractive indices), optic angle and sign, color and its variation with direction in the crystal, and the relative orientation of optical directions and crystallographic directions. One will need to know some elementary crystallography to get very far with these matters; a review will be found in Chapter 14. Other physical properties, such as crystal form, habit, and density may also prove useful in identification.

After the requisite data have been noted down, the identity of the crystal is sought in one of several standard tables. Most of these are arranged by refractive index. One often makes a few measurements to narrow down the candidates to a manageable number, then consults the tables, and finally makes some more diagnostic measurements to complete the identification. Problems may arise if the tables are incomplete, the natural crystals variable, or the optical properties not absolutely diagnostic. For many purposes, however, the optical method is the quickest and simplest of rigorous identification methods. It may at times need to be supplemented by others, such as electron microprobe analysis (which gives chemical composition but not crystallographic information), X-ray diffraction, or spectroscopic examination.

Most polarizing microscope practice beyond the introductory level involves a large measure of sight recognition, or examination of substances about which one has some prior knowledge. In this case the search through standard tables is simplified and unlikely candidates easily set aside.

OPTICAL PRINCIPLES

Certain elementary principles of optics apply directly to the use of the polarizing microscope and to the study of the optical properties of non-opaque solids.

Familiar explanations of optical principles deal with materials like undisturbed air or water in which the velocity of light is equal in all directions of propagation. This is true of isometric (cubic) crystalline substances, called *isotropic* substances, but in all other substances, called *anisotropic* substances, the velocity of light varies with the direction of propagation in the crystal, and light vibrates with its maximum amplitude only in certain planes. Therefore, we shall be concerned largely with light that vibrates in planes and is propagated with velocities dependent on the orientation of these planes.

LIGHT

The physical disturbance that stimulates the nerves of the eye is called light. It is transmitted to our eyes from objects we see. The optical properties of light[1] may be effectively explained by considering light

[1] See Peckett (1992) for a treatment of the propagation of light in three equivalent representations using trigonometry, vectors, and complex exponential functions. A lucid account of the physiology of vision and the perception of color is also given.

to be a transverse wave motion consisting of rapidly alternating electrical and magnetic disturbances in all planes that include the direction of advance of the disturbance. This disturbance is a *light wave.* A line from the source of light to a point at which the light is perceived is called a *ray*. One plane through the electrical disturbance or vibration associated with a ray of light is illustrated in Fig. 1-1. The sine curve represents the magnitude of the electrical disturbance at a given instant at all points in a plane that includes the line. Fig. 1-1 is confined to the plane of the paper, but the wave vibrates in all planes containing the propagation direction *XY*.

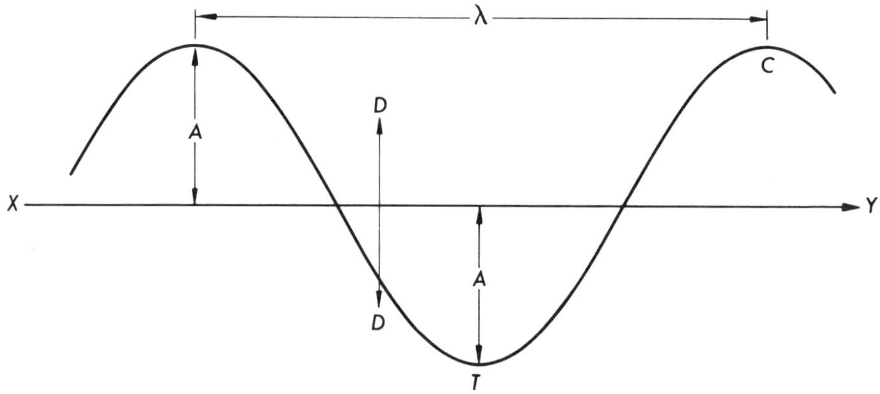

Fig. 1-1. A section of the wave disturbance along the path of a light ray XY.

The figure illustrates:

- $X \rightarrow Y$ = the propagation direction, the direction of advance of the disturbance.
- D = the vibration direction. All other directions normal to XY are also vibration directions.
- C = crest, the point of maximum relative upward displacement.
- T = trough, the point of maximum relative downward displacement
- A = amplitude of the wave, half the distance between crest and trough.
- λ = wavelength, the distance between wave crests.

The amplitude of the disturbance at every point along the wave changes with time. As a result, the wave crest migrates with time from

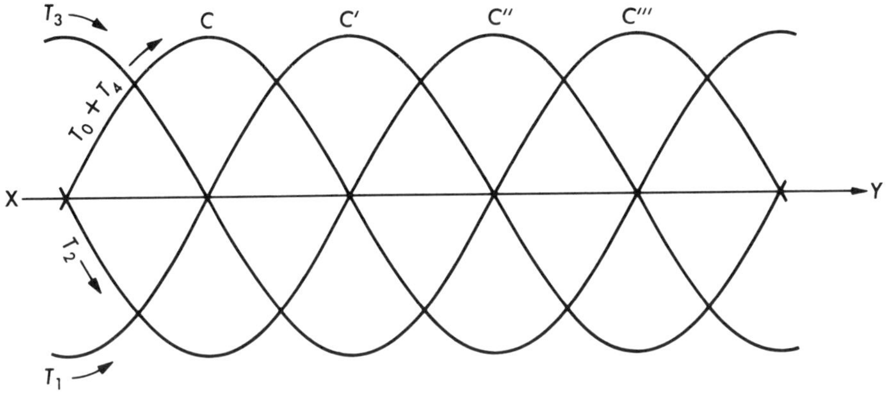

Fig. 1-2. Amplitude of wave illustrated in Fig. 1-1 at equal time intervals $T_0, \ldots, T_4$. Crest C migrates toward Y with time, reaching C''' at T_3.

X toward Y, which is the direction of wave propagation (see Fig. 1-2). The velocity of the wave, v, is the distance, d, traveled by the wave crest in unit time, t.

$$v = \frac{d}{t} \tag{1-1}$$

The wavelength is the measure of the distance, so

$$v = \frac{\lambda}{t} \tag{1-2}$$

The number of vibrations per unit of time is the *frequency of vibration*, f.

$$f = \frac{1}{t} \tag{1-3}$$

Combining equations 1-2 and 1-3,

$$f\lambda = v. \tag{1-4}$$

The *intensity* of light varies directly as the square of the amplitude:

$$I \propto A^2. \tag{1-5}$$

Plane Polarized Light

The phenomenon of polarization was discovered by Christiaan Huygens (1690) in experiments on the double refraction of the mineral calcite, often called Iceland spar or calc spar when transparent. Light that vibrates perpendicular to the propagation direction of the light wave in a single plane is called *plane polarized light*. Light may also be elliptically or circularly polarized. It is, however, simpler and for most purposes adequate to consider circularly and elliptically polarized light as consisting of two components polarized in planes at right angles to each other. This is the practice followed in this book. There is therefore no ambiguity introduced if plane polarized light is referred to more simply as polarized light, as long as it is recognized that on occasion this statement may be taken as shorthand for elliptically polarized light.

Polarizing Devices (Polars)

By using prisms or sheets of certain anisotropic substances, polarizing devices or *polars* may be constructed that transmit polarized light vibrating chiefly in a single plane, called the *vibration direction* or by some writers the *allowed direction*. When two polars are placed at appropriate locations in the optical system of a microscope, materials may be examined in polarized light.

A simple optical system of two polars may be arranged so that the vibration directions are parallel. This case will permit passage of a light wave vibrating parallel to the vibration directions without change in amplitude (except a small change due to absorption in the polarizing materials, which we shall ignore hereafter, treating polars as though they had a transmission efficiency of 100 percent). If, however, one polar is rotated about the propagation direction so that its vibration direction makes some angle with that of the other polar, the vibration direction of the transmitted light is that of the last polar traversed, and the amplitude is decreased, as shown in Fig. 1-3. The plane of vibration of light transmitted by each polar is shown in the figure by the dashed lines VV' and WW'. Light is propagated from O to O'. M is the amplitude of the wave leaving P', and N is the amplitude of the wave leaving P''. The amplitude of the vibration of the transmitted light may be

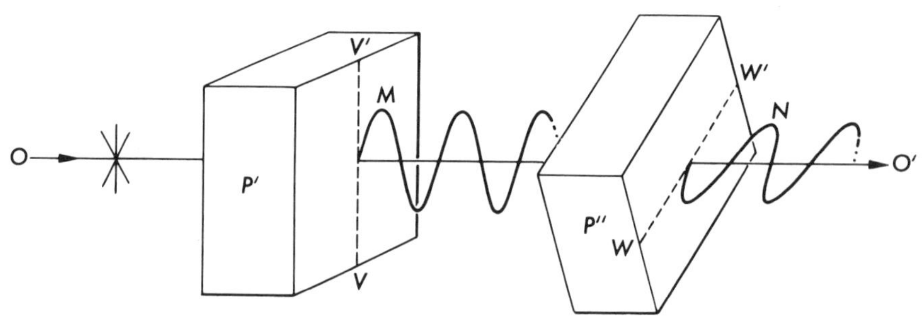

Fig. 1-3. Amplitude of polarized light passing through two polars P' and P''.

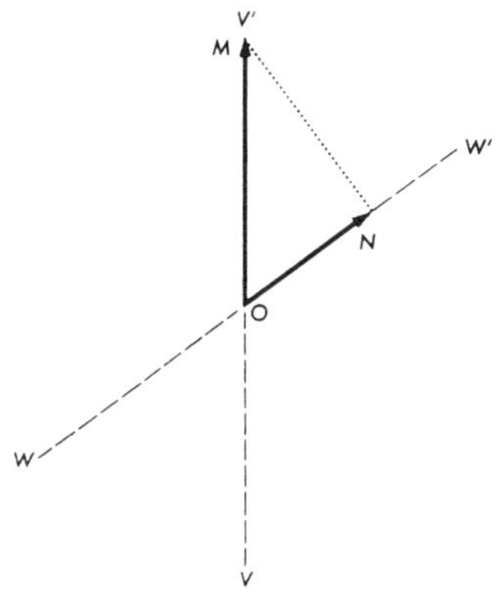

Fig. 1-4. Graphical determination of amplitude of light leaving polar P'' in Fig. 1-3. Symbols are as in Fig. 1-3.

determined by using simple vector principles such as are employed in the resolution of forces. This is shown graphically in Fig. 1-4, where *MN*, the normal to *WW'*, defines the vector *ON*, the required vibration direction and amplitude.

> This effect is familiar to many owners of Polaroid sunglasses, and may be demonstrated by superimposing two lenses of sunglasses and rotating one relative to the other while looking at a source of light. The sunglasses are made with the vibration direction vertical, to cut off the horizontal component of light crudely polarized by reflection from surfaces such as a road or water. The vibration direction may be identified by observing, through the sunglasses or polar, light reflected from a nonmetallic surface such as a table top. This light is partly polarized in the plane of the surface, and when the reflection is most attenuated on rotation of the polar or lens, the vibration direction of the latter is perpendicular to the plane of the reflecting surface. An account of the discovery by Malus in 1808 of polarization due to reflection is given by Hartshorne and Stuart (1970).

By a graphical construction similar to that of Fig. 1-4, it can be shown that no light will be transmitted by a simple optical system consisting of two polars whose vibration directions are perpendicular to one another. This condition is illustrated in Fig. 1-5. The wave entering the second polar has zero amplitude in the vibration direction of the second polar, and with zero amplitude the intensity, by equation 1-5, is zero.

CRYSTALLINE SUBSTANCES; CRYSTALS

Crystalline substances are those in which the atoms have an orderly arrangement. Minerals are naturally occurring inorganic crystalline

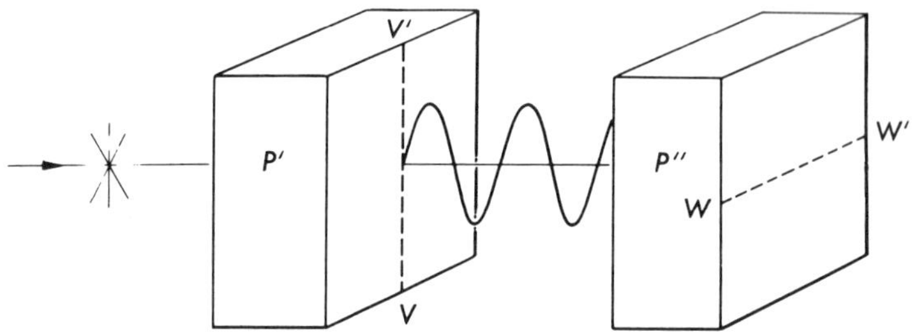

Fig. 1-5 A system of crossed polars, showing how light does not pass through. The polars P'' and P' are crossed when WW' is normal to VV'.

substances. A *crystal* is a polyhedral form, bounded by plane faces and possessing an orderly internal structure.

The term crystal is frequently, in this book and elsewhere, used in a broader sense to include crystalline substances whether or not bounded by crystal faces. The reader unfamiliar with the elements of crystal form and symmetry should refer to one of the many standard texts, for knowledge of elementary crystallography is a prerequisite to identification of crystalline substances with the polarizing microscope. A brief review of symmetry is given in Chapter 14.

Optical Classification of Crystals

To recapitulate, crystals are classified, on the basis of their optical behavior, as isotropic or anisotropic. *Isotropic crystals* transmit light with the same velocity in all directions. *Anisotropic crystals* do not transmit light with the same velocity in all directions. Light entering an anisotropic crystal in all but one or two special directions is resolved into two plane polarized components, the vibration planes of which are normal to each other. Vibrations within the planes are normal to the directions of wave propagation. The polarized light in each of the two planes travels at a different velocity through the crystal.

As suggested before, there are in anisotropic crystals one or two directions of wave propagation in which the velocities of both plane polarized components are equal. These directions are the *optic axes.* Whether a crystal has one or two optic axes is the basis for subdivision of anisotropic crystals into *uniaxial* and *biaxial crystals. Uniaxial* crystals have one optic axis and *biaxial* crystals have two. The optic axes and other optical directions of any anisotropic crystal have specific relationships, dictated by crystal symmetry, to the crystallographic directions.

The optical character of a crystal is related to its crystal system. Isotropic crystals are isometric (cubic). Uniaxial crystals are hexagonal or tetragonal. Biaxial crystals are orthorhombic, monoclinic, or triclinic. Some exceptions are noted in later chapters devoted particularly to isotropic, uniaxial, and biaxial crystals.

The plane of polarized light is rotated during passage through some crystals. This phenomenon is the property of *rotary* (or *rotatory*) *polarization*. Crystals that possess this property are said to be *optically active*. Since in almost all instances optical activity can be identified using the microscope only with thicker sections of material than are commonly studied, the property is of no importance in the identification of crystals with the petrographic microscope. The definitions and discussion here refer to substances that are not optically active, but, in effect, the discussion applies to all crystals because the modifying effects of rotary polarization are not visible in common practice.

RAYS AND RAY SURFACES

The direction along which light energy is propagated is called a ray. Rays may be parallel, as those from a distant source such as the sun, or nonparallel, as those emanating from a point source. The speed at which light energy travels along a ray is called the *ray velocity*. A *ray surface* is the front of the advancing light energy from a source, or in other words the envelope connecting the extremities of the rays after a given time has elapsed. The ray surface may be a plane, as with light from a distant source, or a curved surface, such as the spherical envelope at the extremities of rays from a point source in an isotropic medium. We shall now consider only the type of ray surface associated with a point source of light.

The ray surface in an isotropic medium is a sphere because the ray velocity is equal in all directions. In an anisotropic substance, light from a point source is polarized in two mutually perpendicular planes. In all but some special directions of travel (optic axes) the two polarized rays have different velocities, so that the ray surface of light emanating from a point is a double surface. Because the velocities depend upon the internal structure of the crystal, hence upon crystallographic directions, the ray surfaces are specifically oriented with relation to the crystallographic axes.

WAVE FRONTS

A *wave front* is a surface that is tangent to a collection of ray surfaces. For parallel rays in an isotropic medium it is a plane identical with the ray surface. For rays emanating from a single point source in an isotropic medium it is a sphere identical with the ray surface. In anisotropic substances, however, wave fronts and ray surfaces are not generally identical, nor are they for a collection of nonparallel rays in an isotropic medium.

The direction normal to a wave front is the *wave normal direction,* often simplified to *wave normal;* it is also the *wave propagation direction.* In the simple cases cited above, the wave propagation direction coincides with the ray direction. The two do not always coincide in an anisotropic substance.

Huygens' Construction

The position of a wave front can be found if its previous position and the ray velocities of the medium are known. The graphical solution to this problem is known as *Huygens' construction*: let all points on the old wave front be considered new point sources of light; construct new ray surfaces about a sufficient number of these points, and draw the common tangent to all the new ray surfaces. This is the new wave front.

This trick of considering any point to be a new point source was Christiaan Huygens' fertile and original contribution to the subject of wave motion. His *Treatise on Light*, completed in 1678 but not published until 1690, was reprinted in English translation by Dover (1962).

The graphical construction is illustrated for an isotropic medium in Fig. 1-6. *AB* is the wave front of light traveling from bottom to top in

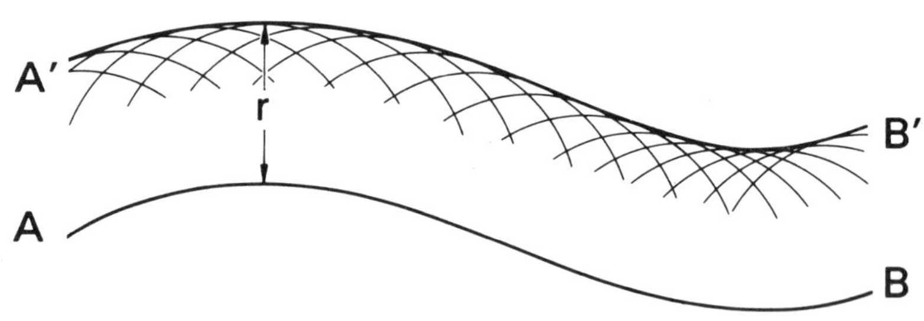

Fig. 1-6. Huygens' construction of a wave front.

an isotropic substance. It is required to find the wave front after the wave has traveled the distance r, which represents the ray velocity. The desired wave front is $A'B'$, the common tangent to all the arcs of radius r. In three dimensions, AB and $A'B'$ would be surfaces, and the circular arcs would be spherical segments.

Wave Fronts in Uniaxial Crystals

Huygens' construction can be applied to the two ray surfaces of a uniaxial crystal to show the relationship between ray surfaces, rays, wave fronts, and wave propagation directions. This is illustrated in Fig. 1-7, which is a special section through the uniaxial ray surfaces. The section is a plane of symmetry through the ray surfaces, so chosen because, with it, all the construction lies in the plane of the paper. The initial wave front intersects the page in AB, advancing toward the top of the page. The ray surfaces of light in this crystal after unit time are known to be of the size and orientation shown by the circles and ellipses around the arbitrarily chosen point sources C, D, and E. Two wave fronts, FG and HI, arise by Huygens' construction. The wave propagation directions are the wave normals DJ and DK, and they coincide. The ray directions are DJ for one of the waves, and DL for the other, since J and L are the only points lying on both a ray surface and a wave front. The ray directions are the directions in which light energy is propagated in each of the waves.

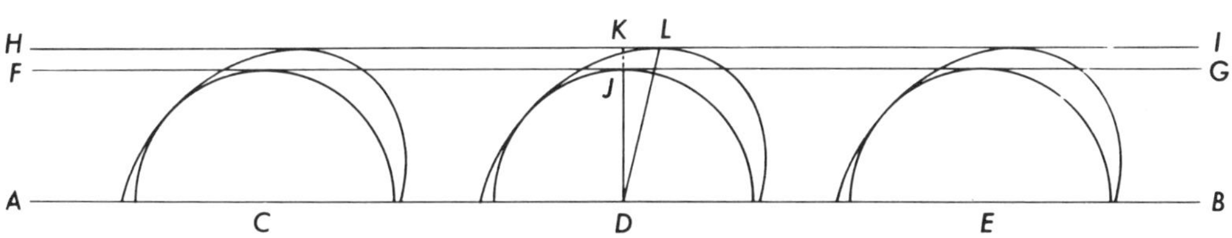

Fig. 1-7. Wave fronts in a uniaxial crystal.

The wave propagation direction in an anisotropic substance may or may not be coincident with the ray direction. Further, the ray direction does not lie in the plane normal to the wave front, the plane of the paper, except in special instances or for one of the rays as in Fig. 1-7. It follows from these generalities that the wave velocity in an anisotropic substance (indicated by the length of the wave normal) is not in all instances the same as the ray velocity (indicated by the length of the ray). Fortunately, most of the optical principles discussed in the succeeding chapters can be treated adequately as phenomena related to waves and wave velocities, so confusion between ray and wave velocities need not arise.

The divergence between the ray *DL* and the wave propagation direction *DK* in Fig. 1-7 is grossly exaggerated with respect to the actual divergence in most uniaxial crystals. In the common uniaxial mineral calcite, however, the ray velocity differences are so great as to generate a readily visible effect. The effect can be analyzed in terms of Fig. 1-7, but in other ways as well. The analysis can be instructive in several ways, and because the effect is spectacular and thought-provoking, it merits discussion in some detail.

Calcite Experiment. This is a version of the experiment in which Huygens discovered the polarization of light. A clear rhombohedral cleavage block of calcite is placed over a black dot on a piece of paper, or on a transparent sheet if an overhead projector is to be used. As discovered in 1669 by Bartholinus (see Born and Wolf, 1964, p. xxii), two images of the dot will be seen through the calcite. If the calcite is rotated on the paper, one of these images remains stationary, while the other describes a circle around the first. The dot itself corresponds to point *D* in Fig. 1-7; one is looking down from the top of the figure. The stationary image corresponds to point *J*, and arises from the wave front tangent to spherical ray surfaces, sections of which appear as circles in Fig. 1-7. The rotating image corresponds to point *L*, and arises from the wave front tangent to ellipsoidal ray surfaces, sections of which appear as ellipses in the figure. The experiment demonstrates that the light energy of this wave front is transmitted along rays that are not normal to the wave front. If the dot were viewed through an isotropic material, no displacement would occur, but no visible displacement would occur for most uniaxial crystals, in which the velocity differences are small and the ellipses only slightly different in shape from the circles. It is interesting to speculate how slowly the understanding of crystal optics might have developed in the absence of calcite.

Having established the behavior of the two images, one can do some more tricks. If the calcite rhomb is tilted in such a way as to lessen the distance between the two images, it will soon become clear that if the dot were viewed along the short body diagonal of the rhomb, only one image would appear. Total reflection prevents seeing this situation directly unless the rhomb is cut and polished at the corners where the short body diagonal emerges. The short body diagonal thus being a unique direction showing a single dot, it must be a direction along which light travels with the same velocity irrespective of vibration direction. This propagation direction, as the reader may have guessed, is the optic axis of calcite.

If an overhead projector is used to view the rhomb down the

short body diagonal (optic axis), or if one squints to see the outline of the rhomb projected onto a single plane, it will be observed that the outline is a hexagon. Calcite belongs to the rhombohedral division of the hexagonal crystal system, and the short body diagonal is the trigonal rotation axis, analogous to the hexagonal rotation axis of the hexagonal division, and is of course the *c* axis. This part of the experiment therefore shows the relationship between trigonal and hexagonal symmetry, and more important for current purposes, it demonstrates that the optic axis is the *c* axis. This is true for all uniaxial crystals.

Return the rhomb to one of its flat sides and again identify the rotating and the stationary images. Place a sheet of polarizing material (polar) on top of the rhomb. (Huygens didn't have polars; he used another crystal of calcite instead.) Now rotate the polar instead of the rhomb. Each of the images will disappear in turn. It may be recalled from Fig. 1-5 that when the vibration directions of two polars lie perpendicular to one another, no light passes through the second polar. It will also be recalled that light passing through a uniaxial crystal (except parallel to an optic axis) may be considered resolved into two mutually perpendicular plane polarized components. The polar interacts with each of these in turn arising from the calcite crystal, just as it does with light polarized by another polar. The disappearance of alternate dot images takes place with 90° increments of rotation of the polar, showing that the vibration directions in the calcite are indeed mutually perpendicular, and that the images of the dot are transmitted by waves having different vibration directions as well as different velocities.

The specific vibration direction associated with each of the dot images can be identified as follows. First ascertain the vibration direction of the polar by viewing a reflected glare from a table top, and turning the polar so that the glare is minimal. The vibration direction is now vertical. Make a scratch parallel to the vibration direction. Place the polar back on the rhomb and turn it until one of the dot images disappears. The vibration direction in the calcite associated with that dot image is now normal to that of the polar. The same can be done for the other dot image. It will be discovered that the vibration direction in the calcite for light that causes the *stationary* dot image is a direction *normal* to the short body diagonal, and conversely, that the vibration direction causing the *rotating* dot image is *parallel* to the short body diagonal, which is the *c* axis or optic axis. We shall not expand upon this discovery at this point except to name the dot images as follows: *ordinary* for the stationary one and *extraordinary* for the rotating one. The reader may wish to return to the experiment after a more detailed discussion of uniaxial crystal optics has been presented. The directly observable phenomena in calcite can then be seen in relation to the more abstract models used in the general discussion.

REFRACTIVE INDEX

When light passes from one substance into another it is either speeded up or slowed down due to the differing atomic structures of the two substances. The refractive index, also called the index of refraction, of a substance is defined in terms of wave velocity:

$$\textit{Refractive index} = \frac{\textit{Wave velocity in vacuum}}{\textit{Wave velocity in substance}} \tag{1-6}$$

To the extent that the velocity of light varies with direction of wave propagation or vibration direction in an anisotropic substance the index of refraction also varies.

A more general form of equation 1-6 relating the index of refraction of one substance to that of another is:

$$\frac{n_2}{n_1} = \frac{v_1}{v_2} \tag{1-7}$$

in which n is the index of refraction, v is the wave velocity, and the subscripts 1 and 2 refer to two substances. The index of refraction of a vacuum is arbitrarily taken as one. Air has an index of refraction of 1.00029 at 15° C and 760 mm Hg pressure; for practical purposes it can be considered to be one. Equation 1-6 may be derived from equation 1-7 by letting n_1 equal one when subscript 1 refers to air or vacuum.

The refractive index (index of refraction) is usually indicated by n, less often by N or R.I. Special notation is used to indicate the values of the refractive indices of anisotropic substances along certain chosen optical directions. In mineralogy or when not otherwise noted, the index of refraction is stated with reference to air.

The range of refractive indices of nonopaque inorganic solids is indicated by the highest and lowest indices reported in standard compilations: Na_2SiF_6, anisotropic, lowest refractive index = 1.3089; stibnite (Sb_2S_3), anisotropic, highest refractive index = 4.303. The range of indices for pure liquids is at least from 1.323 (methyl alcohol) to 1.84 (ethyldiiodoarsene) and can perhaps be extended in both directions. Regarding the indices of refraction for nonopaque minerals, Larsen and Berman (1934) noted that 54 percent of the minerals in their tables have an intermediate refractive index between 1.475 and 1.700. This value remains essentially unchanged (about 56 percent) in the more recent compilation by Fleischer et al. (1984).

All slices of anisotropic crystals except those cut perpendicular to an optic axis have two refractive indices in the plane of the slice. One is associated with one plane polarized component, the other with the other component. Since there are an infinite number of different values depending on the orientation of the anisotropic grain it is conventional to record only the principal values, which are the greatest, the least, and in biaxial crystals, a particular intermediate value. This intermediate value is that of light propagated along either optic axis.

Relation of Refractive Index to Density

Refractive index and density are related to the kind and position of the atoms in a substance. Fairbairn (1943) points out that in ionic compounds refractive index is dependent upon ion radius, valence, ion coordination, and atomic number. A relationship between refractive index and density, the empirical rule of Gladstone and Dale, is:

$$n - 1 = K\rho \qquad (1\text{-}8)$$

in which n equals refractive index, K is a constant of the material, the specific refractivity of the substance, and ρ equals density. For any compound the value of K is the sum of the individual specific refractivities of the components, k, multiplied by their amounts in the compound. Values of k have been calculated by many authors, particularly Larsen and Berman (1934), Jaffe (1956), and Mandarino (1976) according to several criteria having to do with the crystal structure of the compound. Tables of these values are given by Jaffe (1988), who provides an extended discussion and review of the method. A few examples are listed in Table 1-1.

Table 1-1
Some Values of k, the Specific Refractivity Constant

Compound	k	Compound	k
F	0.043	Al_2O_3	0.204
BaO	0.125	MgO	0.200
UO_3	0.134	SiO_2	0.206
Na_2O	0.178	CO_2	0.217
FeO	0.185	V_2O_5	0.366
K_2O	0.186	TiO_2	0.398
P_2O_5	0.190		

For a compound whose composition can be expressed as a sum of oxides, the value of K is the sum of the value of k for each constituent oxide times the weight percentage of the oxide. From the list it is clear that, for compounds containing appreciable amounts of TiO_2 and V_2O_5, K will be larger than 0.2, the approximate mean value of k for all oxides. For substances containing UO_3, BaO, or F, the value of K will be smaller than 0.2. Thus, barium compounds will have a lower mean refractive index and titanium compounds a higher mean refractive index than the average compound of the same density.

In the discussion of refractive index related to density above, it has been tacitly assumed that the substance is isotropic. For anisotropic substances it is the mean refractive index that shows the relations discussed. The mean refractive index is the average value of the refractive indices in three perpendicular directions in the anisotropic substance.

Other formulas relating refractive index to density include those of Lorentz and Lorenz, and Allen (1956), as discussed by Jaffe (1988). Since the refractive index depends on the manner of aggregation of the atoms as well as their kind and amount, there is not perfect agreement between calculated values and the true refractive indices using any of these formulas. The rule of Gladstone and Dale seems, however, to be as good an approximation as any of the other relationships that have been suggested. Jaffe (1956) states that in calculating the refractive index using the rule of Gladstone and Dale a deviation of not more than 0.020 in the mean index of refraction is to be expected, and the error will be much smaller if the appropriate constants for the specific crystal structure are used (Jaffe, 1988).

Variation of Refractive Index with Temperature

Since the refractive index varies directly with density, and density normally varies inversely with temperature, it is not surprising that the refractive index normally varies inversely with temperature. The temperature coefficient of the refractive index is the change of refractive index with a change in temperature of one degree Celsius, sometimes written $-dn/dt$, °C. The refractive indices of both liquids and solids become lower as the temperature is raised, hence the negative sign of the temperature coefficient. The temperature coefficient of liquids is appreciable. This is of importance to the microscopist who uses liquids as immersion media for refractive index comparison. The temperature coefficients of most of the liquids used as immersion media in the range of refractive index of 1.45 to 1.74 are from −0.00035 to −0.0007 per degree Celsius. The temperature coefficients of solids are very small. For fluorite, for example, the refractive index is lowered 0.00001 for each degree rise in temperature. This is an order of magnitude less than the limit of accuracy in the determination of refractive indices of solids with the microscope. The change of refractive index with temperature is not used in identification of crystalline substances.

Refraction—Snell's Law

Refractive index is defined in terms of velocity, but as the name indicates, it is related to the refraction (change in direction) of the wave propagation direction of light passing from one substance into another. Such refraction results if the velocity of light in the two substances is different, and if the angle of incidence is other than 0°. The law of refraction, *Snell's law*, expresses the relationship:

$$\frac{n_2}{n_1} = \frac{\sin i_1}{\sin r_2} \tag{1-9}$$

where n is the index of refraction, i the angle of incidence, and r the angle of refraction. The angles i and r are always measured from the boundary surface normal, as shown by Fig. 1-8, to the incident or refracted wave propagation direction. The subscripts 1 and 2 refer to two substances.

The Dutch scientist Willebrord Snell (Snellius) discovered the law of refraction in 1621, but died in 1626 without publishing. According to the sources used by Born and Wolf (1964), the law was first published by Descartes in 1637 without an acknowledgement to Snell, whose manuscript he is believed, however, to have seen. This apparent plagiarism is perhaps mitigated by the fact that Descartes published anonymously.

As shown in Fig. 1-8, a wave, with wave front normal to the page, intersecting the page in *OP*, is passing from isotropic substance S_1, in which the wave velocity is represented by *PQ*, into isotropic substance S_2, in which the wave velocity is represented by *ON*. *ST* is the boundary between S_1 and S_2, and *FOG* is normal to *ST*. The wave propagation direction in S_1 is *AO* or *BP*, normal to *OP*. At the end of unit time *P* has reached *Q* and *O* has reached some point on the ray surface around *O* which is a circle of radius *ON*. The wave front in S_2 intersects the page in *QN*, a line from *Q* tangent to the circle about *O*. The wave propagation direction in S_2 is *NX*, the normal to the wave front. The angle of incidence is *AOF*, and the angle of refraction is *NOG*.

If S_1 is air, the refractive index of S_2 is, by Snell's law,

$$\frac{\sin AOF}{\sin NOG} \tag{1-10}$$

and equation 1-9 simplifies to

$$n = \frac{\sin i}{\sin r}\,. \tag{1-11}$$

The general cases of the refraction of a wave propagation direction at the boundary between an isotropic medium and a second medium in which the wave velocity is less are diagrammed in Figs. 1-8 and 1-9. In Fig. 1-8 the second substance is isotropic; in Fig. 1-9, it is anisotropic.

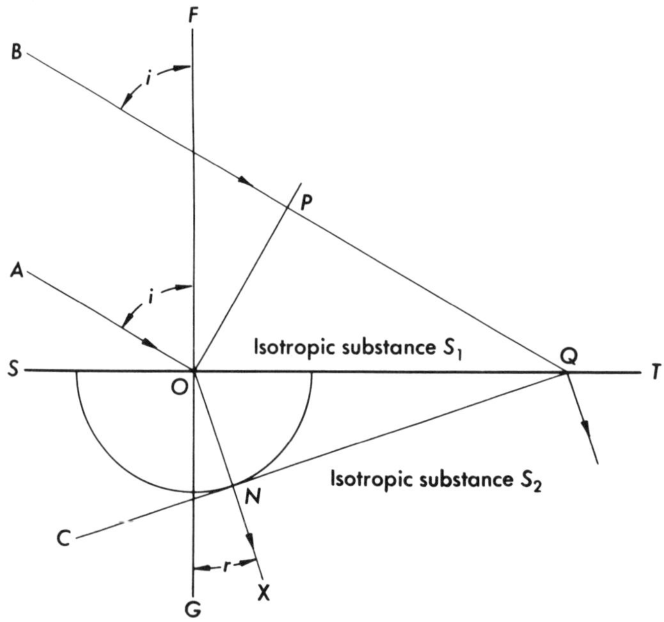

Fig. 1-8. Refraction of light passing from one isotropic substance into another.

In Fig. 1-9, a wave front *OP* is passing from S_1, in which the velocity is represented by *PQ*, into S_2, in which the ray surfaces after unit time are known to be of the size and orientation of those drawn around *O*. The wave propagation direction in S_1 is *AO* or *BP*, normal to *OP*. After unit time, *P* has reached *Q* and *O* has reached points on the ray surfaces around the original *O*. The wave fronts in S_2 are *CQ* and *EQ*, the tangents to ray surfaces through *Q*. The wave propagation directions are *OX* and *OY*, the normals to these tangents. The angle of incidence is *AOF* and the angles of refraction are *XOG* and *YOG*. If S_1 is air, the refractive indices of S_2 for wave normals *OX* and *OY* are, by Snell's law:

$$\frac{\sin AOF}{\sin XOG} = \frac{\sin AOF}{\sin YOG} \tag{1-12}$$

Figures 1-8 and 1-9 both illustrate that on entering a substance in which the velocity is less, the wave propagation direction, if incident other than normal to the boundary, is refracted toward the perpendicular. This generalization is often stated in terms of the refractive index: light entering a substance of higher refractive index, incident other than normal to the boundary, is refracted toward the normal. This statement follows from equation 1-9. The reverse of the generalization is also true, that light entering a substance of lower index of refraction,

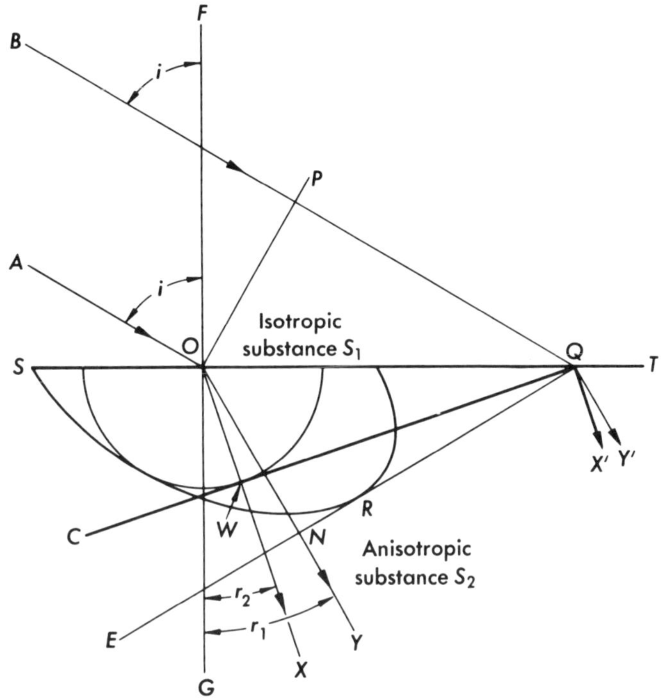

Fig. 1-9. Refraction of light passing from an isotropic substance into an anisotropic substance.

incident other than perpendicular to the boundary, is refracted away from the perpendicular.

If light is incident perpendicular to the boundary of an isotropic or an anisotropic substance, as in Figs. 1-10 and 1-11, the wave propagation direction of the incident light is not refracted despite the change in velocity.

Figure 1-10 shows the wave propagation direction $AB = DE$ normal to the boundary ST between the two substances. The ray velocity in S_2 is represented by $BC = EF$, and the new wave front in S_2 is CF, tangent to the ray surfaces represented by circles drawn around B and E. The wave propagation direction in S_2 is $BC = EF$, the same as in S_1; there is no refraction of light at the boundary, ST.

In Fig. 1-11, the wave front in S_1 is parallel to the boundary ST. The ray surfaces in S_2 are known to be of the size and orientation of the circles and ellipses drawn around points B and E. The wave fronts in S_2 are the tangents to the ray surfaces, CF and PR. The wave propagation direction in S_2 is therefore $BCM = EFN$, which coincides with that in S_1 ($AB = DE$). No refraction of the wave propagation direction occurs at the boundary. Snell's law cannot be used if light is incident perpendicular to a boundary.

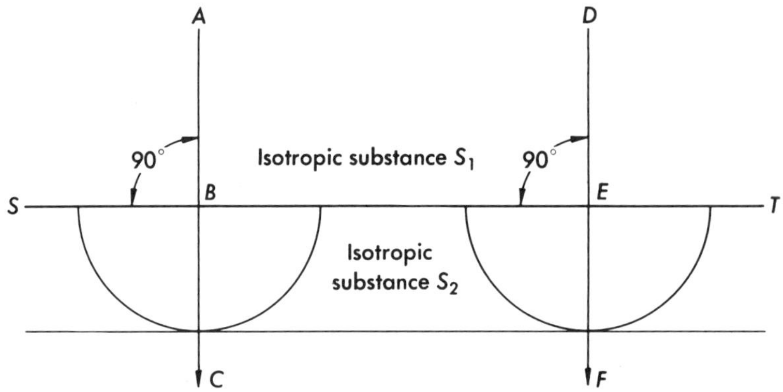

Fig. 1-10. Passage of light at normal incidence between two isotropic substances.

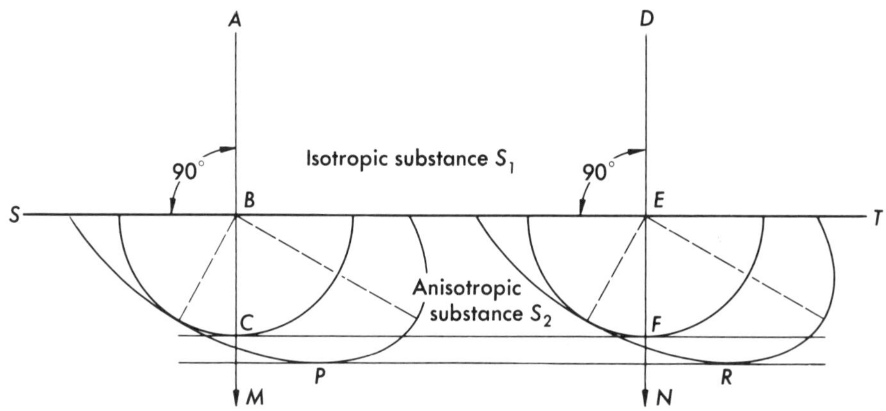

Fig. 1-11. Passage of light at normal incidence between an isotropic and an anisotropic substance.

Application to Microscopic Study. In the microscopic study of crystals, the fragments or sections for study are usually chosen with essentially plane parallel top and bottom surfaces, and light through the microscope is for most procedures incident normal to the object. Anisotropic material under these conditions can be rather simply studied despite the complexities of anisotropic optics. To the extent that light is incident normal to an anisotropic crystal plate with plane parallel surfaces:

1. Waves will be propagated through the plate without deviation of the wave propagation directions.
2. The vibration directions of the polarized components, being

normal to the wave propagation directions, will be parallel to the surface of the plate.

3. The vibration planes of the polarized components being normal to each other, the vibration directions in the plane of the plate lie in these planes and are normal to each other.
4. Wave propagation directions of both components being coincident, vibration directions, being normal to each other, can be studied separately.
5. Each vibration direction is related to a propagation direction and this in turn to a refractive index so that the vibration directions may be correlated with refractive index.

For these reasons, **attention most often need only be directed to the vibration directions in anisotropic material,** seldom to the ray directions or the wave propagation directions. We shall concentrate on the vibration directions.

Double Refraction and Birefringence

Anisotropic crystals exhibit *double refraction,* which refers to the fact that each ray of light in such crystals (and the wave propagation directions except when light enters with perpendicular incidence) is refracted by a different amount. The calcite experiment (p. 12) illustrates double refraction. Birefringence, abbreviated as B or δ, is the measure of double refraction in terms of the difference in refractive index.

The *birefringence of a substance* may be defined as in equation 1-13, where the subscripts h and l signify "highest"

$$B = |n_h - n_l| \tag{1-13}$$

and "lowest." The birefringence of a substance (or crystal species) is therefore the maximum difference in refractive indices. This expression is also valid for the partial birefringence of any section or fragment of a crystal, in which case n_h and n_l refer to the higher and lower refractive indices, each associated with one of the two vibration directions normal to a specified wave propagation direction. In practice, this propagation direction is usually incident normal to the crystal boundary and parallel to the axis of the microscope, so the two vibration directions lie in the plane of the microscope stage. Since n_h and n_l for a random section are generally not the same as those for the substance itself, the birefringence of a section is generally less than that of the substance. Because waves propagated along an optic axis all travel with the same velocity, some sections of anisotropic crystals are not birefringent.

Reflection

When light reaches the boundary between two substances some of the light does not cross the boundary, but is reflected (sent back into the first substance), and the angle of incidence, *i*, equals the angle of reflection *i'*, as illustrated in Fig. 1-12.

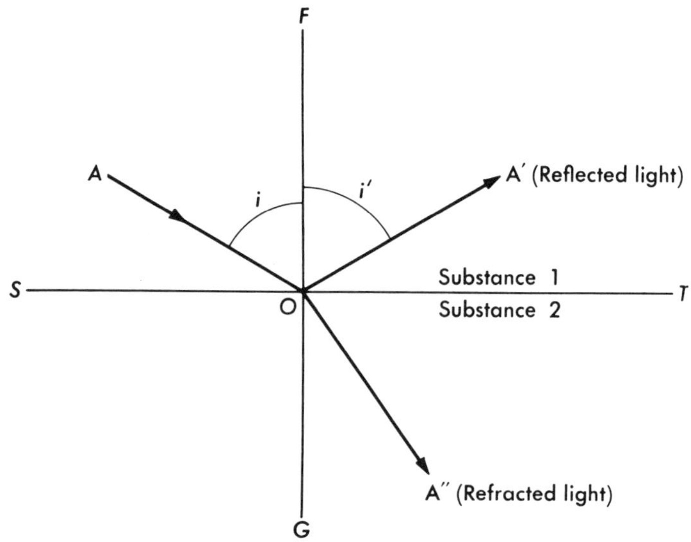

Fig. 1-12. Reflection. The wave propagation direction *AO* is reflected at the boundary *ST* so that the angle of reflection *i'* equals the angle of incidence *i*.

Critical Angle

It has been stated that light passing into a substance of lower refractive index at an incident angle other than 0° is refracted away from the perpendicular to the boundary surface. At some angle of incidence the light will be so greatly refracted that it will just graze the surface as shown by angle *BOB'* in Fig. 1-13. This angle of incidence, which is the largest angle at which light can enter a substance of lower refractive index, is the *critical angle.* At angles of incidence greater than the critical angle all light is reflected.

In Fig. 1-13, substance 2 has the lower refractive index. *FG* is normal to *ST*. Wave propagation direction *AO* is refracted to *A'*; direction *BO* is incident at the critical angle and is refracted at grazing angle *OB'*; direction *CO* is incident at an angle greater than the critical angle and is reflected to *C'*.

The numerical value of the critical angle of a high index substance

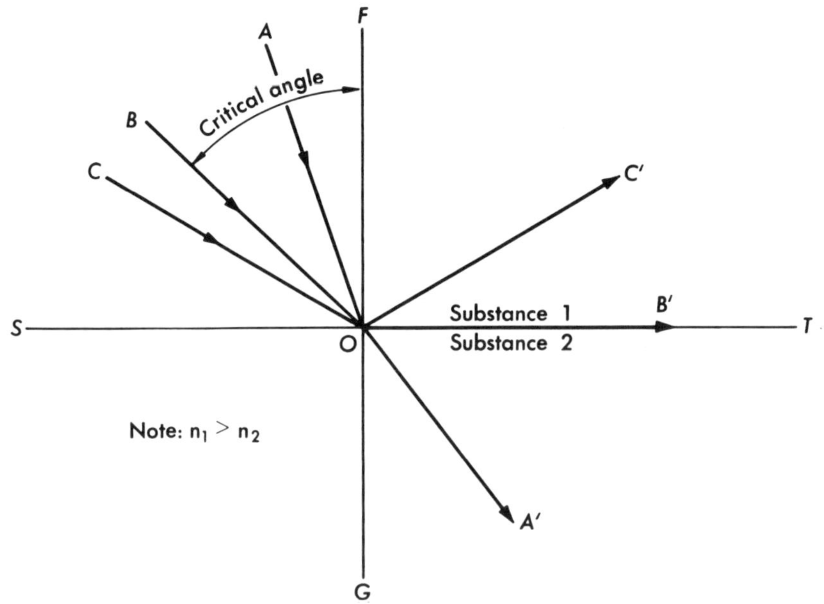

Fig. 1-13. Critical angle.

with respect to a low index substance can be determined from a simple relationship derived from Snell's law (equation 1-9). Let the critical angle of incidence be designated c. At the critical angle, the angle of refraction, r, is 90°, and $\sin r = 1$. Snell's law now becomes

$$\sin c = \frac{n_l}{n_h} \tag{1-14}$$

When the substance of low refractive index is air, n_l may be taken as 1, and equation 1-14 simplifies to

$$\sin c = \frac{1}{n_h} \tag{1-15}$$

If c is measured, n can be calculated, and this relation is the basis for the Abbe refractometer (p. 40), with which the refractive index of a liquid can be measured.

In anisotropic substances the critical angle associated with each wave propagation direction may be calculated using the associated value of n.

Because light passing into a substance of higher refractive index is refracted toward the perpendicular to the boundary surface, light incident at any angle can enter a substance of higher refractive index.

COLOR

The color of light depends upon the frequency of vibration of the light reaching the eye. The frequency does not change when light passes from one substance to another. By equation 1-4, therefore, since the velocity changes, the wavelength must also change. For a given color of light, velocity, frequency, and wavelength are sufficiently constant in air to permit reference to air as a standard medium for most optical purposes. For this reason, the *color in air*, where light is most commonly observed, may be defined by the wavelength of the light. Typical wavelengths for some standard colors are given in Table 1-2. The sensitivity to color of the foveal cones of the human eye peaks at about 550 nm and tails off to zero near 400 and 700 nm.

Table 1-2
Typical Wavelengths for Some Standard Colors

Color in Air	Typical wavelength in nm (10^{-9} m)
Violet	410
Blue	470
Green	530
Yellow	580
Orange	610
Red	670

Monochromatic Light

Light is said to be *monochromatic* if it includes only a small range of wavelengths (which the eye sees as a single color), or if the intensity of light of a single wavelength is very much greater than that of all the other wavelengths present.

White Light

White light is a mixture of light of many wavelengths from all parts of the visible range or all colors of the rainbow, for the rainbow is sunlight separated into its component colors. White light is the light from clouds in a bright sky.

Dispersion

When monochromatic light passes from one substance to another, the frequency remains fixed, but the velocity and wavelength vary directly with each other and the light is refracted. The components of white light are refracted unequally in passing from one substance to another, depending on their velocity, and inversely on their wavelength. The effect, made famous by Newton's experiment in 1666, can be seen if

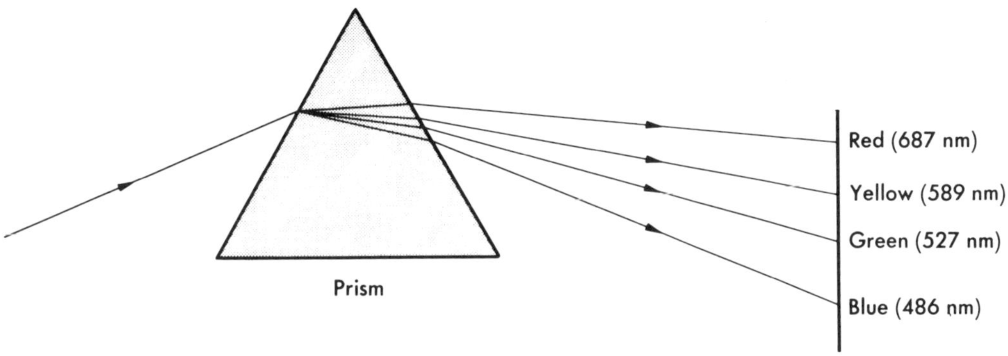

Fig. 1-14. Spectrum due to refraction of white light passing through a glass prism. The numbers are wavelengths in nanometers (nm).

white light is passed through a prism, as shown in Fig. 1-14. The refractive index of glass is greater for blue light than for red, so by Snell's law the angle of refraction is also greater for blue. The unequal refraction of the different colored components of white light is termed *dispersion*. All crystals and liquids show dispersion in this sense, sometimes termed *dispersion of the refractive index*, to differentiate it from another class of phenomena found in anisotropic crystals. All these types of dispersion are discussed in more detail after the following consideration of spectra.

Spectrum

When light is dispersed into its component wavelengths, the resulting pattern is called the *spectrum* of the light. A spectrum is usually formed with a prism or grating, often using a *spectroscope*. Fig. 1-15 shows the optical system of a simple spectroscope. A source of light is focused on a slit and passes through the slit to a prism. On passing through the prism, the different wavelengths, associated with different velocities, are refracted differently, and the position of each may be observed. If a photographic film or electronic sensing system is substituted for the eye, the instrument is termed a *spectrograph* or *spectrometer*. The spectrum obtained may be due to *emission* of light by the source or *absorption* of light by intervening material.

The several types of spectra depend on the nature of the source.

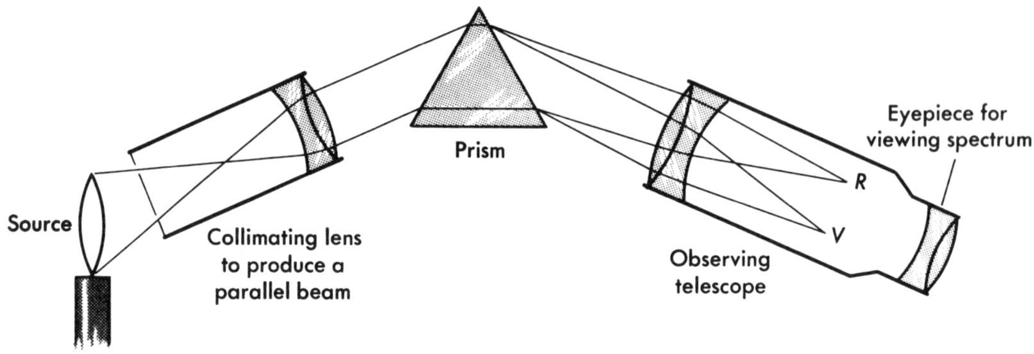

Fig. 1-15. Plan of a spectroscope. *R* and *V* stand for red and violet light, refracted different amounts by the prism. Divergent paths for *R* and *V* within the prism ignored.

Line spectra result when atoms are excited in an arc or flame. Light of distinctive wavelength is emitted when electrons undergo a transition from one quantum level to another. A collection of discrete wavelengths results from the various transitions common to a given species of atom (chemical element), and thus each element generates its own unique line spectrum when excited. The spectrum is a collection of separate *lines, images of the slit,* whose positions in the spectrum indicate their wavelengths. Line spectra are dominated in their intensity by those transitions whose probability of occurrence is greatest, and thus a dominant color results that is characteristic of each atomic species. This dominant color is responsible for certain diagnostic "flame tests" for elements:

Lithium, Li, red flame, chief wavelength 6708 Å
Sodium, Na, yellow flame, chief wavelengths 5890 and 5896 Å
Thallium, Tl, green flame, chief wavelength 5351 Å

(Spectrographers conventionally designate wavelengths in Ångstrom units; 10 Å = 1 nm.)

Spectra from glowing solid bodies form a continuous band and are called *continuous spectra. Absorption spectra* are dark lines or bands, the positions of which depend upon the absorbing substance between the source of light and the slit. The spectrum of the sun's atmosphere is an absorption spectrum. It consists of dark lines superimposed on the continuous spectrum of the radiant body of the sun. The dark lines result from absorption, by the sun's atmosphere, of wavelengths of light that are characteristic of the gaseous elements in the sun's atmosphere. This dark line spectrum is known as the Fraunhofer spectrum. Important *Fraunhofer lines* are designated by letters and are used as standards of wavelength:

F 4861 Å in the blue region, due to hydrogen

D 5890 Å and 5896 Å in the yellow region, due to the two sodium lines, which are so similar in wavelength that they are not separated in instruments of low resolution

C 6563 Å in the red region, due to hydrogen.

Josef Fraunhofer discovered the dark lines in 1814–17 and they were later (1861) recognized as absorption lines by Bunsen and Kirchhoff. The eventual recognition of absorption and emission as phenomena unexplained by Maxwell's electromagnetic theory gradually led to the birth of quantum physics, which therefore has its roots in Fraunhofer's discovery (see Born and Wolf, 1964).

Dispersion of the Refractive Index

The difference in the refractive indices for different wavelengths is expressed by indicating the numerical difference for a specified range of wavelengths. The range commonly used is from F to C ($n_F - n_C$). The range from the green line of thallium to the red line of lithium ($n_{Tl} - n_{Li}$) is sometimes employed. Dispersion may be described as strong or weak, which refers to large or small numerical values of dispersion.

Figure 1-16 shows the refractive index of alpha-chloronaphthalene,

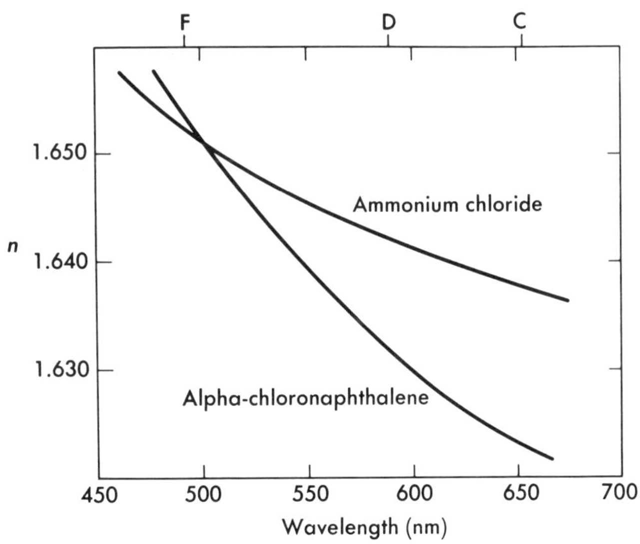

Fig. 1-16. Dispersion curves for alpha-chloronaphthalene (liquid) and ammonium chloride (solid).

a liquid used for immersion mounts, plotted against wavelength. The value $n_F - n_C$ equals 0.0308 for this substance, according to Butler (1933); the index of refraction of light for the wavelength 486.1 nm is 0.0308 greater than for light of 656.3 nm. The dispersion curve for the solid ammonium chloride (the mineral salammoniac), $n_F - n_C = 0.0149$, is also plotted in Fig. 1-16. Dispersion curves for both liquids and solids are, like the ones chosen, not linear but concave upward; conversions to straight lines are discussed in Chapter 16. These examples of dispersion curves for a representative crystal and liquid illustrate the generalization that, for solids and liquids having refractive indices in the same range, the dispersion of liquids is stronger than the solid; the liquid curve is steeper. It is also generally true that the dispersion of liquids, and of solids to a lesser degree, is stronger for those of higher refractive index than for those of lower refractive index.

Refractive index data are complete if they refer to a specific wavelength and, because refractive index varies with temperature, to a specific temperature. A subscript indicates the wavelength by any of the conventional designations and a superscript indicates the temperature in degrees Celsius, for example: $n_D^{20°} = 1.560$. If index of refraction is written without reference to the temperature or wavelength, room temperature and the D line of sodium, the accepted standard wavelength for refractive index measurements, are understood. A standard temperature is most important in calibrating the refractive index liquids used to compare with indices of crystals; 25° C is most commonly used.

Dispersion of Other Optical Properties

Anisotropic substances may show dispersion related to the several optical directions and parameters. Included are the dispersion of the refractive indices, the birefringence, the directions of vibration of the polarized components, and the optic axis directions. Knowledge of the dispersion may occasionally be of critical importance in crystal identification. Dispersion of the refractive indices of liquids relative to solids, as discussed above, is fundamental to some of the tests made in determining refractive index.

Color of Crystals

The color of a crystal in transmitted light depends upon the wavelengths of the light emitted from the light source and the properties of the crystal lying between the light source and the eye. Only such wavelengths of light as are able to pass through the crystal reach the eye. The wavelengths not transmitted are said to be absorbed by the crystal. This is the property of *absorption*. The amount of light absorbed varies with the thickness of the crystal so that the color of thin fragments, for example those employed for microscopic examination, is of lighter intensity than the color of thicker specimens. Many minerals such as fluorite that may be colored in hand specimen show no appreciable color in thin fragments.

In anisotropic material having absorption, the wavelengths absorbed are dependent upon the direction along which light vibrates in the crystal. The color of the crystal in different vibration directions is therefore different. Such crystals exhibit the property of *pleochroism*, change of color for different vibration directions. This phenomenon can be useful in crystal identification, as discussed in Chapters 9 and 12.

Origin of the Vibration Directions. This is a topic for the curious; it is not essential to further progress in this book, but may be of interest to some readers sooner or later. As shown by the calcite experiment (p. 12) and mentioned many times in this chapter, there are in general two mutually perpendicular vibration directions for any wave propagation direction in an anisotropic crystal. The question reasonably arises, why are these directions always mutually perpendicular, when the crystal structure is both anisotropic and randomly oriented with respect to an arbitrary wave propagation direction? The quick but incomplete answer is that polarized light is elliptically polarized with vibration directions given by the major and minor axes of the ellipse, necessarily at right angles. The rest of the answer is that the optical properties of the crystal depend on the integrated electrical properties of the atoms along the path of light, and are dictated by the long-range symmetry of the crystal.

To be brief, light is composed of both an electric and a magnetic vector, but its interaction with crystals can be described for most purposes in terms of the electric vector. The vibration directions of our treatment are the electric vectors of crystal physics. The interaction of these electric vectors with the anionic electric field of a nonopaque crystal governs the retardation of light as it passes through the crystal. For any given section of the crystal, there is always some vibration direction whose interaction with the electric field of the anions is least, whose retardation is minimum, and therefore whose velocity of propagation is maximum; this is the "fast" or low index vibration direction. There is also, conversely, a "slow" or high index vibration direction, and the question is, why should this be normal to the first? For a uniaxial crystal it is easy to show that one of the vibration directions lies in the unique circular section of a spin ellipsoid, normal to the *c* axis of the crystal. The other vibration direction is constrained to lie in the plane *containing* the *c* axis (the optic axis) of the crystal, for if it did not, the crystal could not have the tetragonal or hexagonal symmetry that it must have to be uniaxial. Therefore the vibration directions must be mutually perpendicular.

An extension of the same reasoning can be used to argue the case for biaxial crystals, and so for anisotropic crystals in general. Briefly, the actual location of the vibration directions relative to one or two optic axes can be determined for any crystal by the Biot-Fresnel law, discussed on page 165 in connection with interference figures. The relative identities (high or low index) of the vibration directions are discussed under the structural basis for the optic sign, at the end of Chapter 7. Mathematical derivations of the origin of the vibration directions are given in Chapter 14 of Born and Wolf (1964), and a general treatment of the physical basis of mineral optics may be found in Strens and Freer (1978).

The moral of the story, then, is that the locations of vibration directions are constrained by crystal symmetry, and conversely that no optical orientation may violate the highest symmetry displayed by the holosymmetric class of the crystal system. It is possible to invert this statement by pointing out that the optic orientation, which can often be determined with the petrographic microscope, contains useful information about the possible crystal symmetry. This topic is therefore of practical importance in the identification of crystals with the microscope.

2

Equipment and Preparation of Materials

For the routine identification of nonopaque materials a polarizing microscope is used. The petrographic or polarizing microscope is essentially a compound microscope with the addition of optical devices that polarize light. A simple variety for qualitative observations can be made from almost any compound or stereo microscope and two sheets of polarizing filter. Most quantitative measurements require special features (rotating stage, slots for accessory plates, diaphragms, etc.) that are incorporated into the design of commercial polarizing microscopes.

The chief functions of the polarizing microscope are to magnify an object and permit investigation of its optical properties in polarized light. To describe the magnifying function, it is pertinent to describe the compound microscope in its simplest form.

THE COMPOUND MICROSCOPE

The compound microscope is a system of lenses mounted in a tube arranged so that an enlarged image of an object is formed on the retina of the eye. A light source is placed under the microscope. The light passes through a condenser and then illuminates an object resting over a hole in a platform or stage. In the lower end of the tube, mounted above the stage, an objective lens is placed so as to form a real image of the object in the upper end of the tube when the height of the tube is adjusted so that the object lies just below the lower principal focus of the objective. This real image is magnified by an eyepiece at the upper end of the tube. An enlarged virtual image is seen by the eye. When the stage and tube and also, therefore, the objective and eyepiece are in such a position that the object is just outside the lower principal focus of the objective, the image is clear and the microscope is said to be "in focus."

THE POLARIZING MICROSCOPE

In a polarizing microscope the condenser, objective, and eyepiece usually consist of several lenses each, and the entire system is more elaborate than the simple optical device described above. The following descriptions of special features of the petrographic microscope are pertinent.

Ocular (Eyepiece)

Oculars are generally of the negative or Huygenian type, with crosshairs provided in the plane of the real image. The crosshairs divide the field of view into four quadrants. For convenience, the crosshairs, quadrants, and any directions within the field are identified by names of compass directions; by convention, north is away from the observer.

Oculars in common use magnify from 5 to 10 times; an intermediate value of magnification is commonly employed. Special scales may be inserted, usually at the factory, in the plane of the crosshairs. A common micrometer scale is 1 mm long divided into 20 divisions engraved on a glass reticule. It can be calibrated for each lens system with an object of known size. Other special scales include square rulings of exact dimensions (grid oculars), useful in counting different types of grains.

> Several aspects of the use and care of oculars deserve comment. The ocular can usually be lifted out of the tube or may, indeed, fall out if the microscope is not properly carried. A detent is furnished on the ocular housing and will fit into either of two slots in the tube, to allow the crosshairs to be oriented either at the zero or 45° position. A knurled ring allows the crosshairs to be focused to suit the observer's eye. This can best be done by removing the ocular and rotating the ring while observing a well-lit blank sheet of white paper. The observer is likely to be either left- or right-eyed, and may wish to test which eye will give the best service with least strain. Eye and muscle strain can be minimized if both eyes are kept open and the dominant eye used at the ocular. Beginning practitioners find it helpful to cover the subdominant eye with the hand or a card while learning to observe with both eyes open. Wearers of eyeglasses should determine from experience whether, depending on the correction needed for the eye, glasses will best be left on or taken off. Astigmatism cannot be corrected by focusing the microscope, whereas near- and farsightedness can. If glasses are worn, special "high-point" eyepieces with rubber cups can be helpful in maintaining the correct working distance. Many people

use glasses successfully without such eyepieces, but eventually find that scratches appear on the eyeglass lens from contact with the frame of the ocular. (The eventual degradation of the lens may well coincide closely with the need for a new prescription, in any event.) If glasses are not worn, it is commonly found that oil from the skin or eyelashes accumulates on the surface of the ocular lens. When this occurs it will blur the field of view, and should be corrected by careful wiping with optical grade disposable tissue.

Magnification

The magnification of the optical system is approximately equal to the nominal magnification of ocular times objective. A table of magnifications for different combinations of lenses is furnished by some manufacturers. The magnification of any combination can be discovered by observing an object of known length with a micrometer ocular. The object is usually a stage micrometer, furnished by most microscope manufacturers. Such a micrometer is commonly divided into tenths and hundredths of a millimeter.

Objectives

Three or four objectives of different magnifications provide a practical complement for use in crystal identification. Magnifications (printed on the objective body) of about 3, 10, 25, and 50 are useful. Such magnifications are often referred to as low, medium, medium-high, and high power.

Objectives can be centered on the optical axis of the instrument by means of built-in devices, commonly manipulated either with special wrenches or by means of knurled rings on the body of the objective. Accurate centering is especially important with high-power objectives, or the object of interest may stray from the center of the field on rotation of the stage. One should learn how to center objectives of the particular model of microscope to be used. It is convenient to focus on a speck of dust or small crystal fragment, bring it manually to the crosshairs, rotate the stage until it is most offcenter, and adjust the lens to bring the speck halfway to the crosshairs. Several repetitions of this entire procedure should suffice to center the objective.

Very high magnifications are occasionally needed for examination of fine particles. The highest (up to 100×) are best obtained with oil immersion objectives, which have a short working distance and require oil between objective and specimen to minimize unwanted reflection and refraction.

As a rule, working distance decreases as the magnification of the objective increases. Special objectives that combine high power with large working distance are available for use with the universal stage.

Numerical Aperture. The working distance is related to the angle of the cone of light entering the objective from an object at the principal focus. This angle varies with the refractive index of the intervening medium, for example, oil or air, and with the design of the lens. Most lenses are marked with a number, called the *numerical aperture* (N.A.), from which the cone angle can be derived by the relation

$$\text{N.A.} = n \sin \mu, \tag{2-1}$$

where n is the refractive index of the medium between object and lens, and μ is the *half-angle* of the cone of light. Knowledge of the numerical aperture is useful in comparing the performance of various objectives, especially in relation to interference figures, described in Chapters 8 and 11. As a rule, the N.A. increases with the magnification of the objective. Common values are N.A. = 0.65 for a 40× objective, and 0.85 for a 50× objective.

The Stage

The stage of a petrographic microscope differs from that of most compound microscopes by being circular, capable of rotation, and calibrated in degrees. Various holes are cut or tapped in the stage to provide for the use of stage clips, mechanical stage, a spindle stage, or a universal stage. Stage clips are seldom needed if the stage is horizontal, and they can be a nuisance if they touch the immersion oil or flip off a cover glass.

Mechanical Stage. This device holds a glass object slide and has two motions at right angles by means of which the slide may be translated in any direction across the stage. A finger knob attached to a rack and pinion is responsible for each motion, and calibrated scales are provided for measuring or recording the amount of translation. The mechanical stage permits systematic examination of a preparation, measurement of various distances, and easy relocation of a region of interest on the slide by means of the coordinate system. A *point counting stage* may be either a manual or motor-driven mechanical stage that advances in jumps along selected rows to lay down a grid of points over the slide. Such a device permits modal analysis of the relative abundances of minerals in a slide by means of identifying and tabulating the identity of a mineral grain at each rest point of the stage.

Spindle Stage. The spindle stage is a device that permits rotation of a crystal fragment about a horizontal axis while the crystal is immersed in a suitable liquid. It allows measurement of all refractive indexes and certain other optical properties upon a single crystal fragment. Its techniques and applications are discussed in Chapter 15.

Universal Stage. A device of great versatility that is seldom needed or used in routine crystal identification, the universal stage permits the object slide to be rotated on from three to five axes, depending on the complexity of the particular model. Many orientations of a crystal on the slide may be observed, and identifiable optical or crystallographic directions can be made coincident with an axis of the universal stage. The device is frequently used for accurate measurement of angles between optical and crystallographic directions, or between two optical or two crystallographic directions. The reader is referred to Chapter 15 for a further description of the universal stage.

Illumination

Most polarizing microscopes of recent design are furnished with a built-in light source having a tungsten or krypton arc bulb and a blue, ground-glass filter to simulate the spectral qualities of daylight. A monochromatic source such as a sodium vapor lamp or monochromator can be used instead. Monochromatic light is required for refractive index determinations of high accuracy.

Various adjustments to the illumination system may be possible, depending on the flexibility of the microscope design. Centering of the source, and attenuation by means of a diaphragm or rheostat are desirable features. Space is commonly provided for additional filters, such as neutral density or daylight-correcting filters for use in photography, or ground glass screens used to diffuse the light. Adjustments to illumination are described in this chapter following the condenser and Bertrand lens descriptions.

Polarizing Devices

Polarizing devices (polars or polarizers) are the distinctive features of polarizing microscopes. These devices permit passage of light vibrating in only one plane. Most models utilize transmission polarizers, but older models and some research or super-polarizing microscopes utilize prisms of calcite. Several types of calcite prisms have been used; the original one was designed in 1828 by the Scottish natural philosopher William Nicol (1768–1851). Since this invention permitted the modern development of crystal optics and petrography, "nicol" became a generic name for polarizing prisms, hence such terms as "crossed nicols" are seen in the older literature. With the advent of transmission polarizers, the broader term "polar" has come into general use.

Transmission polarizers, developed by the American inventor E.H. Land (1909–1991), are familiar in simple form as Polaroid sunglasses and the screens that make liquid crystal displays visible. These devices are stretched films of mutually aligned intensely pleochroic long-chain polymers that have the power of absorbing light strongly in one direction (strictly dichroism, a variety of pleochroism). Such devices are described in detail by Shurcliff (1962).

There are two polarizing devices in the optical system of a polarizing microscope. Light passes through one polar before reaching the condenser, and through a removable polar after leaving the objective. The lower polar is often called the *polarizer*, and the upper one the *analyzer*. The vibration directions of polarizer and analyzer are at right angles to one another. When the analyzer is in the optical path, the object is said to be viewed with *crossed polars*. Since the 1970's it has been the standard convention among manufacturers for the polarizer direction to be east–west, but for largely historical reasons, in many institutions all the microscopes are adjusted to have the polarizers north–south. Unless otherwise stated, the north–south convention is assumed in this book. It is of utmost importance to know which orientation is present in the microscope being used.

Orientation of Polars. The orientation of the polars should be examined to ensure proper adjustment. If the polars are crossed at exactly 90°, the field will be black if there is no object on the stage. If the field is not black, one of the polars must be rotated until blackness results. In most microscopes this will be the polarizer, since the direction of the analyzer is fixed. When blackness results, the rotation angle of the polarizer should read zero, and the vibration direction of the polarizer should be parallel to a crosshair. If this is not the case, a factory adjustment of the analyzer may be in order.

The vibration direction of the polarizer (whether N–S or E–W) may be ascertained by removing the polarizer in its housing and viewing through it light reflected from a nonmetallic horizontal surface such as a table top. Such reflected light is partly polarized with its vibration direction parallel to the reflecting surface. The reflected light appears attenuated when the vibration direction of the hand-held polarizer is perpendicular to the plane of the observed surface.

Substage Condenser

Below the stage, and above the polarizer, is the substage condenser. The lower set of lenses in this condenser has a long focal length so that an object on the stage is illuminated by a narrow cone of nearly parallel rays. The most oblique rays can be eliminated, and the angle of the cone limited so that most rays are parallel, by adjusting the diaphragm at the lower principal focus of the condenser just below the polarizer. A swing-out converging lens is provided just below the stage for observations in strongly converging light. In a few models this lens is permanently in place.

Bertrand Lens

The Bertrand lens is mounted below the ocular and can be inserted in the optical path at will. In some microscopes the lens is equipped with an iris, and in some its position in the tube can be varied vertically by rack and pinion or movable sleeve, and laterally by tangent screws. These two movements allow focusing and centering.

The Bertrand lens is a converging lens that permits observation of the image at the back focal plane of the objective. This is the image of the source of light, which can be seen unmagnified by removing the ocular and looking down the tube.

The use of the Bertrand lens in conjunction with the substage swing-out converging lens converts the microscope to a *conoscope* and permits the observation of *interference figures* (Chapter 8). *Conoscopic observation*, in which the image of the source of light is observed, is thus distinguished from the more usual *orthoscopic observation*, in which the image of the object is observed.

Adjustment of Illumination

It is helpful to have the illumination well adjusted to achieve the best balance between contrast and resolution.

The condenser should be centered so that when an intermediate objective is used with the Bertrand lens, the pinhole of light seen when the substage iris diaphragm is fully stopped down is centered on the crosshairs. If this is not the case, the condenser should be centered with its adjusting screws, or, if these are not available on the instrument, the adjustment should be made by the manufacturer.

In some student microscopes the substage condenser can be moved up and down, for example by means of a knurled knob driving a rack and pinion. This condenser should ordinarily be at the upper limit of travel, at which position the uppermost swing-in converging lens should not disturb the slide on the microscope stage when it is moved in or out of position. In many other microscopes, the position of the condenser is fixed at the factory and its position is presumed to be well adjusted for good illumination. The most practical adjustment that can then be made is to reduce the stray illumination from the field of the objective. Using a medium power objective, observe the back of the objective with the Bertrand lens (or by removing the ocular). A sharp image of the substage diaphragm (often called the aperture diaphragm) should be visible when the iris is stopped down, and the iris should then be opened to the periphery of the objective field, so that the back lens of the objective is just filled with light. This procedure will minimize glare and produce good contrast and resolution when the Bertrand lens is then removed and the preparation viewed in the normal way.

Köhler Illumination. If photography is to be attempted with the microscope it is essential to have the field evenly illuminated. This occurs with Köhler illumination, if it can be achieved with the microscope to be used. The Köhler adjustment requires a diaphragm on the light source (often called the field diaphragm) and a substage condenser that can be moved up and down to focus the image of the field diaphragm. Any ground glass filter below the stage should be removed in order to make the adjustment. The objective lens should be focused on the specimen. The Bertrand lens is inserted to view the back focal plane of the objective. The substage condenser is focused to provide a sharp image of the partly closed field diaphragm. Now with the substage converging lens in place, the field diaphragm is opened to a position just beyond the margin of the field. The substage iris diaphragm (aperture diaphragm) is then stopped down to leave the rear lens of the objective appearing to be filled about three-quarters full of light. Normal illumination can then be restored for observation under optimal conditions for photography. Specific directions for achieving Köhler illumination should be obtained from the manufacturer when possible.

Accessory Plates

To cause retardation of the light after it leaves the objective, accessory plates are inserted at will into the microscope tube via a slot. They are discussed in Chapter 6.

STANDARD LIQUIDS FOR REFRACTIVE INDEX DETERMINATION

A standard set of refractive index liquids (generally oils) to be used as immersion media is essential for the routine examination of crystalline material. The set should include liquids throughout the range n = 1.45–1.78. Liquids or melts of higher refractive index are also quite useful. Liquids of lower refractive index are less frequently employed. The standardized refractive index of a liquid is printed on the label. The dispersion characteristics of commercial liquids are also commonly printed on the label, usually as the function $V = (n_D - 1)/(n_F - n_C)$. Temperature coefficients ($-dn/dt$) are also commonly given.

Properties. Some knowledge of the materials used in refractive index liquids and of the optical characteristics of standard liquids is prerequisite to their intelligent use and proper care.

Several requirements should be met by a standard set of liquids to be used for routine refractive index determination with white light. The standard liquids should not be deeply colored, poisonous, harmful to the skin, or solvents for the materials to be examined. Their temperature coefficients should be as low as possible. The refractive indices should stay reasonably constant during use and storage, which demands either that each liquid be a chemically stable compound, or that a series of liquids be made by mixing chemically stable miscible compounds of similar vapor pressures. It is most convenient, in estimating intermediate values of refractive index, to have a set of liquids in which there is an equal interval between each pair of liquids in the set. This requirement can only be met by using mixtures. The standard liquids should have slight to moderate dispersion continuously related in the successive liquids of the series. This requirement (the reason for which becomes apparent when the methods used in index comparison are considered fully) is also met only by mixing liquids in various proportions. A series of pure liquids, with equal intervals between liquids in the series, each with dispersion progressively different from the liquid adjacent in the series, is not known. Most standard sets of liquids consist of several series of liquids. Each series is composed of two end members, and intermediate members that are mixtures of the end members in the required proportions. There is then a regular variation of dispersion throughout the series and a convenient standard interval between liquids, both obtained without sacrifice of stability. The desired range should be covered using as few end members as possible so that the maximum number of liquids will share the same progressive change in properties. Melts may be used for refractive index standards in the region above 1.91. These are less easy to use and less satisfactory than liquids.

Dispersion is very strong and they are strongly colored. For Merwin's dispersion method, liquids with relatively high dispersion may be desirable.

In caring for liquids, tight caps on the bottles are of paramount importance. Differential evaporation of the components of the mixture will change the refractive index. Liquids in the range below 1.74 remain quite stable. After being in use for a year, changes of the order of 0.0001 to 0.003 are reported. Liquids at 1.70 and above 1.74 are less stable.

Most refractive index liquids are organic, but water-based liquids that may not dissolve organic materials are sometimes necessary. Materials that are useful for making refractive index liquids are listed by Hartshorne and Stuart (1970).

Direct Measurement of Refractive Index of Liquids

Instruments with which to standardize newly prepared liquids or to measure the value of old ones are required from time to time. A simple device for refractive index measurement is the Abbe refractometer. This instrument measures the angle at which light emerges from a prism. The light from a liquid of an unknown index enters the prism at grazing incidence and is refracted at the critical angle (as in B′B of Fig. 1-13). For this to be possible the refractive index of the prism must be greater than that of the liquid. If white light is to be used as the source, the instrument is commonly provided with a Bertrand–Amici prism, which compensates to the sodium line. The settings of this prism can then be used to determine the dispersion of the liquid. The accuracy of the conventional Abbe refractometer is about ±0.0002, according to Tilton (1942), but is probably better in more modern instruments. Newer instruments may have prisms with refractive index as high as 1.81 that are not seriously attacked by corrosive high-index liquids, but older instruments may be more vulnerable. Precise directions for the use of the Abbe refractometer are given by Bloss (1981).

A high-accuracy Abbe refractometer without a Bertrand–Amici prism, to be used with a monochromatic light source, is capable of determining refractive index to better than ±0.0001. Equations for the use of this instrument are given by Su et al. (1987).

The Jelley refractometer, manufactured by Leitz, is a simple instrument allowing rapid determination of the refractive index of liquids to an accuracy of ±.001. The principle of the Jelley design is essentially that of perhaps the oldest manifestation of refraction known to man, the stick-in-the-water effect. Just as a stick appears to bend when thrust into water, so light from a slit, when passing through a prism, appears to come from some other direction (as would be the case in Fig. 1-14).

It is said that the eminent Cavendish Professor Lord Rutherford, when asked what first attracted him to physics, remarked that when young he thrust a stick into the water and observed that it appeared to bend, "which gave one furiously to think."

In the Jelley refractometer, a glass prism is made into a compound prism by inserting a drop of the liquid to be measured, and the apparent source of light is located by sliding a hairline along a scale until it coincides with the image of the slit. Thus the angle of refraction is indirectly measured. The linear scale is ruled off in values of refractive index, which may then be read directly. The Jelley refractometer, though less accurate than the Abbe type, may be used with liquids up to $n = 2.35$, and is less expensive.

The most precise instrument for measuring refractive index consists of a hollow prism with plane parallel and optically flat sides, mounted on a spectrometer. The spectrometer is used to measure the *angle of minimum deviation* of light passing through the hollow prism when filled with the liquid.

The angle of minimum deviation is found by rotating the prism filled with liquid on an axis parallel to its solid angle and simultaneously swinging the telescope. The refractive index, a function of the angle of minimum deviation, is calculated from data thus obtained. Light passing through the prism filled with liquid is refracted in an amount that depends upon the refractive index of the liquid, the orientation of the prism, and the angle between the prism faces. The angle of minimum deviation is the smallest angle that one can obtain between the emergent beam and the incident beam.

The liquid whose refractive index is to be measured is placed in the hollow prism, and the height of the prism is adjusted so that the light beam from the collimator tube of the spectrometer passes through the liquid. Monochromatic light is allowed to enter the liquid at any angle as illustrated in Fig. 2-1(a). The prism is rotated, and during rotation the emergent beam is followed by the observer by swinging the telescope. As rotation progresses an angle of rotation is reached beyond which the emergent beam starts to move in the other direction, as illustrated in Fig. 2-1(c). The angle of rotation at which the direction changes [Fig. 2-1(b)] is recorded. The liquid is removed from the prism and the telescope is swung to receive the beam that passes directly through the empty prism. This angle is also recorded. The difference in the two readings is the angle of minimum deviation. At the position of minimum deviation the prism is so oriented that the light makes equal angles with the prism walls, angles *a* and *b* in Fig. 2-1(b). At this position of rotation, the position of minimum deviation, the refraction in the prism walls is equal and can be ignored.

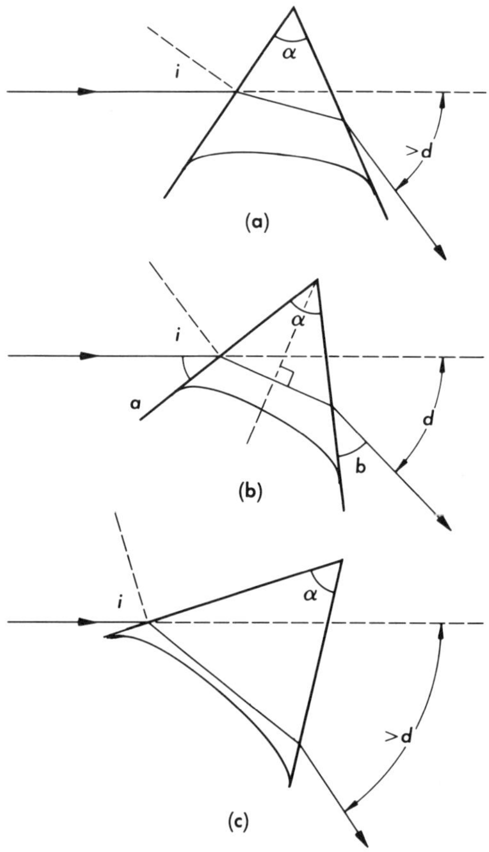

Fig. 2-1. Path of light through a prism at several positions of rotation. Angle α is the prism angle; *d* is the minimum deviation angle, at which angles *a* and *b* are equal.

It is a geometrical consequence of Snell's law that

$$n = \frac{\sin 1/2\,(\alpha + d)}{\sin 1/2\alpha} \tag{2-2}$$

or

$$\tan\left(\frac{\alpha}{2}\right) = \frac{\sin (d/2)}{n - \cos (d/2)}\,, \tag{2-3}$$

where n is the refractive index of the liquid, α is the prism angle, and d is the angle of minimum deviation. Geometric proofs of these equations are given by Ford (1932). The prism angle α can either be measured directly or computed by using distilled water in the hollow prism. The accuracy of the calculated value for an unknown liquid depends upon the precision of the spectrometer and knowledge of α and the temperature. It is not difficult to achieve an accuracy of ± 0.0002, and if a spectrometer precise to ± 10 seconds of arc is used, an accuracy of ± 0.00005 can be attained.

SLIDES AND COVER GLASSES

Crystals studied with the polarizing microscope are usually mounted on a glass slide. Standard slides are 45 × 26 × 1.2 mm. Longer slides frequently used for biological preparations are less convenient when used with a rotating stage. Cover glasses of various dimensions are employed. The thickness is not critical with objectives of low magnification. With high magnification it is desirable to use size 1 (0.15 mm thick) rather than the thicker sizes 2 or 3. Size 1 most nearly corresponds to the cover glass thickness of 0.16–0.18 mm for which objectives are usually corrected. The areal dimensions of cover glasses vary from nearly the size of the slide, in the case of thin sections, down to about 8 mm square for crushed grains. The latter may require a special order, but are convenient because with skill, they allow six immersion mounts to be made on one slide.

PREPARING MATERIALS FOR EXAMINATION

The method of preparing materials for examination and the particular mounting medium used is governed by the purpose of the investigation, the material available, and the fact that grains less than approximately 0.02 mm or greater than 0.5 mm thick are difficult to study. The optimum range of thickness is 0.03 mm to 0.1 mm. Some optical properties may be determined for crystals thinner than 0.02 mm, but difficulties are encountered in examination. If crystals thicker than 0.5 mm are to be studied, a microscope slide with a shallow depression may be used.

Methods of rock sample preparation are succinctly described by Mange and Maurer (1991).

CRUSHED GRAINS

Preparation

If it is necessary to reduce the crystals or aggregates to convenient size, it will be found useful to crush the material until it passes an 80- or 100-mesh (180- or 144-micron) screen and to save for examination that which lies on a screen of 150 to 200 mesh (96 to 72 microns). Crushing promotes breaking along cleavages. Grinding the samples is to be avoided because of the possibility of inducing twinning in the crystals. Close sizing will increase the percentage of fragments that have approximately the same thickness. This is advantageous if one is to estimate the birefringence from interference colors, because if the thickness is controlled, the birefringence and interference color vary together. In a homogeneous sample a few grains, perhaps a dozen or so, may suffice. Small screens made of bolting cloth are useful and easily cleaned to avoid cross-contamination of samples.

All the minerals in the rock can often be rapidly identified by experienced workers using crushed grains. No special preparation is needed; a chip of rock may be crushed, and the grains sized as described below and immersed in several diagnostic liquids that quickly reveal the commonest minerals. This procedure is quicker and at times more definitive than making and using a thin section, which would, however, be needed for textural examination.

As a shortcut for rapid work, screening may be avoided. Instead, a small amount of the crushed material can be spread on a sheet of paper and elutriated by tapping with a finger. The coarser grains will roll off faster, leaving finer sizes behind. When the right size is reached, the desired grains are caught on a small piece of paper, and the finest grains reserved or thrown out with the paper, if the sample is plentiful.

If a truly clean and well-sized sample is desired, the grains can be washed in a beaker of clean water. After agitation, the fines are poured off, leaving the desired grains behind. When these have settled, the remaining water can be decanted, and the sample can be rinsed with an appropriate solvent such as ethyl alcohol and dried quickly on a filter paper under a lamp.

Mounting

The medium used for mounting is chosen primarily because of its refractive index, although ease of manipulation, lack of solvent action on the material being studied, and value as an impregnating substance may affect the choice. Liquids and various resins are employed as mounting media.

Temporary mounts in liquids or melts of known refractive index are always used when refractive indices are to be determined. Temporary immersion mounts have the advantage of being quick and easy to prepare. They are conveniently prepared if the immersion medium is a liquid, by placing a few fragments on a clean slide, and covering with a small cover glass. A drop of the liquid of desired refractive index from a standard set of liquids is placed on the slide as near as possible to the edge of the cover glass. Capillary action will draw the liquid under the cover glass. If the entire area beneath the cover glass is not covered with liquid, more may be added. Excess liquid and excess solids are both to be avoided. If several mounts are made on the same slide or different slides using liquids of different refractive index, it is well to preserve their identity by placing cover glasses at different angles to the edge of the slide and making a sketch map of them on the worksheet. Slides should be preserved until an examination is complete. For several mounts the same grains may be used in a mount with a liquid of different refractive index if they are washed in xylene or acetone, but if enough material is available it is simpler to use other fragments in the other mounts. If permanent mounts are also desired, more of the same sample may be used or the identical fragments may be used after being washed in xylene and allowed to dry. Slides that have served their purpose may be cleaned in ethyl alcohol or in xylene.

Slides covered with a gelatin film are effective as a modification of the usual temporary immersion mount (Fairbairn, 1943). Grains are embedded in the gelatin coating, which has been softened with a drop of water. The gelatin is set by drying before placing the cover glass on the slide and introducing the liquid. The immersion liquid can be removed by washing with carbon tetrachloride, and a liquid of different refractive index can then be applied without disturbing the grains. The gelatin also lessens the tendency of grains to lie in preferred orientations as, for example, on cleavages or crystal faces.

Resins of various kinds may also be used as viscous mounting materials. Grains may be sifted onto the partially softened resin. Commercial resins such as the Cargille Meltmount Series are available in various refractive indices for the convenient examination and identification of mineral grains.

DETRITAL GRAINS

Detrital mineral grains of high density ("heavy minerals") in sedimentary rocks are the subject of study by the sedimentary petrologist. Minerals of sedimentary rocks are separated from the rock mass by methods discussed by Milner (1962), Krumbein and Pettijohn (1938), and Mange and Maurer (1991). The shape, size, and abundance of the several varieties of minerals in a sedimentary horizon are recorded from a representative portion of the heavy minerals removed from the sample. The grains are mounted without crushing so that the size, shape, and relative number of each kind in the slide will be representative of the sample.

A permanent mounting medium is usually chosen that will separate the different crystals strategically into two groups. In one the refractive index is greater than the mounting medium and in the other it is less.

Closer determination of refractive index is not usually necessary, but if it is, refractive index liquids may be used. Examples of different minerals mounted in resin are shown in color plates by Mange and Maurer (1991).

Heavy minerals may also be recovered from igneous and metamorphic rocks after crushing. Larger crystals may be broken during crushing, but smaller ones often survive to show characteristic external forms or growth zones. Among those that often survive intact are zircon, garnet, and rutile. Igneous and metamorphic heavy minerals are generally studied in immersion mounts or under alcohol, rather than in resin mounts.

CRYSTALLINE PRECIPITATES AND MILL PRODUCTS

The polarizing microscope may be used to identify crystalline precipitates formed by microchemical reactions, following the procedures of Short (1940) and Staples (1936). In these and other instances in which the products of chemical reaction are to be identified, the mother liquor is a convenient mounting medium, at least for preliminary examination. Refractive index liquids can also be used if they do not dissolve the substance to be examined.

To the extent that the crystals are small enough for convenient study, crushing, which will alter the size, shape, and crystal form, may be avoided so that these features can be observed, recorded, and used as an aid in identification. Products resulting during mineral beneficiation processes have been studied by Faust and Gabriel (1940) without further crushing in permanent mounts. The minerals present may be identified, and frequently the percentage of each of the minerals present and the degree to which the beneficiation processes have separated the constituents is estimated quantitatively. A permanent mount immobilizes the grains to permit quantitative estimations and provides a record of the investigation.

CRYSTAL AGGREGATES OR ROCKS

Thin Sections

Aggregates of crystals are commonly studied in thin section. A standard thin section is a slice of the material 0.03 mm thick, usually about 10–20 mm by 20–30 mm in horizontal dimensions, which has been mounted on a glass slide with material of known refractive index near 1.54.

Henry Clifton Sorby (1826–1908) of Sheffield, England, was the first to make thin sections of rocks (in 1849), and his descriptions of what he saw with the microscope were at first greeted with ridicule, skepticism, and neglect. Nevertheless, he professed to take no notice of such matters and was soon vindicated by enthusiastic followers in Germany, particularly by Zirkel. He had a long and illustrious career as microscopist, mineralogist, and geologist, and became president in succession of all three societies of these disciplines. Late in life he was honored as the "Father of Microscopical Petrography" (see Judd, 1908). His 1858 remark, "I argue that there is no necessary connexion between the size of an object and the value of a fact" (see Rosenfeld and Chase, 1961) suggests that he was not always completely serene about his doubters.

Thin sections have the advantage of being ground to standard thickness. This aids in the determination of absorption characteristics and the interpretation of the interference colors of anisotropic crystals because both interference and absorption are functions of thickness of the section. Thin sections of an aggregate are useful since they provide information, which may aid in identification, as to the relationship of individual crystals. Thin sections are particularly useful to the geologist for whom identification of the crystalline constituents may be only one of the purposes of the examination. The geologist is also interested in the abundance, size, and shape of the minerals, their relationships to other minerals, and the orientation of the optical directions in the mineral grains.

Preparation of Thin Sections. Making a thin section is a technical process demanding time and skill; a description of the method is given by Humphries (1992). Good instruction and practice are essential, but some people are more gifted than others at this art. Some of the most successful make thin sections commercially, and their services can be found through the advertising pages of professional journals.

The procedure involves slicing the rock with a diamond saw, trimming the piece to suit the dimensions of the slide, grinding one side flat, cementing it to the glass slide, slicing off all but a millimeter or so of the cemented piece, and grinding the cemented piece to the standard thickness of 30 microns. The final stages of grinding are done by hand on a glass plate, to achieve uniformity of thickness. This is checked by the maximum interference colors of known minerals under crossed polars. For example, the maximum interference color among grains of quartz should be a faint straw yellow.

During slicing and rough grinding, the glass slide is commonly held in a vacuum chuck. All sawing and grinding are done wet. Some machines process a number of samples automatically, in sequence. Thin sections to be used for reflected light or for other purposes such as electron microprobe analysis are polished in successively finer grits and left uncovered. For examination of delicate features without the disturbance of light due to grinding and surface roughness, double polishing is used. This more difficult procedure introduces a polishing step before the rock slice is finally cemented to the glass slide.

Opaque Material. Any thin section may be studied in reflected light by conventional methods for opaque crystals if the upper surface of the thin section is polished and left uncovered. Methods for polishing and examining opaque minerals are described by Craig and Vaughan (1981) and Peckett (1992).

VIDEO STATION

The provision of a video work station greatly accelerates the learning process in crystal identification. A video camera on a petrographic microscope can be mounted with a monitor on a split-level trolley, and wheeled to any vantage point in the laboratory. There a seated group can watch the procedure and learn what to expect to see in the field of the microscope. This mode of demonstration is particularly effective with color fringes and interference figures. It can also be used to teach petrography and the textures of thin sections.

3 The Immersion Method

The immersion method permits measurement of refractive index by comparison of the unknown crystal with liquids of known refractive index. A determination of refractive index usually suffices for the identification of isotropic crystals. Refractive index determinations are also fundamental in identification of anisotropic crystals, and the techniques of this chapter are applicable to these as well. Anisotropic crystals, however, have either two or three principal refractive indices, and determination of these requires alignment of principal vibration directions with the polarizer, as well as certain other techniques described in succeeding chapters. Anisotropic crystals also possess other optical and crystallographic characteristics such as optic angle and extinction angle, which are usually needed for identification, and may be determined at the same time as refractive index.

RELIEF

The relative index of refraction of one substance adjacent to another affects the refraction and reflection of light at the boundary. The contrast between a colorless solid grain and the surrounding medium is termed *relief*, by analogy to topographic relief.

The appearance of relief in immersion mounts is due to the presence of a dark line at the boundary of the mineral when it is in focus. This is caused by total reflection of some of the light at the border of the crystal. The relief is a rough measure of the difference between the index of the crystal and the surrounding medium because the greater the difference, the more often the critical angle is exceeded, the more rays are totally reflected, and the more marked the dark border. Relief is greater when the difference between the index of the crystal and that of the mounting medium or adjacent crystal is greater.

Relief is described as high, moderate, or low, depending upon whether the grain in the mount is conspicuous, moderately prominent,

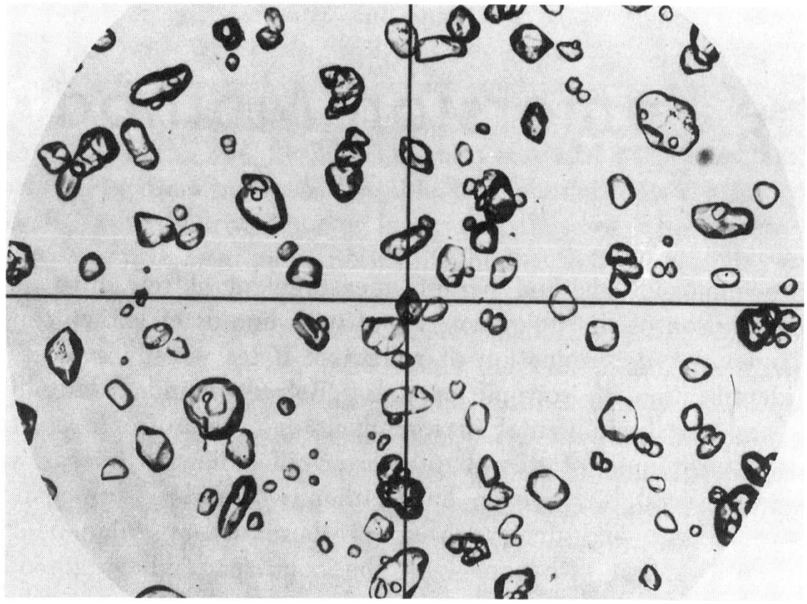

Fig. 3-1. High relief. Quartz grains (*n* around 1.54) in a liquid of $n = 1.40$; 10× objective. Compare with Fig. 3-2, which shows low relief.

or barely noticeable. High relief is shown in Fig. 3-1. Low relief is shown in Figs. 3-2 and 3-3.

A difference in refractive index of about 0.04 or less between grain and mounting medium is characteristic of grains with low relief. A difference of about 0.04 to 0.12 is moderate, and above 0.12 is high.

Relief is also described as being *positive* or *negative*. If a grain has a higher index than the surrounding material it is said to show positive relief. If it has a lower index it is said to have negative relief. Positive relief is distinguished from negative relief by the oblique illumination or Becke line tests described below. The positive or negative nature of relief is an important characteristic of crystals in an immersion mount, for it guides the observer to the next correct step in the experiment. It is also important in the identification of minerals in thin sections, and it forms the basis for many determinative tables, including Appendix A in this book.

STRATEGY USED IN THE IMMERSION METHOD

The goal of the immersion method is to find a liquid (or "oil") of known refractive index that matches the refractive index of the unknown crystal. Most of this chapter is devoted to the technique of discovering whether a given liquid has a refractive index greater than, smaller than, or equal to that of the unknown crystal. Given this competence, one makes a series of immersion mounts, each of a different group of crystal fragments with a different liquid. Ideally, each mount contains a liquid closer to the crystal in refractive index than the last, until a match is found. The liquids are chosen strategically so that the smallest possible number of mounts will lead to a match, or will result in bracketing the crystal between two adjacent liquids. To this end, it is efficient to approach the refractive index of the crystal from both above and below rather than from one direction alone. The first liquid chosen should lie in the range of most frequent indices, between 1.5 and 1.6. If the crystal is found to be considerably higher than the liquid, the next liquid should be about halfway toward the high end of the standard series. Each succeeding choice should similarly reduce the range by one-half, thereby requiring preparation of a minimal number of mounts.

By noticing a change in relief from mount to mount, it is often possible to judge whether the next mount should be made with a near or distant liquid. Low relief dictates the use of a nearby liquid, perhaps much less than half the range away. If the relief is positive, the next mount needs to be made with a liquid of higher refractive index, and conversely for negative relief.

The range of refractive indices to be considered in the preceding strategy may be narrowed by some prior knowledge of the unknown. If the immersion method is being used simply to confirm a prior tentative identification, an appropriate liquid can be chosen and a single mount may suffice to produce a match.

If a crystal falls between two liquids in the standard series, and if further refinement of refractive index is desired, an intermediate liquid can be made by merging an equal number of drops of each of the two adjacent liquids under the cover glass. A more precise refinement can often be made by varying the wavelength of light until a match is found, and calculating the refractive index of the liquid (and matching crystal) by the dispersion method.

Liquids used in routine identification are commonly standardized at the D line at 25° C. If the working temperature at the microscope is known to differ considerably from the standard value, an appropriate temperature correction should be made to the labeled value of the matching liquid. Daylight or filter-corrected tungsten light is normally used, but sodium (or filtered monochromatic) light may be used to check the final match.

COMPARISON OF LIQUIDS AND CRYSTALS IN IMMERSION MOUNTS

Either of two tests may be applied to determine the relative refractive index of crystal and liquid in the immersion mount. These are the *Becke line or central illumination test* and the *oblique illumination test*. Both depend on refraction of light at the boundary between substances of differing refractive index, or on reflection of light at the boundary when the critical angle is exceeded. Both may be used under some circumstances to judge qualitatively *how far* the crystal and liquid are from matching as well as in *which direction* the ultimate match will lie. Both may be used with either monochromatic or white light, but the oblique illumination test is more sensitive near matching conditions in white light. Both are often used concurrently, since not all grains give rapid, unequivocal answers with one test alone.

In making either test, one is asking whether the crystal has a refractive index higher or lower than the surrounding liquid. The question is framed in terms of the crystal, the object of study, rather than the liquid being used as a reference material. To minimize confusion it is advantageous to use a single formulation of the question exclusively.

BECKE LINE OR CENTRAL ILLUMINATION TEST

Practice

A crystal fragment is observed with an objective of low or medium power, preferably around 10×. The polars are not crossed. The substage diaphragm is partly closed to remove the more oblique rays from the cone of light. The focus of the microscope is *raised* (stage *lowered)* to bring it just above the crystal being observed. As this is done, a bright line parallel to the crystal boundary will appear to move either into the crystal or out into the liquid. The bright line is the *Becke line.*

If the crystal index is lower than that of the liquid, the Becke line moves out into the liquid. Such a case is illustrated in Fig. 3-2, where faint bright lines lie just outside the borders of all the crystals in the field of view. The effect of motion seen while raising the focus is more pronounced to the eye than that of mere position as recorded by the photograph.

If the crystal index is higher than that of the liquid, the Becke line moves in toward the center of the crystal. Such a case is illustrated in Fig. 3-3, where faint bright lines lie just inside the darker borders of all the crystals in the field of view. Again, the effect of motion seen

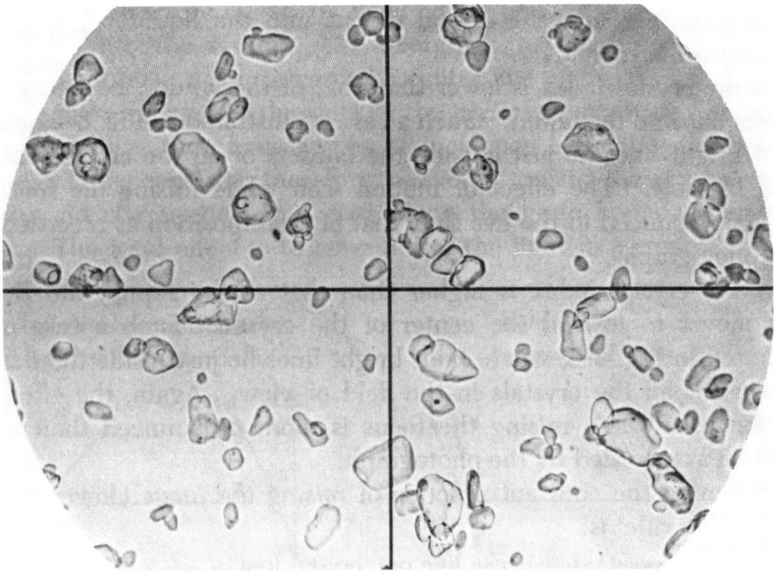

Fig. 3-2. Becke line out, crystal low. Quartz grains (n around 1.54) in liquid of $n = 1.56$; 10× objective; central illumination.

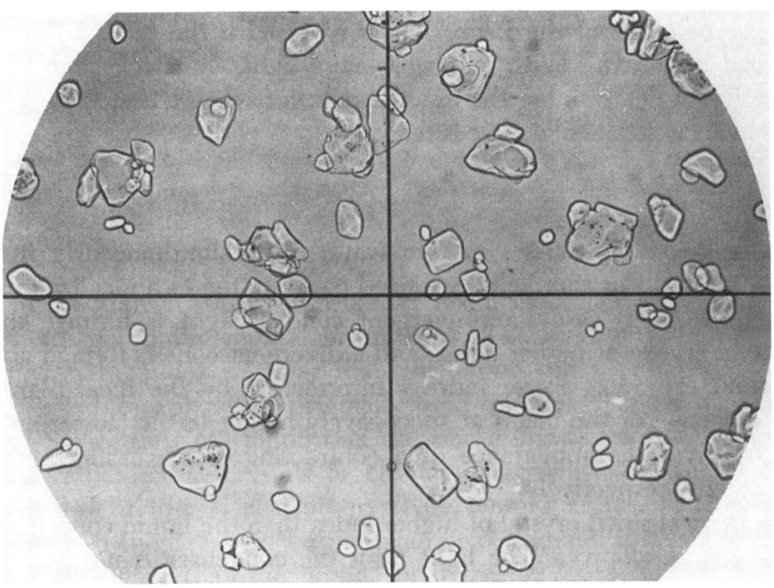

Fig. 3-3. Becke line in, crystal high. Quartz grains (n around 1.54) in a liquid of $n = 1.52$; 10× objective; central illumination.

while raising the focus is more pronounced than mere position as recorded by the photograph.

Assuming the constant practice of *raising the focus* (lowering the stage), the rule is:

Becke line out, crystal low
Becke line in, crystal high,

where "low" and "high" refer to refractive index relative to the surrounding medium.[1]

The behavior of the Becke line is reversed if the focus is lowered instead of raised. With monochromatic light, the Becke line test is effective in determining the exact match, for when a match in refractive index is attained, no line is seen.

Principle

The Becke line may arise in two ways, often simultaneously in the same grain. For simplicity, we need to consider the two idealized cases alone. In both cases, a convergent cone of light is formed above a round crystal of higher index, and a divergent cone is formed above a round crystal of lower index. Intersection by the focal plane of the objective of the cones at successively higher levels (lowering the stage) gives the appearance of a contracting or expanding annular Becke line, respectively.

A vertical-sided crystal of higher index than the liquid concentrates the light as follows. The light from the condenser is not perfectly parallel, as suggested in Fig. 3-4. Those rays inclined into the sides of the crystal are refracted by Snell's law toward the normal to the interface, into the crystal. On emerging, they are refracted away from this normal, again toward the center line of the crystal. Those rays slightly inclined to the sides of the crystal from within are totally reflected, since they have angles of incidence exceeding the critical angle. These rays too, therefore, converge toward the center of the grain. Strictly vertical rays pass along the grain boundary undeviated. The total effect is to concentrate the light in a conical fashion above the grain. Similar effects may result even if the grain boundaries are not quite vertical or smooth. The opposite effect results if the crystal index is lower than the liquid index (Fig. 3-5).

This is the explanation originally used by Becke (1893) for the bright

[1] For those who read German, the time-honored rule is the "3H Rule": Beim *H*eben des Fokus wandert die *h*elle Linie ins *h*öher brechende Medium.

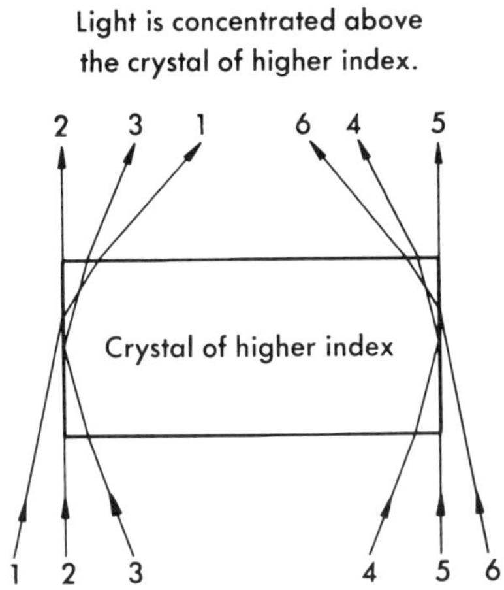

Fig. 3-4. Becke line in, crystal high. Wave normals 2 and 5 are undeviated; 1 and 6 are refracted, and 3 and 4 are reflected.

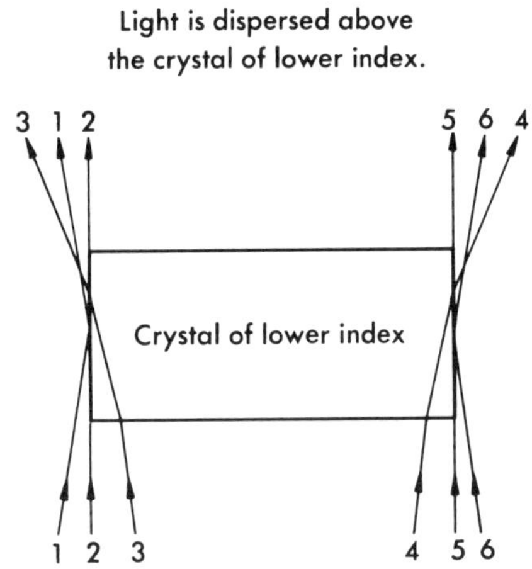

Fig. 3-5. Becke line out, crystal low. Wave normals 2 and 5 are undeviated, 1 and 6 are reflected, and 3 and 4 are refracted.

line. Saylor (1935) found that the Becke line test is most sensitive when vertical-sided crystals or cleavage fragments are used.

> Friedrich Becke in 1893 first called attention to the effects of total reflection at the boundary between two crystals in thin section, and showed that they could be used to determine relative refractive index. He applied the method ingeniously to the determination of the plagioclase feldspars by reference to quartz in thin section. The Becke principle was later carried over to the immersion method, which allows a much simpler determination of refractive index by reference to standard oils. Becke also contributed greatly to the interpretation of interference figures (see Chapter 8).

The bright line is also observed on lensoid grains, as suggested in Figs. 3-6 and 3-7. Such grains can be considered small converging lenses (if higher than the liquid) or diverging lenses (if lower than the liquid). Once again, the focus when raised will intersect successively smaller segments of a conical concentration of light if the crystal is higher than the liquid in index, and successively larger segments of a divergent cone if the crystal is lower.

Becke Line with Colored Crystals

There is some difficulty in applying the Becke line test when crystals are deeply colored. A line moving into such a crystal will often not be

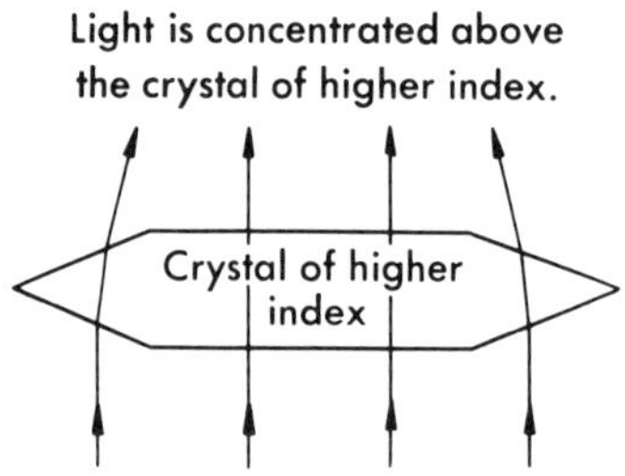

Fig. 3-6. Central illumination, crystal high.

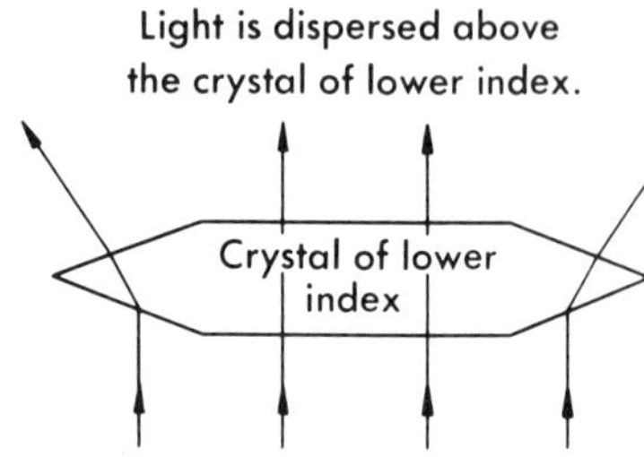

Fig. 3-7. Central illumination, crystal low.

clearly visible, so the line must be made to move out from the colored grain by lowering or raising the focus as required.

Becke Line Test If Refractive Indices Nearly Match

If crystal and liquid indices match (are equal) for any wavelength in the middle part of the visible range, there will be two Becke lines. One moves out as the other moves in. The color of the two lines depends on the wavelength for which the refractive indices of grain and liquid match. The line moving out is a combination of colors for which the refractive index of the crystal is lower than that of the liquid: blue, blue-green, or green. The color of the line moving in is a combination of the colors for which the refractive index of the crystal is higher than that of the liquid: brown, orange-red, or yellow. The relative intensity of the two colored lines varies. This also depends upon the wavelength for which the refractive index of liquid and crystal match. If the wavelength for which there is a match is the standard yellow, approximately in the middle of the visible spectrum, the intensity of both lines is approximately equal. A match for yellow, the standard wavelength, is therefore indicated when one line (going out) is blue-green and the line going in is orange-red and both are of approximately equal intensity.

Distinction between lines of similar color and estimation of the relative intensity are highly subjective, so that the decision that a match for yellow has been obtained is not free from error. When the liquid and crystal have refractive indices so nearly equal that two Becke lines are seen, the color fringe test described below is a more effective indicator of match. The colors of the fringe can be identified more certainly than those of the Becke lines.

OBLIQUE ILLUMINATION TESTS

Practice

The image of a crystal will be shaded on one side if rays leaving one side of the object are prevented from forming an image, or if the crystal is illuminated by oblique rays from only one side. To create the shadow, the light is cut off by a stop introduced into the optical system from one side. Which side of the crystal will be shaded depends upon the refractive index relative to the surrounding medium, the position of the stop in the optical system, and whether the object is below the focus of the condensing lens system used during the test. All the variables except one, the refractive index relative to the surrounding medium, depend upon the technique chosen and the optical system employed.

The image of the crystal and surrounding material may be shaded by introducing a stop in the path of the light above the object. The frame of an accessory plate or the frame of the analyzer may be used as a stop by inserting it into the tube until half the field is shaded. A low- (about 3 to 4×) or medium-power (10×) objective is employed. For consistent results the crystal being examined should be in the center of the field and the substage diaphragm should be partly closed.

As an alternative method of performing this test, one may illuminate the object with oblique rays from only one side. A stop, consisting of a card or other shading device, may be inserted below the converging lens or below the substage assembly, or the finger may be placed over part of the light source. The positions where the stop may be placed are either approximately at the back focal plane of the objective or at a conjugate focal plane below the stage. Which of these methods of preventing light from reaching one side of the observed image will be easier to use, and will at the same time give good results, depends upon the construction of the microscope. If the location of the analyzer or the accessory plate is appropriate for this test, the placing of a stop in the tube will probably be the most convenient method of performing the oblique illumination test.

Having chosen a technique appropriate to the optical system being used, one should examine crystals of higher and lower refractive index than the medium in which they are mounted and note on which side of the crystal the shadow appears. Rules formulated from these observations may then be applied to determine, for the technique used, the refractive index of the grain relative to surrounding material. General rules for the oblique illumination test are not trustworthy because the many variables in the technique and optical system offer too many chances of reversal of the effect. Nevertheless, the rules below will apply in a great many cases and will be useful after making sure they apply to the optical system being used.

With many microscopes in which the accessory plate proves to be a useful stop, a 10× objective is employed and the upper converging lens of the condenser is not employed. The object thus lies below the focus of the lens system of the condenser. The stop is inserted so as to shadow approximately half of the field of view, and one notes the direction from which the stop was introduced. If the crystal index is lower than that of the surrounding medium, the side of the crystal opposite the direction from which the stop was introduced will be shadowed, and the other side of the crystal will remain bright, as illustrated in Fig. 3-8. If the crystal index is higher than that of the

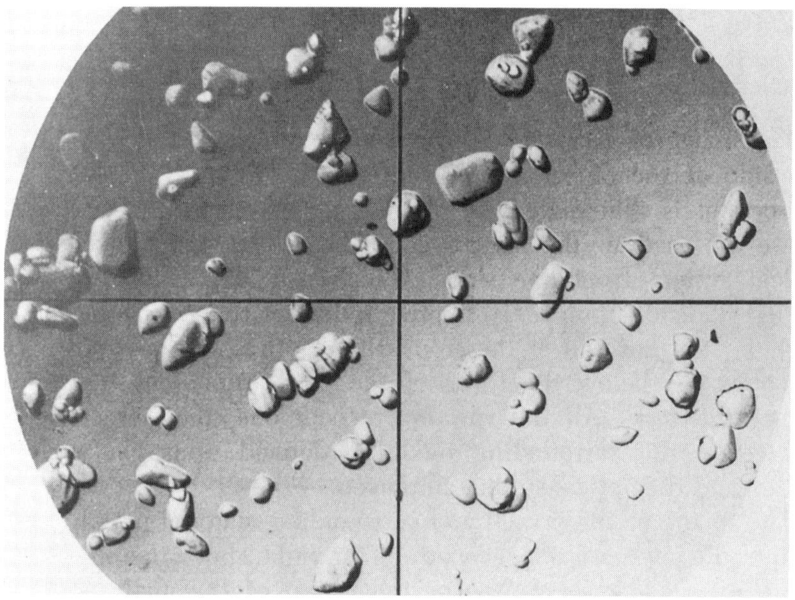

Fig. 3-8. Shadow opposite, crystal low. Field shadowed from NW by an accessory plate, as discussed in text. Quartz grains (n around 1.54) in liquid of n = 1.56. 10× objective.

surrounding medium, the side of the crystal toward the direction from which the stop was introduced will be shadowed, as illustrated in Fig. 3-9. Assuming these practices are followed, the summarized rules are:

Shadow opposite, crystal low
Shadow near, crystal high

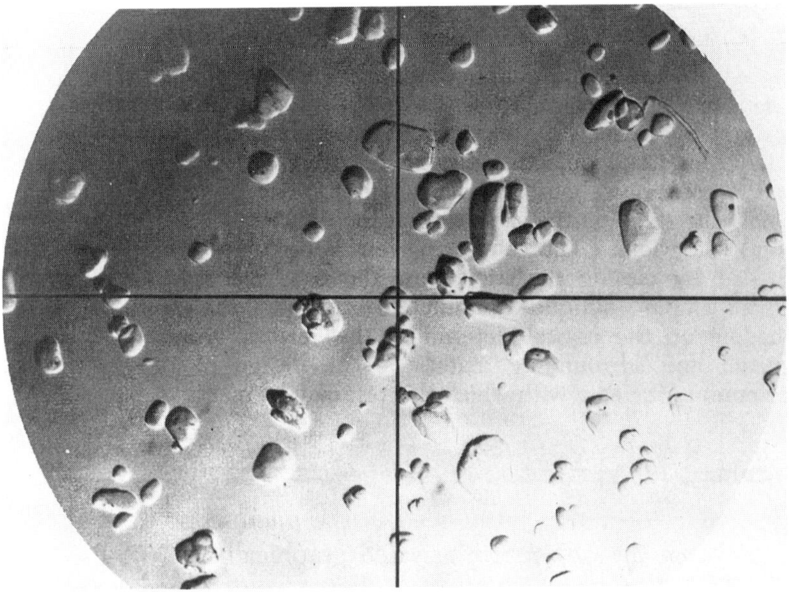

Fig. 3-9. Shadow near, crystal high. Field shadowed from NW by an accessory plate, as in Fig. 3-8. Quartz grains (*n* around 1.54) in liquid of $n = 1.52$. 10× objective.

where "shadow" refers to the shadowed crystal border, "opposite" and "near" refer to the direction from which the stop was introduced, and "low" and "high" refer to the refractive index relative to the surrounding medium.

In the formulation above, emphasis has been laid on the direction from which the stop was introduced, which in this case is also the direction from which the shadow appears to move across the field. The same rules may also then be valid for a stop introduced below the stage, even though the *field* shadow and its motion are reversed, as illustrated in Figs. 3-10 and 3-11.

Principle

The effect of a stop introduced in the microscope tube upon rays of light from the object is illustrated diagrammatically in Fig. 3-10. The partly shaded image that results is due to convergence or divergence of light from the object being studied as compared to the light from the surrounding material. As illustrated in Fig. 3-6, a lens-shaped grain (thinner on the edges than in the center) that is of higher refractive index than the surrounding material will converge the light from the

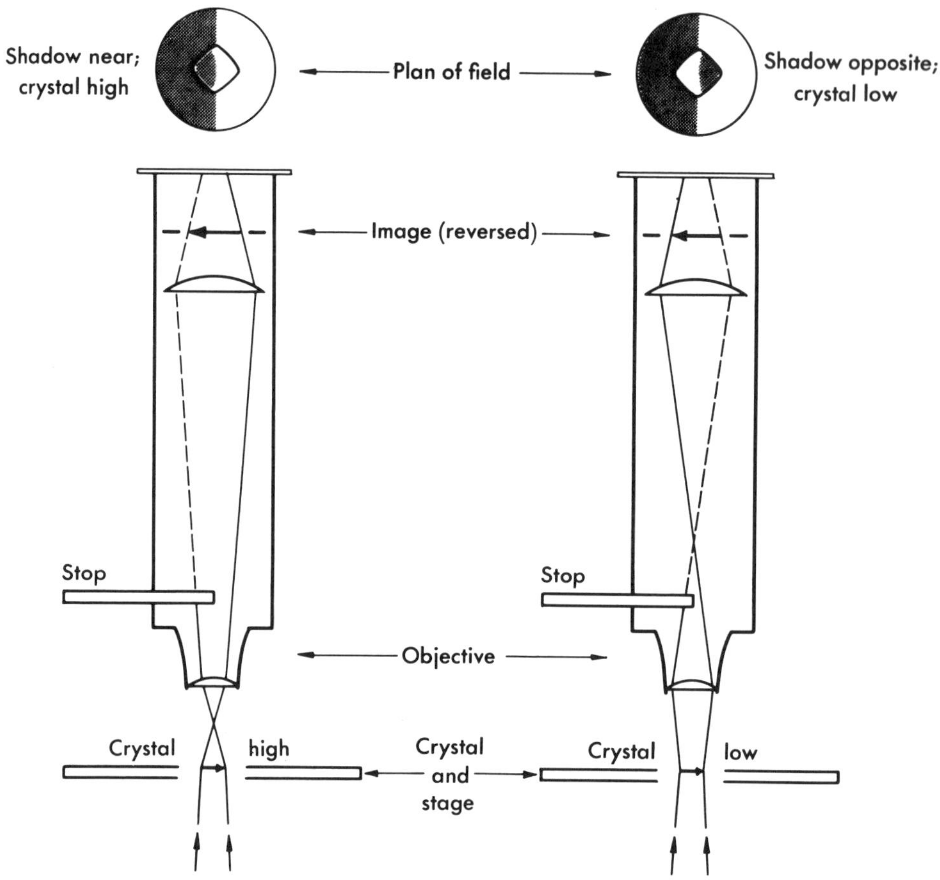

Fig. 3-10. Oblique illumination, stop above objective. See text for precise meanings of "near" and "opposite."

condensing lens. If the slot for accessory plates lies above the focus of the convergent light, as it does in many microscopes, a stop inserted here will prevent the passage of light from the side of the grain opposite the stop upward along the tube. In the image of the grain, which is reversed, the shadow is on the same side as that from which the stop was introduced. An object of lower refractive index than the surrounding material is shaded on the side far from the stop. Light diverges on leaving the object and comes to a focus at a point in the tube above the stop. Passage of light from the same side of the object as that from which the stop is introduced is prevented by the stop. The reversed image of the object is, therefore, shaded on the far side.

A schematic diagram of a stop inserted below the stage is shown in Fig. 3-11. A crystal of lens shape and higher refractive index than the

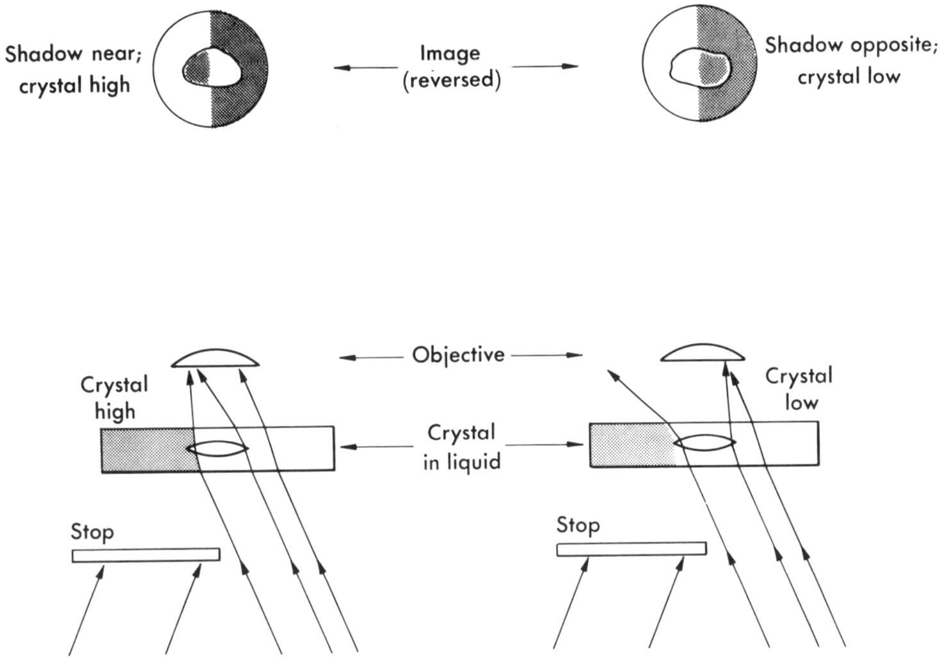

Fig. 3-11. Oblique illumination, stop below stage. See text for precise meanings of "near" and "opposite," which in this case are not as seen in the plan views.

surrounding material converges the light, much as in Fig. 3-6. With oblique illumination, half the oblique light is prevented by the stop from reaching the crystal. Light from the side of the grain opposite the stop is refracted so that it does not enter the objective; this side of the grain appears dark. Light from the side of the grain near the stop is refracted into the objective, and this side therefore appears illuminated. The illuminated side of the crystal, in the *reversed* image, is the side *away* from the actual stop position. By analogous arguments, the opposite effect is produced when the crystal has a lower refractive index than the surrounding material. Such a crystal diverges the light from below. Light from the side of the crystal near the stop is refracted out of the optical path, and this side appears dark. Light from the side of the crystal away from the stop is refracted into the objective, and this side appears illuminated. The dark side of the crystal, in the reversed image, is the side away from the actual stop position.

It will be noted that the same principles govern the behavior of both the Becke line and the shadows produced in oblique illumination. Oblique illumination effects can also be produced and explained with the vertical-sided crystals of Figs. 3-4 and 3-5. Furthermore, just as two Becke lines result when the crystal matches the liquid for some visible wavelength, so colored shadows (color fringes) result with oblique illumination. These are illustrated in the frontispiece.

COLOR FRINGES

If the index of the crystal is so near that of the liquid that, for some wavelength of visible light, the crystal index and the liquid index are equal, the shadows resulting from oblique illumination become colored and are called *color fringes* because they occur chiefly on the borders of the crystal. The fringes are differently colored on the sides of the crystal toward and away from the side from which the stop was introduced. The exact color of the blue, green, or violet fringe need not be observed because of the difficulty of distinguishing the colors in this range of the spectrum. Attention is directed instead to the fringes that are brown, orange-red, or yellow. If the fringe is brown the refractive index of the crystal (for the standard wavelength yellow) is lower than the refractive index of the surrounding medium. If the fringe is yellow the crystal index is higher, and if orange-red the refractive indices of the crystal and surrounding medium are approximately equal. There are also intermediate shades of color that may be distinguished, as for example a bright red, which indicates that the crystal is a very little bit low, and a yellow-orange, which indicates that the crystal is a bit high. In making the observations, it is important for consistency that the field be half shadowed and that the crystal lie at the crosshairs. Conveniently summarized, the important distinctions to be made with reference to the fringe that is not blue or green are:

Yellow, crystal high
Brown, crystal low
Orange-red, crystal matches the surrounding medium

One can also determine relative index from color fringes without observing the specific color. The intensity of the color fringe on the opposite side of the stop is either equal to, less than, or greater than that on the near side. If the color fringe on the side *opposite* the stop is:

More intense, crystal low
Less intense, crystal high
About equal in intensity to the fringe on near side, crystal matches

Color-blind people can make use of color fringes by means of these rules.

Principle

The appearance of color fringes under the circumstances described is essentially the Christiansen effect discussed by Wright (1911). Two generalities form the basis for the explanation of this effect in the oblique illumination test: the refractive indices of both liquid and crystal are greater for blue light than for red, and the dispersion of a liquid is greater than that of a crystal, as discussed in Chapter 1.

Several simple considerations explain the origin of the fringes. If the index of a crystal is higher than that of the surrounding liquid for all colors of the visible spectrum (not equal for any wavelength in the visible range), oblique illumination allows white light (not color fringes) to appear on one side of the crystal. But if the index of this crystal were higher only for some of the colors of white light, it is reasonable that light only of those colors, and not white light, would have appeared on one side of the crystal. This condition is illustrated in Fig. 3-12, a composite sketch analogous to the two sketches of Fig. 3-11 (compare also Figs. 3-6 and 3-7). In Fig. 3-12, white light is dispersed by the crystal into various components of different wavelength (color). The

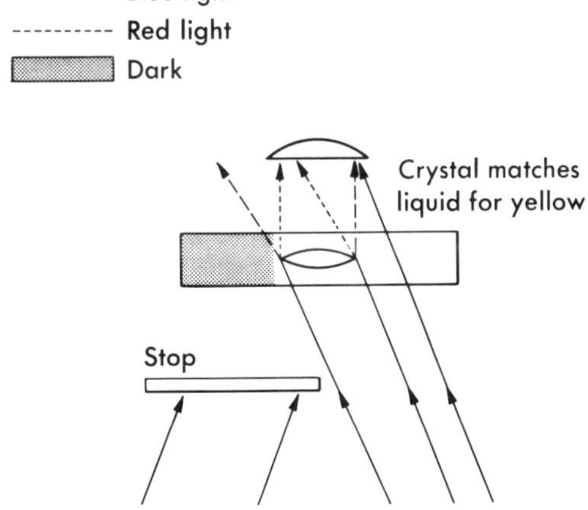

Fig. 3-12. Crystal obliquely illuminated by red and blue light. A match occurs for yellow light and color fringes appear because the crystal is low for blue and high for red.

crystal index matches that of the liquid for standard yellow light, and this ray is undeviated by the crystal. The crystal index is higher for red light and lower for blue light, and these wavelengths are accordingly converged and diverged respectively. Red light illuminates one border of the crystal, and blue the other (see frontispiece).

Three possible cases arise when crystal and liquid have an index equal for some wavelength in the visible range. Examples of each of these will be examined. The examples chosen are those in which the crystal and liquid index match for orange (crystal low), for green (crystal high), and for yellow (crystal on). These conditions are illustrated in Fig. 3-13, after M.J. Buerger (1933).

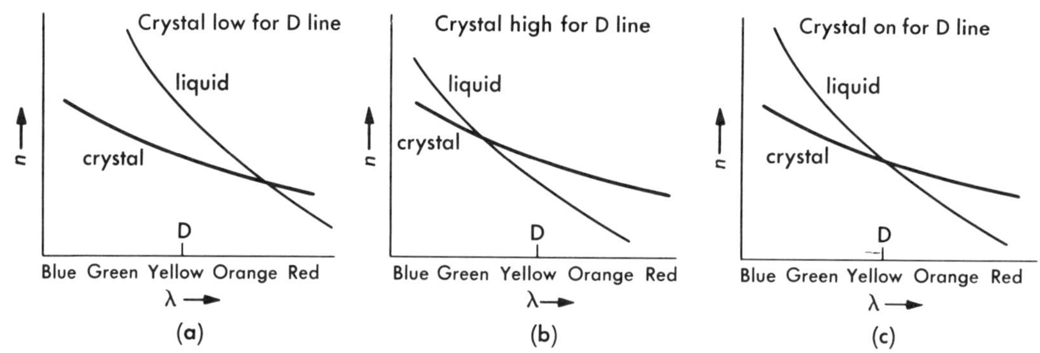

Fig. 3-13. Dispersion curves. (a) Color fringe brown; crystal high only for red light. (b) Color fringe yellow; crystal high for yellow, orange, and red. (c) Color fringe orange-red; crystal matches at D.

When crystal and liquid index match for orange, the dispersion curves for crystal and liquid cross as in Fig. 3-13(a). The crystal index is higher than the liquid index for red and lower for the standard wavelength yellow. When obliquely illuminated with white light, red light, for which the crystal is high, will appear on one edge of the crystal, but the other colors will not. The intensity of the red light will be a small percentage of the total intensity of light used and, since low-intensity red appears brown, the fringe will be brown.

When crystal and liquid index match for green, as in Fig. 3-13(b), the crystal index is higher for the standard wavelength yellow, and also for orange and red. These three colors together will therefore illuminate one edge of the crystal. The resulting color fringe will be yellow or orange-yellow, since yellow has the dominant intensity.

When crystal and liquid index match for the standard wavelength yellow, as in Fig. 3-13(c), the color fringe will be orange-red because the crystal has a higher index for orange and red, and these two colors will illuminate one edge of the crystal. The orange-red color fringe is thus the signal for a standard match between crystal and liquid. Further discussion of this condition appears at the end of the chapter.

Further Tactics

If liquids of refractive index successively closer to that of the grain are being used for a series of immersion mounts, it is useful to know how near the match has been approached when shadows first give way to color fringes. It is found that the span of refractive index values in which color fringes appear is greater in the high index range of refractive indices than in the low range. In the low range the span of refractive index values in which color fringes appear may be smaller than the interval between standard liquids, so that color fringes are not observed in any immersion made with liquids at the standard intervals. The range of liquids in which color fringes appear is dependent upon the generality that differences in dispersion between a liquid and an immersed solid are greater for materials of higher refractive index.

The exact interval in which shadows give way to color fringes depends upon the dispersion of liquid used for immersion and the dispersion of the crystal being examined. Figure 3-14 illustrates two cases: one, the dispersion of a crystal of low refractive index in relation to the dispersion of liquids of similar index; and the other, the dispersion of a crystal of high index and liquids of similar index.[2] In Fig. 3-14 the refractive index of zunyite is equal for some wavelength in the visible range to that of each liquid in the series over a range of index values n_D = 1.605 to 1.585. Color fringes will be seen in immersions made with any of these standard liquids. For sylvite, color fringes will be seen only in the immersion made with standard liquid of n_D = 1.490. As indicated for specific cases in Fig. 3-14, a larger difference between liquid and crystal index is indicated by a color fringe if a crystal of high index rather than one of low index is being examined. It follows that color fringes are more sensitive indicators of match, the lower the refractive index of the liquid employed.

THE ORANGE-RED COLOR FRINGE

Reference Standard

The color that signifies a match between crystal and oil under oblique illumination in white light is a particularly bright, fluorescent orange-

[2] Dispersion data for Fig. 3-14 may be found, for liquids, in Butler (1933); for zunyite, in Larsen and Berman (1934); for sylvite, in Hintze (1915). The liquids are mixtures of kerosene and monochloronaphthalene.

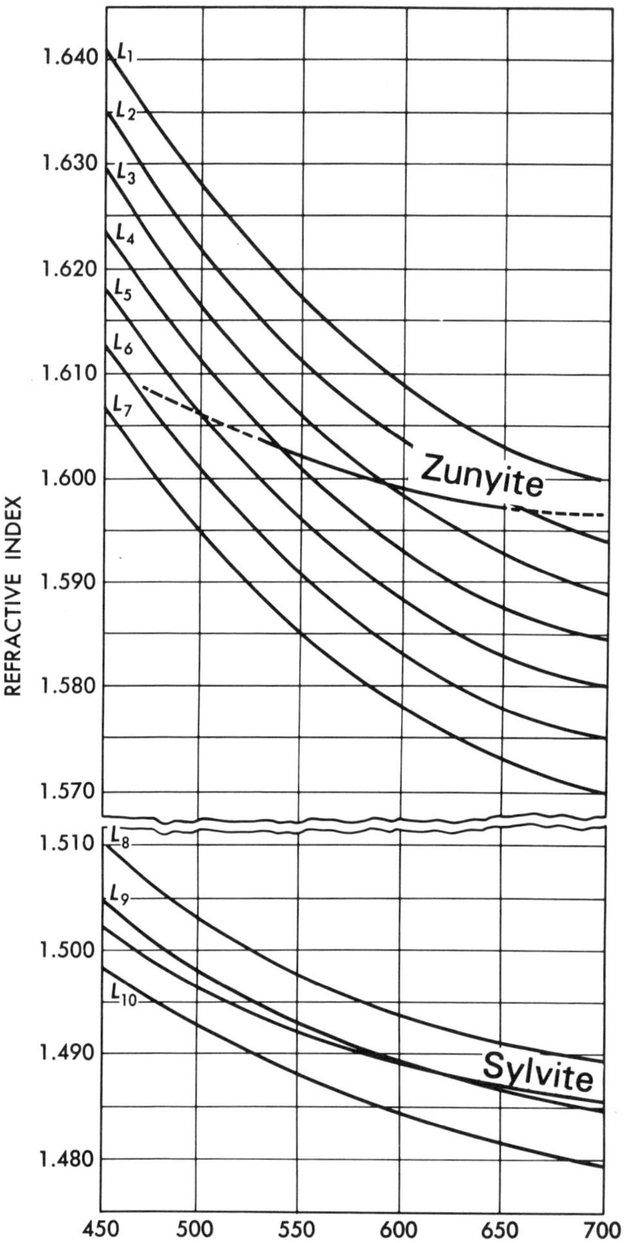

Fig. 3-14. Dispersion curves of selected liquids and isotropic crystals in the middle and low index range. The mineral zunyite has $n_D = 1.6$, and the mineral sylvite has $n_D = 1.49$.

red. One way of learning this color is to observe crystals that very closely match their liquid, for example the lowest refractive index of quartz grains in a 1.544 oil. It is helpful, however, to have a portable reference standard that is close in color to the orange red color fringe. Certain plastic traffic control cones or *pylons* are very close to this color, and have the necessary fluorescence. When chips of this material are available in the lab, the criterion of match between crystal and oil can be described as "pylon color." Pylons used by telephone companies may be more likely to approximate the matching color fringe than the more orange pylons used by other contractors. The desired color is a little more orange than a just-ripe cherry tomato.

The specification used to make one well-matched batch of pylons in 1973 was liquid plastisol mixed with 4 weight percent of pigment D 3534 from the Lauder Chemical Co. This material is slush-molded into cones and baked until it reaches a Munsell color of 10R as determined from ASTM testing code D-1535. The baking induces the fluorescent quality of the color. Scraps from physically flawed cones may be obtainable from manufacturers or local users. Single color chips cut from a successful batch are available from the "Pylon Chip Supply," Department of Geology and Geography, University of Massachusetts, for calibrating the search for larger pieces that may serve as secondary reference standards. Thumb-sized color chips can then be furnished for the laboratory.

Calibration of Color Fringes

Although the recognition and description of colors is a subjective matter, it can be quantified enough to be useful in the measurement of refractive index by the oblique illumination method in white light. Figure 3-15 shows the results of one such investigation. In this study, color fringes were assigned one of ten descriptive names that could be recognized repeatedly by the observer. These colors are given as labels on individual curves showing the matching error as a function of refractive index. Figure 3-15 shows that the sensitivity of match decreases with rising refractive index. This occurs because oils of higher index have higher dispersion slopes, and hence show a given color over a greater range of refractive index. If a red-orange color fringe is observed on a crystal immersed in a 1.625 oil, the crystal is high by 0.002. The same color observed in a 1.5 oil signifies a crystal only 0.001 high. The pylon color used as a reference standard defines a match within ± 0.001 for all but the highest liquids. The pylon color is systematically a bit too orange, and the best match is given by a color that is just slightly redder than the pylon color chip.

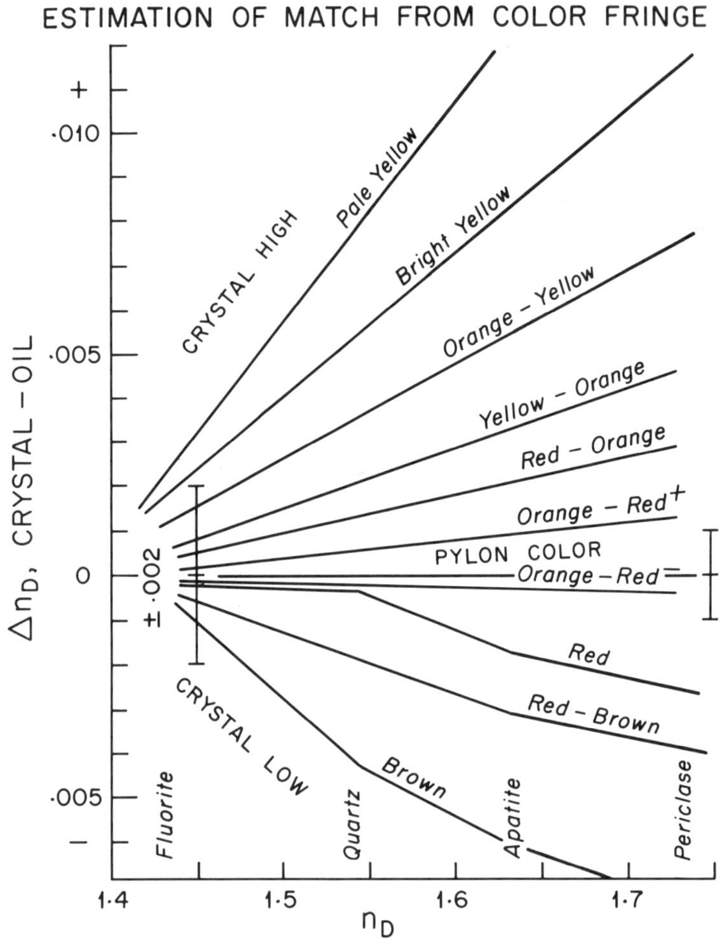

Fig. 3-15. Calibration of color fringes. The four reference minerals are indicated at the bottom of the diagram, and the observed fringe colors are arbitrarily connected by straight lines.

Figure 3-15 can be used to estimate the refractive index of the crystal even if a pylon-color match is not achieved. For a 1.6 oil, for example, if the color fringe observed can be described as bright yellow, the crystal is higher than the oil by about 0.007. If the color fringe is brown, the crystal is about 0.005 low. If a cherry tomato is used for reference, the estimated refractive index will not be more than about 0.0015 low. Such estimates can be used to make judicious choices about the next oil to use in immersion, or can simply be taken as final estimates if low precision will suffice for identification of the crystal.

Figure 3-15 also illustrates that no color fringe will be seen at all if the refractive index is very low and the crystal not within about 0.001 or 0.002 of the oil. If the refractive index is high, the upper limit for a color fringe to be seen is about +0.01, whereas the lower limit is about −0.006.

Figure 3-15 was constructed by observing four reference minerals in many different oils at various incidental stage temperatures, and comparing the color fringe seen with the wavelength of match using a shuttle monochromator. The dispersion curve of the mineral was assumed to be linear in Hartmann space. The refractive index of the oil was calculated with the Sellmeier equation (Chapter 16). The actual difference between crystal and oil could then be calculated for the D line at the conditions of each observation. The minerals used were fluorite (isotropic, $n = 1.434$), the low refractive index of quartz, 1.5442, the high refractive index of apatite (sample KI 3377, 1.6363), and periclase (isotropic, 1.737). All oils used were standard commercial Cargille Refractive Index Liquids. The measurements were made in November, 1978, by S.A. Morse.

DISPERSION STAINING

The term *dispersion staining* has come to be used for the production of color fringes consisting of pure or impure spectral colors, by means of a variety of stops placed at the back focal plane of the objective. The use of such stops is termed *focal screening* or *focal masking*. The word "staining" in this case does not have its usual pigmentary connotation but refers instead to the coloring of the border of the crystal by a pure spectral or complementary color of light. Some of the colors seen are exactly those of color fringes in oblique illumination, which are themselves manifestations of dispersion staining, but are not usually grouped under that term for historical reasons. The chief applications of dispersion staining involve stops that are circular or annular; and hence that produce color fringes around the entire border of a grain rather than just at one side. The technique therefore produces color fringes by central illumination. These are of course related to, and in some cases simply broadened versions of, Becke line colors. Nevertheless, one method of focal screening, called *unilateral* screening, involves oblique illumination and partially splits the spectrum to opposite sides of the grain, so it somewhat resembles ordinary oblique illumination. It seems a reasonable distinction to define dispersion staining as the practice of producing color fringes by focal screening, mainly by central illumination.

A brief review of the history and practice of methods involving stops at the back focal plane of the objective is given here in view of the widespread industrial and forensic applications of dispersion staining. A more explicit guide to practices is reserved for Appendix B, where applications to asbestos are discussed. More detailed accounts of the method are given by Dodge (1948), Cherkasov (1960), McCrone et al. (1977), Bloss (1981), Wilcox (1983), and Fleischer et al. (1984).

With central illumination, color fringes can be produced along the entire boundary of a grain, or of many grains, by various methods of shadowing axial or peripheral areas of the illumination system. These methods are valued for their effect on all of the many fragments in view. For example, one or more of these special methods of screening the field often permits easy discovery of undesirable foreign grains when a mineral sample is being checked for purity prior to chemical analysis. The dispersion staining method is practiced especially in dealing with fibers such as those of the asbestos group of minerals (e.g., McCrone, 1987).

In dispersion staining a medium power objective is modified by placing a screen at the back focal plane. This screen may be an opaque disc at the center (*central screening*) or an annulus around a central hole (*annular screening*). The sizes of the opaque disc and the central hole are roughly comparable, near 2 mm for a 10× objective. Special objectives with flip-in screens are described by McCrone et al. (1977) and may be available commercially for the particular make of microscope at hand.

In annular screening, the undeviated rays at the wavelength of light for which the crystal matches the liquid are transmitted to the eye, and the refracted rays at other wavelengths are blocked by the screen. Examples of such undeviated and refracted rays are shown in Figs. 3-4 and 3-5. In this case, the pure spectral colors seen resemble those of color fringes in oblique illumination. When the refractive index of the crystal matches or is low for sodium light, the colors resemble the yellow to orange-red colors seen on the opposite side of the crystal from which the stop was introduced in oblique illumination. When the crystal is high for sodium light, the colors seen are bluish-green to violet, like the colors on the near side of the crystal in oblique illumination.

In central screening, the image of the crystal is seen against a dark background, in what is called darkfield illumination. The undeviated rays at the margins of the grain are blocked by the central stop, and the nearby rays at wavelengths to each side of the matching wavelengths are combined in the image seen by the eye. These complementary wavelengths, from which the matching wavelength has been subtracted, yield impure spectral colors that are quite opposite in sense to those described above, namely blue to violet for crystal low, and orange-red to yellow for crystal high. It is therefore evident that the use of central screening will require mental agility for those trained on oblique illumination, or else reference will need to be made to an interpretive chart of colors, as for example given on p. 8 of Fleischer et al. (1984).

Focal screening is generally appreciated for its effect on all the grains in the field of view. It is therefore most applicable to crystals that are isotropic or of low birefringence. Groups of crystals of intermediate or high birefringence will give mixed color signals, and only those signals obtained from single grains at extinction will be relevant to one of the two or three principal refractive indices. Given crystal fragments of convenient size, the ordinary methods of oblique illumination will be preferable to dispersion staining for most anisotropic rock-forming minerals.

The practice followed in this book is to give emphasis to a single method of refractive index determination that can be learned as a rote procedure and used effectively with individual grains viewed with ordinary equipment. The use of color fringes, quantitatively developed as described above, fulfills this function, and is easily followed up with observations in monochromatic light for refinement of the refractive indices by objective criteria using the dispersion method (see below). For finely divided material, however, and in certain other applications, dispersion staining has many champions. The methods of investigation will best be chosen to suit the primary needs of the individual laboratory and the preference of the observer.

FACTORS AFFECTING ACCURACY

The accuracy of the immersion method of refractive index determination is affected by the accuracy with which the standard liquids have been calibrated, and their subsequent stability, as discussed in Chapter 2. In addition to the error introduced by liquid instability, error may be introduced because of estimation of refractive index in the interval between the standard liquids. This error may be minimized by reference to the color calibration diagram of Fig. 3-15, described below, or by mixing an intermediate liquid as required from a known number of drops of two adjacent liquids.

Accuracy in Routine Work

The sensitivity of methods of match was investigated by Saylor (1935). In monochromatic light the ordinary oblique illumination method was found least sensitive, the Becke line test using central illumination more sensitive, and the double diaphragm method of oblique illumination the most sensitive of the three. Vertically bounded fragments may be matched more closely than those of irregular shape. The use of low-power objectives increases sensitivity. Inaccuracies in the fourth place were reported for most methods and conditions investigated. The use of continuously and rapidly variable monochromatic light in the dispersion method, discussed below, requires modification of Saylor's findings. The shadow-hopping property of rapid wavelength variation renders ordinary oblique illumination more sensitive than the Becke line method, and the sensitivity can be evaluated objectively for each experiment.

Precision refers to the repeatability of an observation. High precision is necessary but not sufficient for high accuracy (which can be affected by calibration errors, for example). *Accuracy* is a measure of the approach to a true value.

An accuracy of ±0.002 often suffices for the identification of a crystal, and is easily achievable in routine work if the refractive index is less than 1.7. It can be achieved in the range 1.7–1.8 if the index oils have recently been calibrated with a refractometer. An accuracy of ±0.001 is more desirable for crystal identification, and can often be achieved with white light. An accuracy of ±0.0005 is routinely achievable with the dispersion method (see below), and an accuracy of ±0.0001 is attainable with considerable care and effort (Su et al., 1987).

> Accuracy can be tested by measurement of standard materials, such as the low refractive index of quartz, 1.5442, or a standard glass. In 1973 at the University of Massachusetts, ten students in a class having two months' experience with the immersion method were asked to determine as closely as possible the refractive index of a crushed glass, using the color fringe in oblique illumination as a criterion. "Pylon" color chips (described in the section on the orange-red color fringe) were used as a reference standard for the orange-red color. The oils were calibrated with an Abbe refractometer to ±0.0001. The consensus result for this glass was $n = 1.5147_4$, with a standard deviation of 0.0003_4. The range was −0.0006 to +0.0004. It is clear that each of these operators came well within the desired criterion of 0.001. A later test among five students with a natural pyroxene (highest index 1.677) gave a standard deviation of 0.0014, which is probably a good indication of routine precision with high index liquids.

Techniques for Improving Accuracy

Calibration with Standard Glass Powders. If fragments of a uniform, calibrated glass are introduced in the immersion mount along with grains of an unknown crystal, the color fringe of the known glass can serve as an internal reference standard for the unknown crystal. Standard glass powders can be obtained from some makers of refractive index liquids, and reference sets may be obtainable from certain scientific supply houses. If a glass powder is available near the refractive index of the unknown (i.e., within the range ±0.002-0.010), its color fringe can be used in conjunction with Fig. 3-15 to calibrate the color fringe of the unknown. A useful feature of this method is that temperature correction and instrumental calibration of the oil can be avoided. Because the available standard glass powders occur only at fixed intervals, their use in white light is somewhat limited. They are especially effective when used with the dispersion method, described below, because they can then be used to provide a precise calibration of the refractive index of the oil at the temperature of observation (Sueno, 1933; Bloss, 1981; Su, 1992). The range and spacing of standard glasses can also be augmented by the introduction of carefully characterized, homogeneous mineral specimens.

The Dispersion Method of Merwin. The dispersion method introduced by Merwin and Larsen (1912) and refined by Posnjak and Merwin (1922) is a rapid and precise method of refractive index determination. It is particularly well suited to refinement of data obtained in white light, to replicate measurements, or to rapid determination of crystals belonging to a solid solution series. Use is made of the known dispersion curves of both the immersion liquids and the species of crystal being examined.

H.E. Merwin and E. Posnjak were staff members at the Geophysical Laboratory of the Carnegie Institution of Washington, and in company with F. E. Wright they made many advances in the techniques of optical mineralogy. Their work was directed especially to the identification of the experimental run products of high temperature silicate research. E.S. Larsen, Jr., noted for his studies of the volcanoes of the western United States, was a member of the U.S. Geological Survey and a professor at Harvard University.

In the method as originally used by Merwin, the dispersion of the crystal is measured during each determination. The light source used is a monochromator, which furnishes monochromatic light of any desired wavelength in most of the visible range. The wavelength may be changed continuously by the turn of a dial or the motion of a substage shuttle. Liquids are used whose dispersion curves are accurately known, preferably from measurements by the minimum deviation method or by use of a high-accuracy Abbe refractometer (Su et al., 1987). Each immersion mount is examined by means of the oblique illumination or Becke line test over the range of wavelengths until a mount is found that provides a match at some wavelength. A mount is then made using a higher or lower liquid, so as to provide a match between crystal and liquid at some other wavelength. The dispersion curve of the crystal can then be drawn through the intersections of the liquid dispersion curves with the observed wavelength of match. This might be done, for example, with zunyite in Fig. 3-14. From the dispersion curve, the crystal index at the standard wavelength of 589.3 nm can be read off.

In modern practice, dispersion curves are either described algebraically or plotted on a Hartmann net (Chapter 16), which converts the curves to straight lines that closely approximate the actual variation of refractive index with wavelength. Dispersion curves for crystals of a given species can be determined once and for all. A single intersection of a known liquid with the crystal at any visible wavelength will then suffice for a reading, by extrapolation, of the n_D of the crystal. If immersion liquids with large dispersion line slopes are chosen, one mount will often suffice for measurement of a refractive index. The sensitivity of the method is increased, however, by the use of more liquids with flatter slopes.

For precise work by the dispersion method, the temperature of the mount must be monitored with a thermocouple or thermometer attached to the microscope. Since the temperature coefficient, $-dn/dT$, of liquids in common use ranges from about 0.0004 to 0.0007 per degree C, it is desirable for very precise work to know the temperature to ±0.1° C, thereby keeping the temperature contribution to refractive index error well below ±0.0001.

The advantages of the dispersion method stem from its precision and its speed. Since determinations of the wavelength of match can be made, under ideal conditions, as rapidly as one grain per minute, the method is capable of yielding information of statistical significance about sample homogeneity or variability. Since the wavelength can be varied continuously about the matching wavelength, accurate error limits can be placed on the matching wavelength; these may be as small as ±5 nm in the case of crystals with vertical boundaries. If the sample is homogeneous, determinations of the index of as many as 20 grains may yield a precision (expressed as mean deviation) of ±0.0001, according to Morse (1968; 1978).

The dispersion method has been used as a standard determinative method for plagioclase feldspars, as developed by Tsuboi (1934) and revised by Morse (1968) and is suitable for many other crystalline solution series, some of which are discussed in Chapter 16.

Other Special Techniques. It is possible to make refractive index determinations with few mounts by simply varying the temperature of the mount, utilizing the known temperature coefficient of the liquid.

A most entertaining version of this method was a trick used by the late J. F. Schairer at the Geophysical Laboratory, when he wanted simply to see whether his synthetic silicate glass compositions were uniform. He mounted a bit of crushed glass on a slide in an oil of index far too high for the glass, and then laid the slide on top of a nearby furnace for a few moments till it heated up a few tens of degrees above room temperature, at which point the index of the liquid was now lower than that of the glass. Then he put the slide on the microscope and watched as the index of the liquid gradually rose through that of the glass. By using oblique illumination, he was at once able to detect any variability in the glass by noticing any variation in the color fringes from grain to grain. His microscope was never covered; it collected dust on every surface, and it was an object only an owner could love. Nevertheless, as he was fond of pointing out, if but one tiny crystal was present among the pieces of glass, it stood out "like an outhouse on a hill."

For normal measurements with this method, special immersion liquids (Emmons, 1943) that are stable on moderate heating are employed; these are pure liquids rather than mixtures, and they have large temperature coefficients. The thermal variation method is far more time-consuming than the dispersion method because of the relatively large thermal inertia of the special cells used for heating the mount. The temperature of the mount is also subject to uncertainties due to the effects of evaporation when the mount is considerably above room temperature.

The double variation method, in which both temperature and wavelength are varied, utilizes liquids with large dispersion as well as large temperature coefficients. The appropriate liquids, technique, and accessory apparatus for the double variation method have been brought up to date by Su et al. (1987). This method is useful for refractive index determination in a single mount when sample material is scarce, as with the spindle stage (Bloss, 1981).

4

Isotropic Crystal Identification

INTRODUCTION

Isotropic substances are those in which light travels with the same velocity regardless of direction. Isotropic substances include crystals of the isometric (cubic) system, unstrained glasses, both natural and artificial, and most liquids and gases.[1]

Among inorganic crystals that are isotropic, the alums, many oxides, halides, cyanides, and bromates are represented. Of the entries in Winchell and Winchell's (1964) tables of inorganic substances, 29 percent are isotropic. Minerals that are isotropic include halides such as fluorite and halite, some of the oxides such as the spinels, and some silicates including members of the garnet group and the sodalite group of feldspathoids. Larsen and Berman (1934; USGS Bull. 848) reported that 14.9 percent of the entries in their tables of nonopaque minerals were isotropic. Fifty years later, about 13 percent of the entries in Fleischer et al. (1984; USGS Bull. 1627) were isotropic.

PROCEDURE OF IDENTIFICATION

Measurement of refractive index is the basis for the routine optical identification of isotropic as well as anisotropic crystals. The ordinary immersion method discussed in Chapter 3 is used. Mounts of crushed grains are made in a succession of immersion liquids having different refractive indices, until a match between crystal and liquid can be shown to exist by one of the diagnostic tests such as the Becke line test or oblique illumination test.

[1] The crystal system name *isometric* is used here in preference to the approved name *cubic* because it is more precise and less misleading. A "cubic" crystal might be expected to look like a cube, but the name isometric introduces no such expectation.

Once the refractive index is known, the identity of the crystal is sought in one or more of the standard tables, such as Tröger (1979) or Fleischer et al. (1984) for minerals, or Kordes (1960) or Winchell and Winchell (1964) for artificial inorganic substances. Other optical and physical properties may then be used to refine or confirm the identification.

The remainder of this chapter concerns the recognition of isotropic character and the supplementary optical and crystallographic properties that can be used as aids in identification. The scheme of identification, beginning with the determination of refractive index, is applicable in principle to anisotropic crystals as well, and the same pattern of treatment is therefore followed in succeeding chapters on anisotropic crystals.

OPTICAL PROPERTIES

Determination of Isotropic Character

Isotropic character is indicated if a grain does not form an interference figure, the image seen at the back focal plane of the objective. It is sufficient to note that, using crossed polars, converging lens, Bertrand lens, and high-power objective, the field will be almost completely dark. With this same optical arrangement a fragment of anisotropic material will produce an interference figure containing bright areas and dark bands. The types of interference figures are described in subsequent chapters. A simpler test for isotropic character demands that several grains of the same crystal observed between crossed polars remain dark on rotation of the stage. Only if these grains are surely oriented differently and therefore cannot all be normal to an optic axis is this test as rigorous as the first. Different orientation of several grains is to be expected in thin section. In grain mounts, cleavage normal to an optic axis may influence the orientation of the grains so that many may be similarly oriented and be dark under crossed polars. Sections of strongly birefringent crystals cut normal to an optic axis are dimly lighted rather than completely dark on stage rotation, for reasons to be discussed in subsequent chapters. They are not likely to be mistaken for isotropic materials.

Isotropic substances remain dark on stage rotation between crossed polars because light passing from the polarizer vibrates in only one plane and has no component normal to that plane. In passing through an isotropic substance the vibration direction of polarized light is unchanged. The plane of vibration of light that can be transmitted through the analyzer is normal to that of the polarizer and is the plane in which there is no component of vibration, so that no light is transmitted, as in Fig. 1-5. This result is in contrast to the effect when birefringent grains of anisotropic crystals are observed. Unless the birefringent grain is so oriented that the vibration directions in the grain are parallel to the planes of vibration of polarizer and analyzer, there is a vibration component in the plane of transmission of the analyzer as in Fig. 5-7, and light is transmitted.

Pseudoisometric Crystals. A few crystals have isometric form but are not isotropic. These pseudoisometric (pseudocubic) crystals are sometimes described as being optically anomalous because they appear to be isometric but do not remain dark when rotated between crossed polars. They are doubly refracting with very weak birefringence. Observed between crossed polars they show a gray interference color (Chapter 5) at some position of stage rotation. The low first-order interference color, gray, indicates a small retardation, a weak birefringence. The gray interference color is sometimes not noticeable for it is nearly black. Pseudoisometric minerals are commonly considered with the isotropic group. Most pseudoisometric minerals crystallized originally in the isometric system but on cooling have inverted to crystals of one of the anisotropic crystal systems, at the same time retaining their isometric external form. The minerals leucite and cristobalite and some garnets are examples; however, some anisotropic garnets are truly nonisometric.

In pseudoisometric minerals polysynthetic twinning commonly accompanies inversion. In the two parts of the twin the vibration directions of the two plane polarized light components are not parallel. Between crossed polars, the units of the polysynthetic twin appear as bands, segments, or spindle-shaped forms, some of which are gray or white at some position of stage rotation at which the remainder of the crystal is at extinction. At another position of stage rotation the areas formerly white will be at extinction and the other areas will be gray or white.

Color

Color is a diagnostic characteristic of some isotropic crystals but is less intense in microscopic preparations than in larger crystals, a fact of significance to the mineralogist who is accustomed to observing the color of larger crystals. Mineralogists and others using the petrographic microscope sometimes name colors using formal names assigned to the chromas and values of the Munsell color appearance system, described in detail by Peckett (1992).

Relief

Relief in immersion mounts has been discussed in Chapter 3. In thin section, relief is prominent throughout a grain and can be used as a basis for crystal identification, as in Appendix A. In addition to the dark line at the crystal boundary, cleavage cracks cause total reflection and the rough surface created by the grinding causes differences in the angle of refraction of the light, and, therefore, darker and lighter areas on the surface of the section. Colored grains stand out more prominently than colorless grains of equivalent refractive index. Abundant fractures or inclusions also increase the apparent relief. Grains of some minerals in thin section are said to have a *shagreen surface*, a rough, pebbly surface, which is due in part to refraction at the roughly ground upper surface of the grain. Garnet and olivine are typical examples. Doubly polished thin sections do not show this effect. Micas and other sheet silicates also commonly show a *birdseye maple effect*, resembling the variably spaced "eyes" found in some decorative veneers. This effect is due to hollows or strained areas in the grain, induced during preparation of the section.

Refractive Index in Thin Section

The refractive index of minerals in thin sections may be determined to be greater or less than that of the mounting medium or the adjacent crystal using the oblique illumination or Becke line tests described in Chapter 3. The refractive index difference between two substances is roughly indicated by relief. Grains at the edge of the section are usually examined because one becomes practiced in estimating relief with respect to the cement (n desirably and commonly near 1.54) used as a mounting medium.

CRYSTALLOGRAPHIC PROPERTIES

In the identification of isotropic substances the form, cleavage, and twinning are effective as supplements to the optical properties of index of refraction, dispersion, and color. Each of these properties is briefly considered in the following paragraphs.

Form

Some crystals in aggregates, as, for example, in rocks or slags, develop external faces during crystallization. In thin section, straight boundaries of a mineral grain enclosed in an aggregate are usually the traces of crystal faces, but in clastic (fragmental) rocks, they may be cleavage planes. These may be observed particularly well if the surrounding material has different optical properties. The area of a single crystal in the plane of the section is indicated by the area over which cleavage cracks are continuous. Dark borders are also frequent indications of crystal outline when two similar individuals lie adjacent. The form of the cross-section of a crystal will vary with the angle of cut, but observation of a number of sections in the same mount may suggest the form of the crystal.

If euhedral crystals are studied in immersion mounts, faces other than those parallel or perpendicular to the plane of the slide are frequently visible. They may be observed most easily if relief is neither too great nor too small. A favorable amount of reflection at the borders of the grain results when the refractive index difference between crystal and mounting medium is about 0.05. Crystal form is usually destroyed in crushed fragments, but the shape of the fragments may be influenced by cleavage. Detrital grains from sediments have typical shapes pictured in atlases and manuals of sedimentary petrography such as those of Mange and Maurer (1991), Krumbein and Pettijohn (1938), and Milner (1962). These shapes may be diagnostic.

Cleavage

There are three kinds of cleavage in isometric crystals: cubic, octahedral, and dodecahedral. The minerals halite, fluorite, and sphalerite illustrate the three types, respectively. The chemist will find cubic cleavage illustrated by fragments of crystals of NaCl but will find the other types of cleavage very rare.

In a thin section, crystals have diverse crystallographic orientations with respect to the plane of the section, and the traces of cleavage planes in the section form geometric patterns related to the kind of cleavage and orientation of each grain. The traces of cubic cleavage appear as triangles unless the section is normal to a crystallographic axis. The traces of octahedral cleavage form rhombs or triangles and the traces of dodecahedral cleavage form polygons.

Cleavage is visible in thin section because cracks have been developed along the cleavage planes during grinding of the section. Cracks are often most abundant near the margins where the section had less strength and tended to pull apart during the grinding. Cleavage may be more prominently developed in larger crystals than in small crystals of the same substance in the same slide. The cleavage cracks stand out as dark lines when viewed without crossed polars. The dark lines are caused by the total reflection of some rays of the nearly parallel light that illuminates the mineral. Total reflection of some rays at the cleavage crack creates a darker area at the crack, the light being concentrated in the adjacent area. More light will be totally reflected and the dark appearance will be accentuated if the condensing lens is introduced, the substage lowered, and the iris of the substage partly closed so that strongly convergent light of reduced intensity is focused just below the thin section, providing optimum conditions for total reflection at the cleavage surfaces. The amount of substage lowering is fairly critical and of the order of a few millimeters. The amount of total reflection is also influenced by the amount of difference in refractive index of the crystal and the adjacent medium.

In immersion mounts, cleavage may be identified if fragments rather than euhedral crystals are examined. Fragments tend to lie on one of the cleavage planes. Since the upper cleavage surface has not been subjected to the grinding action applied to the upper surface of a thin section, it appears as a flat surface on a grain, usually parallel to another flat surface on which the grain is resting. A straight edge bounding a fragment is also likely to be a cleavage direction. The three types of cleavage in isometric crystals can be differentiated by the shape of the fragments lying on one cleavage that is partly or wholly bounded by straight edges. Since total reflection and refraction make the straight edges more clearly visible they are best observed on a fragment that is mounted so as to show a moderate relief. Cubic cleavage fragments have vertical boundaries at right angles. Octahedral cleavage fragments are triangular or rhombic. Dodecahedral cleavage fragments are commonly four-sided polygons with a flat upper surface. Fragments of a crystal with good cleavage are, however, rarely completely bounded by cleavage planes and some show no cleavage at all. Therefore, the absence of cleavage in an assemblage of fragments does not prove the absence of cleavage in the crystal from which they came.

Twinning

Twinning with composition plane (111) is common in isometric crystals. Since all parts of the twinned individual have the same optical properties, twinning in isotropic crystals can only be determined by observation of the external form, not by optical methods.

OTHER PHYSICAL AND CHEMICAL PROPERTIES

Density

The density of small crystal fragments can be determined quite accurately with the Berman microbalance. Lists of minerals ordered according to density are available for reference (Tröger, 1979). A microscopical technique for estimating density in immersion mounts has been advanced by Bloss (1961) who points out that refractive index media can be so chosen as to furnish simultaneously a suitable range of densities. Observation of whether the immersed fragments sink or float can be made concurrently with refractive index observation. Bloss furnishes reference data arranged in such a way that the density estimate can be used to advantage in identification.

Hardness

A test for hardness may be diagnostic in the identification of a crystal, especially if the sample is very soft or very hard. If soft, it will deform under a needle. If hard, it may scratch glass with more or less ease, and this can be tested by rubbing a few grains dry between two glass slides and looking for scratches on the glass.

Solubility

The solubilities of minerals in standard reagents such as H_2O, HCl, HNO_3, and aqua regia are sometimes helpful in identification. Ease of solution, etching, and gelatinization are effectively studied by use of small fragments under the microscope. If the reagent used produces fumes or could otherwise be harmful to the microscope objective, a large glass screen should be used above the preparation. Solubility information is included in standard reference works such as that of Fleischer et al. (1984).

Microchemical Reactions

Occasionally a specific test for a chemical constituent is helpful in final identification. Although such tests are now more normally made with an electron microprobe, a spectroscope, a spectrograph, or an X-ray fluorescence (XRF) or atomic absorption (AAS) or inductively coupled plasma (ICP) spectrometer, diagnostic microchemical reactions can be rapidly obtained with extremely small amounts of material, and are therefore quite useful. Tests can be made with a few crushed grains or powder scrapings from a larger sample. These are taken into solution in a drop of solvent on a glass slide, and a drop of reagent is then merged with the solution while the resulting reaction is viewed under the microscope. Often the result is a crystalline precipitate, the form and color of which are characteristic. Detailed procedures and descriptions of reaction products are given by Short (1940), Cameron (1961), and Galopin and Henry (1972).

WORKED EXAMPLE OF ISOTROPIC CRYSTAL IDENTIFICATION

Observations in white light will be described first, after which the use of the dispersion method will be discussed and illustrated with an example.

White Light

The sample was first mounted in a 1.542 oil. The following observations were made:

1.542 oil:
Crystal clear—not opaque
Crystal black under x-polars—isotropic
Square and rectangular inclusions—confirm isometric
Becke line in—crystal high
Shadow colored—see color fringe
Color fringe yellow-orange—crystal a bit high
By reference to Fig. 3-15, the yellow-orange color fringe implies that the crystal is high by 0.002. A new mount was therefore made in 1.544 oil.

1.544 oil:
Color fringe orange red—crystal on
Refractive index 1.544—table says halite, NaCl.

Dispersion Method (Monochromatic Light)

An example of crystal identification is shown in Fig. 4-1, based on our worksheet for crystal identification provided with Chapter 16.

The dispersion method is used to determine the refractive index of the crystal at any wavelength and to calculate its value at the D line. The crystal is observed in monochromatic light after a preview in white light to ascertain that the crystal and oil are somewhere near a match. The field is partly shadowed, and the substage diaphragm closed somewhat to give oblique illumination. At any random wavelength, a shadow appears on one side of the crystal and the other side is bright. There is no need to interpret the shadow except in terms of its presence or absence.

REFRACTIVE INDEX WORKSHEET

SAMPLE NO. TS-1

	ALPHA α (NLOW) N	BETA β (OMEGA) N	GAMMA γ (NHI)
Oil:	1.542	1.544	____
Shuttle reading $\bar{X}$	82.3 mm	114 mm	____ mm
Lambda match = λ_0	515±6 nm	594±2 nm	____ nm
k_D, from table	0.4647	-0.0235	____
t match	23.2 °C	23.0 °C	____ °C
dn/dt (bottle)	– 4.1 E-4	– 4.1 E-4	– ____ E-4
$n_D^L = n_{25} + [(t-25)(dn/dt)]$	1.5427_4	1.5448_2	____
Δ^L(F-C) (bottle or graph)	0.0165	0.0166	____

Crystal:

Calculate Δ^S(F-C) = 10↑(5.9 log n - 3.2) or = exp (5.9 ln n -7.368):

Provisional Dispersion, Δ^S(F-C)	0.013	0.013 book value, halite	
$n_D^S = n_D^L + k_D(\Delta^L - \Delta^S)$	(1.5443_7)	1.5447_4	____

(If warranted, recycle with refined values of the crystal dispersion) done here

Book gives 1.5443 but this is clearly a bit greater.

γ - β = 0 B = 0 $Vx \approx \arcsin \sqrt{(\gamma - \beta)/B}$ = ____

$2V_{calc}$ — $2V_{obs}$ — from figure —

Sign — Cleavages — Color, etc. colorless

Optic orientation — Crystal system isometric

Distinguishing features inclusions, cubic shape

Mineral Name Halite + Composition NaCl +

Observer Morse Date 17 Oct. '89

Fig. 4-1. Two worked examples of isotropic crystal identification, using the dispersion method with the worksheet provided in Fig. 16-4.

The monochromator is an interference filter held in a shuttle inside a holder at the base of the microscope. A slit in the holder allows only a narrow range of wavelengths to emerge into the microscope. The monochromator is shuttled back and forth past the slit, varying the wavelength of light until the shadow hops from one side of the crystal to the other as the shuttle moves back and forth. The eye seeks the null position in which the shadow is neither on one side nor the other; the crystal nearly or quite disappears. The shuttle scale is read. The matching operation is repeated and the shuttle read until the operator is satisfied that the reading is being reproduced closely enough, usually within a millimeter or two.

The temperature at the stage of the microscope is noted by laying a sensitive thermometer on the stage and letting it equilibrate for a minute or so.

These two observations, the shuttle reading and the temperature, are the only observational data needed to complete the exercise.

The rest of the work is reduction of data involving external information, either printed on the bottle of oil or taken from tables. The problem is solved in Hartmann space, mentioned in Chapter 3 and described in detail in Chapter 16. In this space, the dispersion curves of liquid and crystal are treated as straight lines (more true of the crystal than of the liquid). The matching condition found in the observation described above defines the refractive index as equal to that of the oil at some wavelength other than the standard, D, 589.3 nm. If the dispersion slope of the crystal is known or assumed, its index at D can be obtained by extrapolation from the wavelength and index of observation.

The calculation can be done by following the worksheet. This is admittedly a "black box" exercise; greater understanding will result from consulting Chapter 16. The black box serves to introduce the method, which will also become more familiar with practice. In the current example, two oils were used, and the problem solved twice, as shown in columns 2 and 3 of Fig. 4-1. As will be seen, one of these would have sufficed; on the other hand, more than two could increase the precision.

The first step in the data reduction is to convert the shuttle reading to a wavelength of observation, λ_0, using the manufacturer's calibration (rendered as a digitized table or parametric equation). For the match made in the 1.542 oil, the result was $\lambda_0 = 515$ nm (green light). The Hartmann wavelength function k_D is then read from Table 16-1, Chapter 16.

The temperature coefficient (always negative) is recorded from the bottle and applied to the observed temperature to find the actual D-line refractive index of the oil at the temperature of observation, $n_{t,D}$. The temperature coefficient of the crystal is taken as zero.

Next, the dispersion of the liquid, Δ^L(F-C), is taken from the bottle or a graph and written down.

It now remains to find the index of the crystal at the D line. We need a provisional dispersion for crystal, and that could be calculated from the general approximation for minerals given in the worksheet and discussed in Chapter 16. The approximation gives 0.008. If we then worked through the problem and identified the mineral, we would learn that the correct value for the dispersion of halite is 0.013. The worksheet has omitted that step for the purpose of illustration, and we take 0.013 directly as the dispersion. If any important uncertainty should arise from the choice of the dispersion of the crystal, we could always make another mount in one or more higher oils and determine the dispersion by solving for the refractive index at two or more wavelengths.

Using the dispersion listed in the worksheet, Δ^S(F-C), The D-line index is calculated from the determinative equation:

$$n_D^S = n_D^L + k_D(\Delta^L - \Delta^S) \tag{4-1}$$

as indicated on the worksheet, Fig. 4-1.

The result for the 1.542 oil was $n = 1.5444$. The same type of calculation was made for the 1.544 oil, with 1.5447 as the result. (The two sets of observations at the respective wavelengths of match define a line giving a value of Δ^S(F-C) $= 0.012 \pm 0.0006$ for the dispersion.) The disagreement between the results for the two oils is 0.0003 in refractive index, which is very small in terms of the accuracy needed for crystal identification. The determination in the 1.544 oil carries more weight because it yielded a match very close to the D line, so the error from the Hartmann linear approximation is entirely trivial. It appears likely that the refractive index of this sample of table salt is very close to 1.5447. Nothing in the mineral tables comes close to this value but halite, for which the average index is 1.5443 (Palache et al., 1951).

The precision of our method allows us to conclude that the refractive index of our sample is greater than that for pure halite by about 0.0004. We note that the sample is table salt, which is commonly iodized in order to prevent iodine deficiency, which can cause goiter in humans. This additive if dissolved in the crystal must increase the refractive index somewhat, because of the much greater atomic weight of iodine than chlorine, and consequently greater density of the crystal. The end of the worksheet accordingly gives the mineral name as "halite plus" and the composition as "NaCl plus". This exercise illustrates the dividends that can sometimes be realized by careful, yet routine, measurement of refractive index.

5 Phase Relationships and Interference

In this chapter we consider the result of light waves traveling the same path from the same source. This consideration will lead in turn to an explanation of the colors of anisotropic material viewed between two polarizing devices (polars) in white light. These colors are known as interference colors.

MONOCHROMATIC PHENOMENA CONFINED TO A SINGLE PLANE

Interference

When two waves of the same wavelength (same color) travel the same path, they may be resolved into a single wave whose amplitude at any point is the vector sum of the amplitudes of the component waves. The combination of the waves is called interference. A wave resulting from interference is shown in Fig. 5-1. The resultant wave has the same wavelength λ as the two component waves, and its amplitude, shown by addition of vectors, is the sum of the original amplitudes at any point along the wave propagation direction.

Retardation

If at any instant the crests of two waves traveling along the same path have not reached the same point, one is said to be retarded with respect to the other. Such retardation, usually designated by Δ, is shown in Fig. 5-1.

Two special values of retardation are of particular interest:

1. Retardation equals any integral number of wavelengths of the light employed:

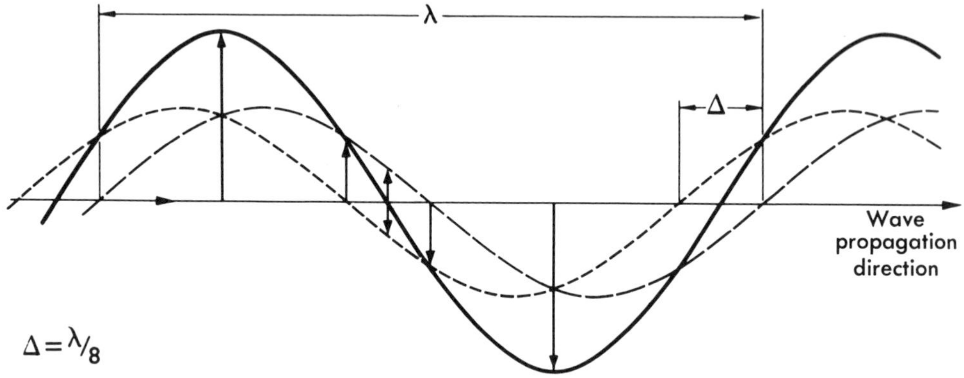

Fig. 5-1. Interference between two waves in the same plane (plane of the page) and traveling along the same path; one is retarded λ/8 behind the other. Individual waves are dashed; resultant is solid.

$$\Delta = n\lambda, \ n \text{ an integer, or}$$

2. Retardation equals any odd number of half-wavelengths:

$$\Delta = n\lambda + \lambda/2, \ n \text{ an integer}$$

These two values are illustrated in Figs. 5-2 and 5-3, respectively.

Δ = *n*λ. When retardation is *n*λ (Fig. 5-2), the two waves are superimposed ("in phase," "constructive interference"), and the resultant wave has an amplitude double that of the original, and shares the same wave propagation direction. The resulting light is brighter than if no

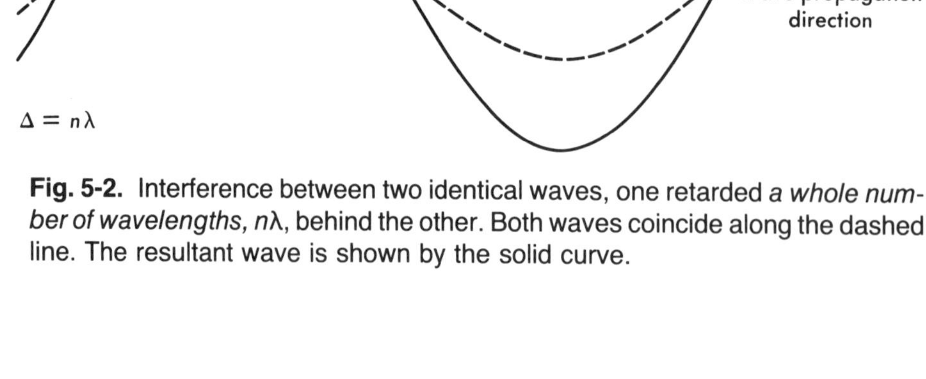

Fig. 5-2. Interference between two identical waves, one retarded *a whole number of wavelengths, n*λ, behind the other. Both waves coincide along the dashed line. The resultant wave is shown by the solid curve.

interference occurred because intensity is proportional to amplitude squared (equation 1-5).

$\Delta = n\lambda + \lambda/2$. When the retardation is $n\lambda + \lambda/2$ (Fig. 5-3), the two waves cancel one another (are "out of phase" exactly, cause "destructive interference"), and the resultant amplitude is zero. Darkness results.

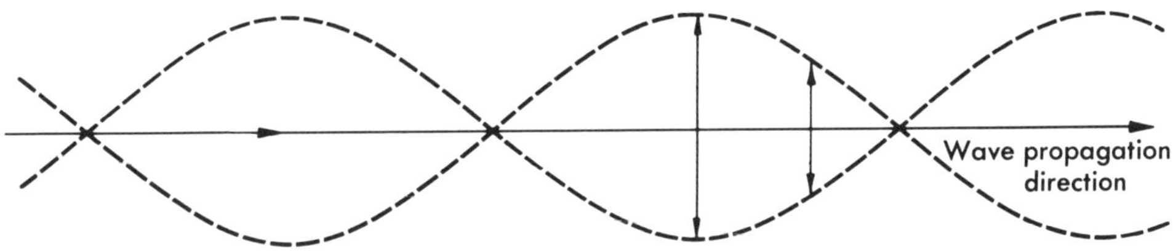

Fig. 5-3. Interference between two identical waves, one retarded $n\lambda + \lambda/2$ behind the other. The sum of the amplitude vectors is everywhere zero, and darkness results.

MONOCHROMATIC PHENOMENA IN TWO VIBRATION PLANES

Retardation of Monochromatic Light by an Anisotropic Crystal Plate

When light enters anisotropic material, it is resolved into two components having planes of vibration that are mutually perpendicular (orthogonal) and also perpendicular to the wave propagation direction. The two polarized components, as they pass through an anisotropic crystal plate, are retarded one with respect to the other, because the atoms in anisotropic material are so arranged that the two components have different velocities. The amount of retardation depends upon the difference in refractive index of the crystal for the two polarized components and the distance traveled through the plate. If light is incident normal to the plate, both components travel the same distance, for their wave propagation directions coincide. This distance is the thickness of the plate. The relationship between the thickness, t, the two refractive indices, n_1 and n_2, and the retardation, Δ, is a form of Fermat's principle:

$$\Delta = t\,|n_2 - n_1| \tag{5-1}$$

Consider the following analogy. Two trains to Boston leave from the same station at the same time, one traveling slower than the other. The *retardation* of the slow train behind the other is a continual function of the distance travelled (the *path length*) multiplied by the *velocity difference*. Applied to wave trains in a crystal, the lag of one wave behind the other continues to grow until the slow wave emerges from the crystal, after which the lag (the difference, the retardation) remains constant and can be observed.

Pierre de Fermat enunciated in 1657 the famous principle of *least time* or *shortest path*, according to which light always arrives at its destination in the shortest possible time. One result of this principle is that the shortest optical path length is given by the product of the physical path length (e.g., thickness, t) and the refractive index n (the inverse of velocity). In the case at hand, where two values of n occur over the same path length, the *difference* in optical path length (our retardation) is therefore the product of the thickness times the difference in refractive index.

Only the absolute value of the refractive index difference is of interest, for the retardation does not depend upon which component has the greater refractive index. Since the birefringence of any section is defined as

$$\mathrm{B} = |n_2 - n_1| \tag{5-2}$$

where the n's refer to the greatest and least refractive index of the section, the retardation may be expressed simply as

$$\Delta = t\mathrm{B}. \tag{5-3}$$

Equation 5-3 has several consequences. If the thickness of the plate is known (as it may be in thin sections of minerals) then $\Delta \propto \mathrm{B}$ and one need only measure the retardation to determine the birefringence. If the birefringence is known (as it is for many mineral species), then $\Delta \propto t$ and a measurement of retardation can provide a very accurate measurement of thickness. The retardation of quartz grains may be used to give the thickness of other mineral grains in a carefully ground bakelite mount. This thickness may then be entered in equation 5-3 to determine the birefringence of the other mineral grains by measurement of retardation (Hess, 1960). Accurately determined birefringence is not often used as a determinative property, but knowledge of it allows the accurate calculation of one refractive index if the other index is known.

Retardation by an Anisotropic Crystal Plate Between Crossed Polars

The preceding section dealt only in a general way with retardation. We must now consider specifically the two separate, mutually perpendicular planes of vibration of polarized light in the crystal. When we consider these two planes of vibration, we find that the relationships between retardation and the amplitude of the transmitted light, as discussed in connection with Figs. 5-2 and 5-3, are no longer valid. Instead, the amplitude turns out to be zero for $\Delta = n\lambda$ and maximum for $\Delta = n\lambda + \lambda/2$, just the reverse of the case for retardation in a single plane. This is because we are now dealing with phase relationships and vector sums in two mutually perpendicular planes. We now consider each of the two limiting cases of retardation.

$\Delta = n\lambda$. In this case, the wave in one plane is retarded a whole number of wavelengths behind the wave in the other plane, as shown in Fig. 5-4. In this figure, the two mutually perpendicular vibration planes in the crystal are arbitrarily labeled D_1 and D_2. Light entering the crystal is polarized north–south, as indicated by the double arrow at the left side of the figure. Because $\Delta = n\lambda$, the crests and troughs in D_1 are side-by-side with those in D_2. This means that the vibration vectors point jointly either up or down (north or south), but never point jointly to either side (east or west). As an example, consider the two crest

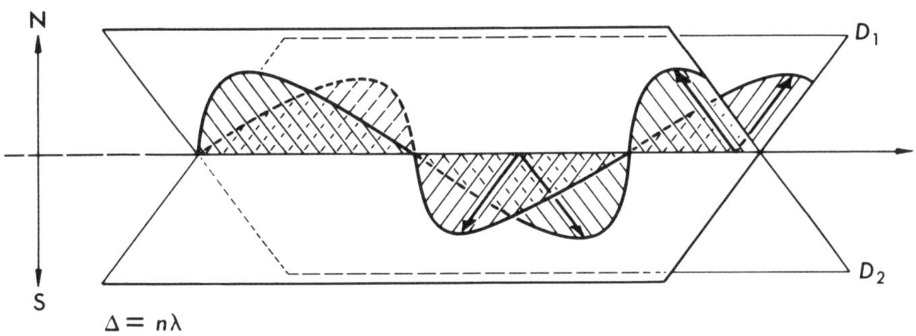

Fig. 5-4. Two waves in mutually perpendicular vibration planes D_1 and D_2 in a crystal have $\Delta = n\lambda$ and therefore have nonzero vector sums north or south if the vibration direction of the polarizer is N–S.

vectors nearest the right-hand end of the vibration planes in Fig. 5-4. These vectors are redrawn in Fig. 5-5, which is a view backward along the propagation direction of Fig. 5-4.

In Fig. 5-5, the N–S polarizer direction is shown by the vertical dashed arrows, the traces of the vibration planes D_1 and D_2 are shown rotated with respect to the polarizer direction, and the E–W analyzer direction is added to emphasize the condition of crossed polars. (The analyzer direction is omitted in Fig. 5-4, but the analyzer is assumed to lie near the right-hand edge of the drawing.) Figure 5-5 confirms that the resultant of the vectors in D_1 and D_2 is nonzero to the north (as in the figure) or south, when $\Delta = n\lambda$. The resultant in the E–W

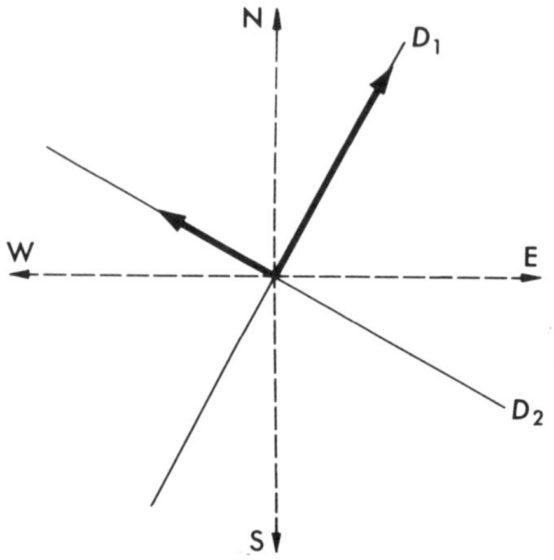

Fig. 5-5. Vibration vectors for $\Delta = n\lambda$, light emerging from page. The first polar is N–S, and vibration vectors D_1 *and* D_2 in the crystal have nonzero sums north or south only.

direction is zero because the vectors are equal and opposite east and west, as can be shown by analysis of the congruent parts of congruent triangles flanking the N–S direction. Accordingly, no light is transmitted through the analyzer. Darkness results.

$\Delta = n\lambda + \lambda/2$. In this case, the retardation of the wave in one plane is exactly a half-wavelength behind that of the wave in the other plane, as shown in Fig. 5-6. In all other respects, the figure is like Fig. 5-4. In Fig. 5-6, the crests in D_1 are side-by-side with troughs in D_2. This means that the vibration vectors point jointly either east or west, but never north–south. Again, the vectors nearest the right-hand side of the drawing are redrawn in a view backward along the propagation direction, in Fig. 5-7. The E–W analyzer direction is again added. The resultant of the vibration vectors is non-zero east or west, never north–

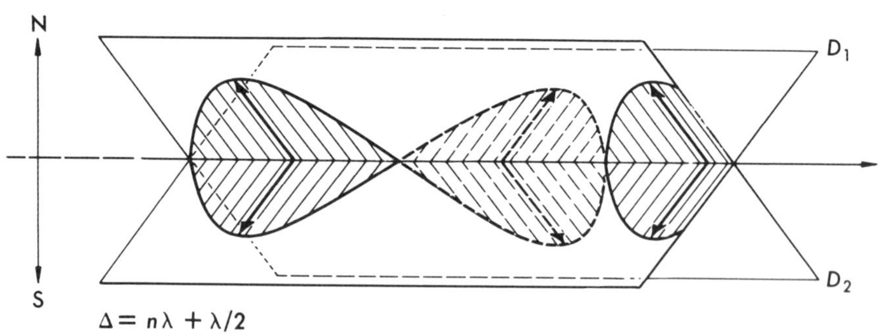

Fig. 5-6. Two waves in mutually perpendicular vibration planes D_1 and D_2 have $\Delta = n\lambda + \lambda/2$ and therefore have vector sums east or west if the first polar direction is N–S.

south. The resulting E–W component is transmitted through the analyzer, and brightness results. This brightness is known as an interference color; it is the color of the monochromatic light used.

Fig. 5-7. Vibration vectors for $\Delta = n\lambda + \lambda/2$, light emerging from page. Vectors D_1 and D_2 in the crystal now do have nonzero sums E–W when the first polar is N–S.

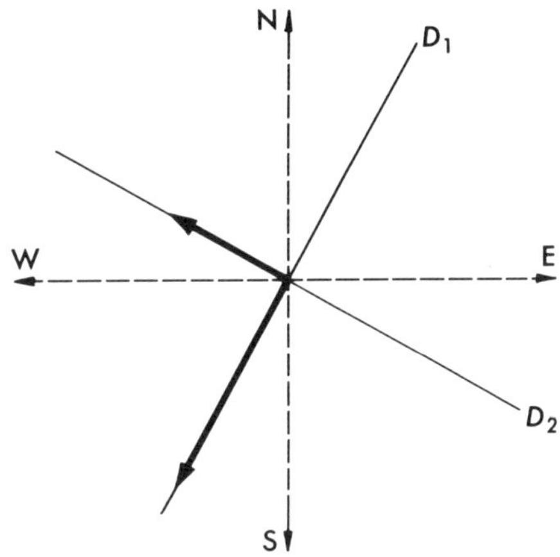

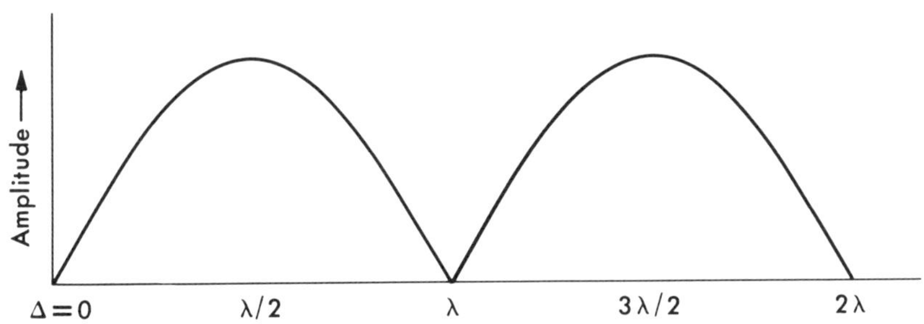

Fig. 5-8. Variation of amplitude with retardation.

There exist, of course, an infinite number of conditions of retardation between $\Delta = n\lambda$ and $\Delta = n\lambda + \lambda/2$. Between these limits there are intermediate values of brightness of light emerging from the system. The brightness is zero only for $\Delta = n\lambda$. The relation between amplitude and retardation is shown by the sine curve[1] of Fig. 5-8. Amplitude (as shown by the ordinate) rises from zero at $\Delta = 0$ to a maximum at $\Delta = \lambda/2$, returning to zero at $\Delta = \lambda$, and so on indefinitely. Intensity, which is proportional to amplitude squared, is similarly related to retardation, as shown by the sine squared curve of Fig. 5-9.

Retardation is more commonly plotted in absolute units of length, such as nm or Å, rather than as fractions of some wavelength. This

[1] That the curve graphed in Fig. 5-8 is a sine curve is shown algebraically in Johannsen (1918, 1969), pp. 343–4.

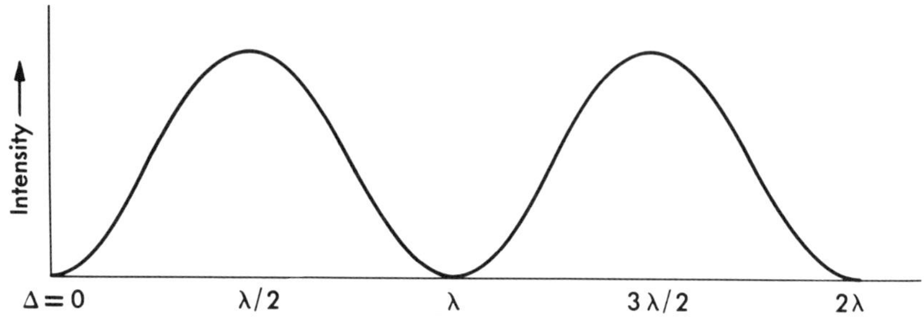

Fig. 5-9. Variation of intensity with retardation

practice readily allows comparison of the intensities associated with several different wavelengths, as shown in Fig. 5-10.

POLYCHROMATIC PHENOMENA

Interference Colors

The phenomena in the preceding sections all result from the use of monochromatic light in the optical system. When white light is used, as is more generally the case, the interference color seen is usually not white, but some other color contained in white light. This is because some wavelengths are suppressed while others are allowed to pass through the system. The wavelengths suppressed are, as the reader may suspect, near or at the wavelength for which $\Delta = n\lambda$.

> Interference colors may be seen by viewing a piece of the white mica, muscovite, between crossed polars. Select a mica sheet about as thick as two or three pages of this book and place it between two sheets of polaroid that have been crossed so that no light comes through them. Look through this mica-polaroid sandwich toward a white light source. The colors seen are interference colors.

The crystal plate considered here has a constant thickness and hence (equation 5-2) constant retardation expressed in units of length (nm) for all wavelengths of light. White light contains all wavelengths of the visible spectrum, so if Δ is constant, there must be some wavelength λ in white light for which $n\lambda = \Delta$. This wavelength is not transmitted by the second polar. It is the wavelength for which Fig. 5-5 was drawn. Similarly, there must be some wavelength for which $n\lambda + \lambda/2 = \Delta$. This wavelength is transmitted as in Fig. 5-7, with maximum brightness. All the wavelengths near that for which $n\lambda = \Delta$ are transmitted with small intensity, that is, they are suppressed. All the wavelengths near that for which $n\lambda + \lambda/2 = \Delta$ have nearly maximum intensity. The color seen represents a combination of those wavelengths transmitted with maximum and nearly maximum intensity.

To illustrate how the intensities of several wavelengths of white light are summed to produce the total interference color, a graphical explanation is given in Fig. 5-10. Six waves are drawn, having the wavelengths of six colors of the visible spectrum. These intensities are plotted as a function of retardation, just as in Fig. 5-9, except that absolute values of retardation in nm are used for the abscissa. Each wavelength has a minimum intensity where $\Delta = n\lambda$ and a maximum intensity where $\Delta = n\lambda + \lambda/2$. The range of Δ in Fig. 5-10 is sufficient to show several maxima and minima for each wavelength.

The total interference color for any value of retardation may be deduced by summing the intensities given by the curves along the appropriate vertical line in Fig. 5-10. The bar graphs for selected values of retardation at the top of the figure summarize the interference color information contained in the curves below. Thus, for example, the bar graph for $\Delta = 400$ nm shows no intensity for violet and increasing intensity for blue, green, orange, yellow, and red. The observed total interference color for such a combination is orange.

Interference Color Scale

If one examines the bar graphs of Fig. 5-10 from left to right, it will be seen that some colors are repeated. These are colors of the second and third orders; the higher the order, the paler they become. A new order begins every 550 nm of retardation. This is the wavelength of yellow light, so the onset of every new order is marked by a band of red. The continuous sequence of interference colors resulting from continuously increasing retardation is called the interference color scale. It "rises" with retardation just as the self-repeating musical scale rises with frequency. Orders of interference colors are loosely analogous to octaves of the musical scale, although since there are no fixed intervals within orders of interference colors, the analogy is not strict unless one imagines a violin string being slurred over several octaves.

Above the fourth order, interference colors are not recognizable except as pastel shades, and interference colors of many orders of retardation have a bright, creamy-white appearance, best illustrated by viewing cleavage flakes of calcite.

> The scale of interference colors may be seen by observing a quartz wedge (with continuously increasing thickness, hence continu-

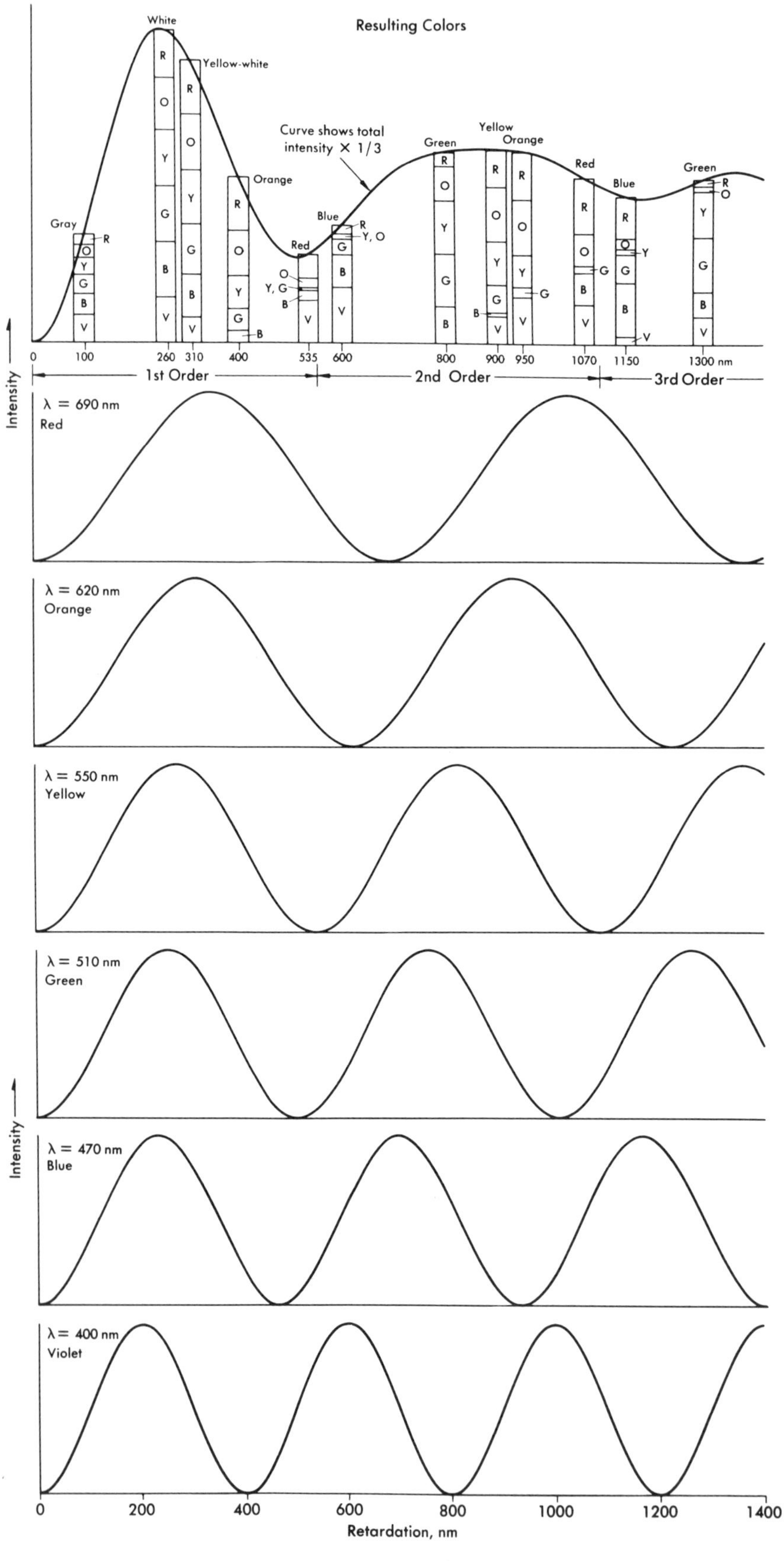

Fig. 5-10. A graphical explanation of interference colors. The interference color seen, for any value of retardation, is the sum of the intensities along a vertical line.

ously increasing retardation) in a polaroid sandwich (with polars crossed) on an overhead projector. By rotation of the quartz wedge within the sandwich, the effects of extinction (see below) can also be seen. If the polars are placed parallel when the quartz wedge is at the 45° position, a different interference color scale will be seen.

The interference color scale (see frontispiece) is depicted photographically in many publications, including some issued by the manufacturers of microscopes. It is conveniently tipped into the back of Tröger (1979). The hand-colored plate in Michel-Lévy (1888), reproduced in Winchell & Winchell (1929), is most accurate in the rendition of colors.

Extinction

The foregoing discussion, and especially Figs. 5-4 through 5-7, has treated only the general case of a crystal plate whose vibration directions are not parallel to the vibration directions of the polars. If the crystal is rotated about an axis coincident with the wave propagation direction until its vibration directions coincide with those of the polars, no light is transmitted, and the crystal is said to be at extinction. Such coincidence is obtained four times upon rotation of the crystal through 360°, since the vibration directions in the crystal and of the polars are at right angles. Extinction can easily be observed by rotating a sheet of muscovite sandwiched between crossed polars.

The reason for extinction when the vibration directions of crystal and polars coincide is that one of the crystal directions is then perpendicular to the second polar direction. The vector sum in the direction of the second polar is therefore zero, and no light is transmitted. The argument can be carried out with the aid of diagrams analogous to Figs. 5-5 and 5-7, as in Fig. 5-11, in which the first polar has a vibration direction N–S (north arrow and dashed line), and the second one, not shown, is assumed to have vibration E–W. D_1 and D_2 are vibration directions in the crystal. In Fig. 5-11(a) brightness results because the crystal directions are not parallel to the polars. In (b) extinction occurs, and in (c) maximum brightness occurs when the crystal directions are at 45° to the polars. Note that upon rotation from one extinction position to another, the amplitude (hence intensity) of the emergent light changes through a maximum, but not the color. The diagrams are valid for monochromatic light or for any interference color produced in the crystal by white light.

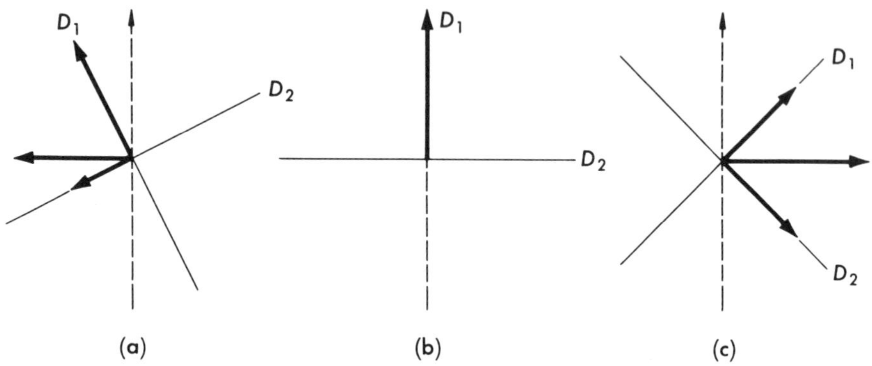

Fig. 5-11. Extinction and brightness in an anisotropic crystal for which $\Delta \neq n\lambda$, at three positions of rotation. The second polar (not shown) is assumed to have an E–W vibration direction.

DIFFRACTION AND BRAGG'S LAW

The interference phenomenon of diffraction occurs in crystals both for visible light and for X-rays, depending on the scale of the periodicity responsible for the diffraction. In either case, diffraction occurs only when the periodicity is close in size to the wavelength of the radiation. The geometrical optics of X-ray diffraction were deduced by the 22-year-old W. L. Bragg (1913; 1975), in what has been called "one of the most prolific ideas in the history of physics" (Lipson, 1970). Bragg's law states the relationship between the wavelength of the radiation, the periodicity of the structure, and the angle of diffraction. Because Bragg's law involves retardation, it is appropriate to discuss its derivation here.

The optical phenomenon of diffraction can readily be demonstrated by observing a monochromatic point source of light such as the reflection of a low-power laser beam. A few metal sieves or screens in the range 40–200 mesh (360–72 microns) will serve as optical diffraction gratings. Holding a sieve at relaxed arm's length, observe the reflected point of light. Its image will be found to be repeated in a square grid. This grid consists of the diffracted images of the source. When other screens are used, it will be found that the size of the diffracted grid is inversely proportional to the mesh size. The reason for this will emerge from the analysis below. If the mesh size is too big, no diffraction will occur. The condition for diffraction is that the spacing of the holes in the sieve must be some near multiple of the wavelength of the light, and that the retardation of light from one hole behind that from the next must be an integral number of wavelengths: $\Delta = n\lambda$. Surprise, surprise! An Optical Transform Kit and laser source can be obtained from the Institute for Chemical Education, Dept. of Chemistry, University of Wisconsin, Madison, WI 53706.

Having noted the existence of optical diffraction, we now treat instead the original problem that confronted Bragg, the diffraction of X-rays by the atoms of a crystal. The explanation can then readily be applied, with slight modification, to the case of the sieves. When a crystal is bathed in X-radiation, its constituent atoms become excited and re-radiate X-rays of the same wavelength. The regularly spaced atoms can then be treated as point sources, and their cooperative effects treated by Huygens' construction (Chapter 1). As a shortcut, however, we will consider only two planes of such atoms, and ignore the actual atom positions. This was the simplification conjectured and then proved correct by Bragg as a young physics student in Cambridge.

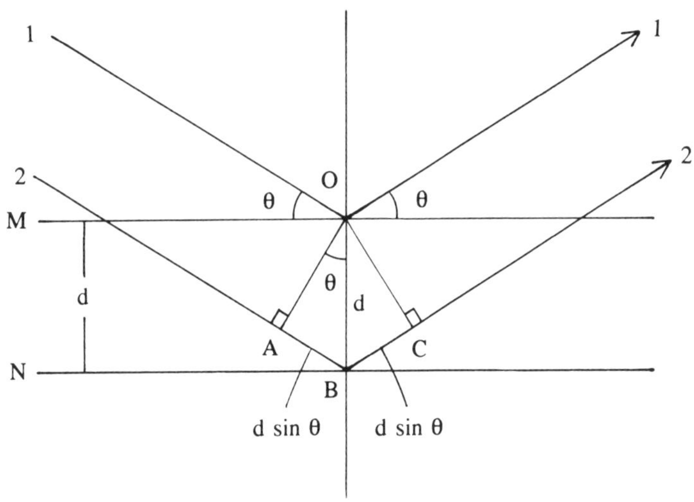

Fig. 5-12. Derivation of Bragg's law. Two rows of atoms, M and N at spacing *d*, are impinged upon by two X-rays, 1 and 2. The retardation upon exit is ABC = 2AB.

The experimental geometry is shown in Fig. 5-12. Two planes of atoms, M and N, are normal to the paper and separated by an interplanar spacing *d*. An incoming wave of X-rays is represented by two rays, 1 and 2. Ray 1 impinges on plane M at O and ray 2 impinges on plane N at B. The initial wave front is OA. The glancing angle (the complement of the angle of incidence in visual optics) is θ. It is stipulated that the condition for diffraction is satisfied, and that diffraction therefore occurs. When this is true the planes of atoms are at minimum deviation (see Fig. 2-1) with respect to the glancing angle. A new wave front is now present at OC.

As ray 1 travels through the point O and beyond, ray 2 must travel the further distance ABC. It is therefore retarded the distance ABC behind ray 1. The distance ABC = 2AB is related to the interplanar spacing by the relation AB = $d \sin \theta$. The retardation, Δ, is therefore evidently $\Delta = 2d \sin \theta$. But the condition for diffraction is $\Delta = n\lambda$, so evidently, if diffraction occurs,

$$n\lambda = 2d \sin\theta \ . \tag{5-4}$$

This is Bragg's law.

William Lawrence (later Sir Lawrence) Bragg was the son of the eminent physicist W. H. Bragg (then at Leeds) and was a research student at Cambridge under the third Cavendish Professor, J. J. Thomson. It is recorded by Lipson (1970) that Bragg could remember the precise moment and place on "The Backs" (place back of the river) in Cambridge when the idea occurred to him that the planes of atoms behaved as mirrors to X-rays. Bragg himself wrote (1975, p. 21), ". . .I studied Laue's photograph further, and it was then that I had the 'brain wave' which led to my maiden effort in research, and which advanced X-ray analysis to its next stage." At p. 23 he describes testing his idea with an X-ray beam incident on a sheet of mica, and "taking my plate, still wet from the fixing dish, down to J. J. Thomson's room and showing it to him. It was very gratifying to see my professor's great excitement." W. L. Bragg determined the crystal structures of many minerals and eventually became, after Lord Rutherford, the fifth Cavendish Professor.

The same reasoning as described above can be applied, usually at normal incidence, to the holes in a grating, spaced d apart. The diffracted wave normals of visible light are constructed from wave fronts found by Huygens' construction on wavelets 1, 2,. . .,n wavelengths behind the first in another hole. Since for monochromatic light, the quantity $2d \sin \theta$ is constant for any integer n, it follows that d and θ are inversely related; hence the behavior of the sieves mentioned above.

Bragg's law is the basis for studying the nature of crystals by X-ray diffraction. If the wavelength of the X-radiation is known (it is characteristic of the metal in the target of the tube), then a solution can be found for d when θ is measured. If the sample is a powder of a crystal, all possible interplanar spacings are represented by a series of lines on a film, or peaks on a graph. If the sample is a single crystal, the diffracted beam occurs only at discrete positions, and the crystal must be rotated to obtain spacings from another set of planes. By such a procedure, the interplanar spacings are obtained, and by collecting the intensities of the diffracted beam, information is gained about the kind and position of the atoms in the planes. This is the basis for crystal structure analysis.

The iridescence sometimes shown by such minerals as peristerite, labradorite, and some amphiboles is a Bragg diffraction of visible light from submicron planes of exsolution within the crystal. In this case, the spacing of the planes is not exactly regular, and the diffraction color is a mixture of wavelengths centered on those diffracted from the average spacing. The color is related to the spacings by Bragg's law, but taking into account the refractive index of the crystal. Substituting the notation *RI* for refractive index, the condition for optical diffraction from lamellae having an average pair spacing $a + b$ is

$$n\lambda = 2\,RId_{a+b}\sin\theta\ . \qquad (5\text{-}5)$$

This type of optical diffraction is reviewed by Ribbe (1983b).

6 Anisotropic Crystals

INTRODUCTION

Certain relationships are true for all anisotropic crystals, whether uniaxial or biaxial. Certain procedures that form part of the routine of crystal identification are therefore generally applicable to all anisotropic crystals.

All anisotropic crystals are alike in that, in general, light passing through them is resolved into two mutually perpendicular plane polarized components. These components define the vibration directions, which lie at right angles to each other and at right angles to the propagation directions. The qualification "in general" above recalls the exceptions: light propagated along a direction called an optic axis is not so resolved, and behaves essentially like light passing through an isotropic substance.

It is convenient to assume that when crystals are examined with the microscope, the light is incident normal to the surface of the fragment and is parallel to the microscope axis.[1] If this is so, the wave propagation directions of both polarized components coincide with the microscope axis, and the two vibration directions in the crystal therefore lie parallel to the stage. Each polarized component has a different velocity, and therefore each vibration direction is associated with a different refractive index. This most important result forms the whole basis of refractive index measurement with anisotropic crystals. To identify a vibration direction is to identify qualitatively the associated refractive index. For this reason vibration directions are named according to the associated refractive index (as high, low, or intermediate). When a crystal fragment is at extinction, its vibration directions lie parallel to the polars, assumed N–S and E–W.

[1] This is not strictly true. It is obviously not true when the strong converging lens below the stage is inserted for conoscopic observation.

NOMENCLATURE

Since each crystal fragment lying on the microscope stage has two vibration directions and two associated refractive indices, the vibration directions can be named according to the magnitude of the refractive index. The vibration directions are therefore referred to as D_h and D_l, or the high-index and low-index vibration directions, respectively. These are not necessarily the highest and lowest refractive indices of the substance, but only the high and low refractive indices for the section of the substance that happens to lie parallel to the stage.[2]

GENERAL PROCEDURES AND CONVENTIONS

The principles mentioned above permit the following kinds of observation in any anisotropic crystal:

1. Recognition of sections cut or broken normal to an optic axis (called "optic axis sections")
2. Location of vibration directions (positions of extinction)
3. Measurement of the "extinction angle," or angle in the plane of the stage between some crystallographic direction and a vibration direction. The crystallographic direction may be revealed by twin planes, cleavage, or external crystal faces
4. The determination of which vibration direction is D_h and which is D_l
5. Measurement of any phenomenon (e.g., color, refractive index) caused by light vibrating in only one of the two allowed vibration directions
6. The estimation of the birefringence of the section being examined.

[2] Some prefer to think in terms of velocity rather than refractive index and to think of the light components as being two rays of different velocity. Accordingly, they identify vibration directions as associated with the faster and slower rays. From the laws of refraction, it is clear that the fast ray (component) causes the lower refractive index, and the slow ray (component) the higher, so "fast = low, slow = high." Since the directly observable and measurable features are the *directions of vibration* and their associated refractive indices, a mental step can be eliminated by naming directions according to the relative size of their associated refractive indices. It is desirable to eliminate as many mental steps as possible.

Several conventions are used in discussing the optical or crystallographic orientation of anisotropic grains with respect to the microscope stage. A grain or section designated by a crystallographic plane or form [a (100) section, for example] lies with that plane parallel to the microscope stage. A grain or section designated by an optical direction (an optic axis section, for example) lies with that direction coincident with the microscope axis.

EXTINCTION

Grains of anisotropic crystals that are not optic axis sections, observed between crossed polars, are dark at four positions in a complete rotation of the stage (see Chapter 5). At such positions of extinction, the two vibration directions are N–S and E–W.

When a grain lies at extinction and the upper polar is then removed, the light passing through the grain vibrates only in the direction permitted by the lower polar. The other component of vibration can be observed by rotating the stage 90°, or more simply, by inserting the upper polar, rotating to the next extinction position, and removing the upper polar to view the grain in plane polarized light.

Identification of Optic Axis Sections

Grains observed between crossed polars that are completely dark at all positions of stage rotation show *complete extinction*. Complete extinction is characteristic of all sections of unstrained isotropic substances and some optic axis sections of uniaxial substances. Other uniaxial optic axis sections remain the same color (gray, for the most part) on rotation.

Imperfect Extinction. Optic axis sections of biaxial crystals and of uniaxial crystals with moderate or strong birefringence do not show complete extinction, but are gray or white rather than black at all positions of rotation. This effect is illustrated by sections of calcite or olivine or of soda niter cut perpendicular to an optic axis. In uniaxial crystals this occurs because, despite the usual assumption to the contrary, light is not all traveling exactly parallel to the optic axis, not all exactly normal to the section. Some light is slightly inclined to the optic axis. In the crystal the oblique light is resolved into two components, the refractive indices of which are not equal, so that to the extent that the section is illuminated by oblique light, it appears to be weakly birefringent. The stronger the birefringence of the crystal, the stronger the birefringence associated with the planes normal to the slightly oblique incident light and the more nearly white the interference color of the section. The lack of extinction is also accentuated if more of the oblique light reaches the eye. This effect is illustrated if the eye of the observer is shifted across the ocular while observing an optic axis section of some strongly birefringent substance.

Undulatory Extinction

Sometimes extinction passes as a wave across a grain as the stage is rotated. Each part extinguishes at a slightly different position, and the effect is that of shadows crossing the grain. This is known as undulatory (undulose) extinction. The effect is due to the slightly different orientations of vibration directions in different parts of the grain. It is commonly exhibited by substances that have been physically deformed (strained), as, for example, quartz in thin sections of many metamorphic rocks. The features produced by this effect are often referred to as *strain shadows*.

Parallel Extinction (Straight Extinction)

When some crystallographic direction such as a trace of cleavage or crystal face lies parallel to a vibration direction (therefore N–S or E–W at extinction), the crystal is said to have parallel or straight extinction. Examples of parallel extinction can be seen in cleavage fragments of the mineral cancrinite (Fig. 6-1a).

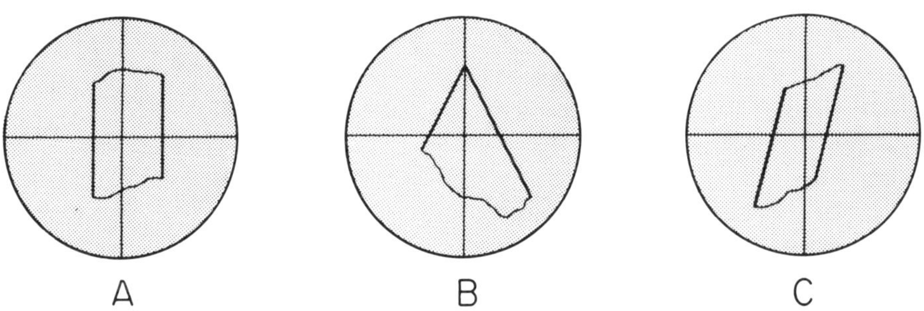

Fig. 6-1. (a) Parallel extinction. (b) Symmetrical extinction, in which the angle between two cleavages is bisected by a crosshair. (c) Inclined extinction.

Symmetrical Extinction

When a vibration direction (crosshair) bisects the angle between two cleavage directions, symmetrical extinction occurs. It can be seen in cleavage fragments of the minerals soda niter or calcite (Fig. 6-1b).

Inclined Extinction

If a single cleavage trace is inclined at any angle to a vibration direction, or if two cleavages are not symmetrically disposed around a vibration direction (crosshair), inclined extinction occurs. Examples are seen in cleavage fragments of the mineral kyanite (Fig. 6-1c).

Effect of Dispersion in Biaxial Crystals. If the vibration directions for red light do not coincide with those for blue light, complete extinction is not obtained with white light. Instead, at the position where the crystal is extinct for blue light, the crystal is reddish brown. At the nearby position where the crystal is extinct for red light, the crystal appears bluish gray. If the interference color of the grain is above first-order white, the effect may be obscured. Dispersion also affects the extinction in sections cut normal to an optic axis in biaxial crystals in which the optic axes for blue and red light are not coincident. Optic axis sections of these crystals may show bluish or brownish colors rather than the usual gray or black. Cleavage fragments of heulandite that show white or gray interference colors exhibit different positions of extinction for red and blue. Some grains of plagioclase in thin section may also show faint colors near extinction. *The effect is best observed when daylight illumination is employed, for it includes more blue wavelengths than most artificial illumination.*

Extinction Angles

When extinction is inclined, the angle (usually acute) between a crystallographic direction and a vibration direction is often measured, and is called the extinction angle. Extinction angles are often recorded in the literature as diagnostic features of monoclinic and triclinic crystals. In such cases, the identity of the crystallographic and optical directions is stated explicitly. The measurement of these angles is part of the broader subject of the determination of optic orientation in biaxial crystals, which is treated at length in Chapter 13.

USE OF ACCESSORY PLATES TO IDENTIFY VIBRATION DIRECTIONS

In any anisotropic grain there are two vibration directions parallel to the stage. They are perpendicular to one another, For light vibrating in one of these directions, the crystal has a higher refractive index, and for light vibrating in the other, the crystal has a lower refractive index. *Accessory plates,* introduced by J. B. Biot, are used to determine whether a given vibration direction in the crystal is the high or low index vibration direction. These plates have known retardation and orientation, and they cause increased or decreased retardation of light in the optical path depending on whether they are oriented similarly to or differently from the grain being studied.

Principle

The accessory plate is a slice of an anisotropic crystal, and therefore has two vibration directions at right angles in the plane parallel to the stage. These are D_h and D_l. They are arranged to lie at 45° to the polarizer-analyzer directions. D_h is usually marked with an arrow on the mounting (sometimes labeled "slow") and is commonly oriented NE–SW.

The crystal slice may become loosened and rotated to a false position within the frame. For this reason, the orientation of round accessory plates should occasionally be checked against a substance of known orientation, such as a quartz wedge, placed upon the stage of the microscope.

The three most used accessory plates cause retardation of one order, a fraction of an order, or many orders of interference color. The first of these was originally made of the mineral gypsum. It is cut so that the retardation is about 550 nm, producing an interference color of red of the first order. The generic name *gypsum plate* still survives because of its convenience, and will be retained here in view of its familiarity and utility, despite the fact that quartz is more commonly used in plates of modern manufacture. Gypsum plates may be identified by the label 1°, λ, or Rot I, or, if older, Gips, which is the German word for gypsum. The mica plate (also called the quarter-wave plate, and marked λ/4 or Glimmer, which is the German word for mica) has a retardation of about 145 nm, λ/4 for yellow light. The interference color produced is white of the first order. The quartz wedge is a graduated wedge, the value of retardation of which depends upon the design. The thickest part usually gives rise to third- or fourth-order interference colors (see frontispiece). Innumerable other accessory plates have been constructed for special purposes, many of which are described by Johannsen (1918, 1968). None is widely employed in the routine examination of crystals.

The accessory plate has its own characteristic retardation (and hence interference color), controlled by the thickness to which it is ground. The 1° or gypsum plate, having a retardation of 550 nm, produces an interference color of first-order red (see Fig. 5-10) when the plate alone is in the optical path between crossed polars.

When an anisotropic crystal (not at extinction) is introduced into the optical path it will modify the retardation of the system, and a new interference color will result. If the crystal lies with its D_h roughly parallel to the D_h of the accessory plate, constructive interference results, and the retardation of the crystal-plate system will appear to have an interference color higher in the scale than that of the plate alone. It will appear to the observer that the crystal has a higher interference color, but it is equally true that the part of the plate subtended by the crystal has a higher color. It is, in fact, the color of the entire crystal-plate system that is important.

If the crystal lies with D_l roughly parallel to the D_h of the plate, the interference is destructive, the retardation of the crystal-plate system is *less* than that of the plate alone, and the color is lower.

One can best understand the interference effects of crystal and plate by reference to Fermat's principle (equation 5-1):

$$\Delta = t\mathrm{B} = t(n_h - n_l) = tn_h - tn_l \;, \tag{6-1}$$

where n_h is the higher refractive index and n_l the lower refractive index of either the crystal or the plate. The retardation of the entire crystal-plate system, Δ_{sys}, is simply the sum of the retardations of the crystal and plate alone, taken with regard to sign. When the high index direction of the accessory plate *(p)* is parallel to the high index direction of the crystal *(xl)*, it can be verified that

$$\Delta_{sys} = \Delta_p + \Delta_{xl} \;. \tag{6-2}$$

This is *constructive interference; retardations add; "colors rise"*. When on the other hand the high index direction of the accessory plate is parallel to the low index direction of the crystal,

$$\Delta_{sys} = \Delta_p - \Delta_{xl} \;. \tag{6-3}$$

This is *destructive interference; retardations subtract; "colors fall"*.

One observes the crystal continuously while inserting the plate, and therefore becomes preoccupied with the apparent effect on the interference color of the crystal. It is well to remember that the important effect is:

how the *crystal* modifies the plate.

Often the crystal is a first-order gray, perhaps with Δ = 150 nm, before the plate is inserted. When a gypsum plate (Δ = 550 nm) is added, the color of the system becomes either second-order blue or first-order yellow. Both are colors higher in the scale than first-order gray. But the blue is the interference color ($\Delta = 550 + 150 = 700$ nm) of the plate (red) **plus** that of the crystal (gray), whereas the yellow is the interference color ($\Delta = 550 - 150 = 400$ nm) of the plate **minus** that of the crystal.

Conventions

If the grain is not flat (the usual case) it may show a range of several interference colors instead of just one, since retardation depends on thickness as well as birefringence. Convenient terms are therefore needed to describe the apparent behavior of all the colors seen instead of one alone.

When the crystal is so oriented that it adds to the retardation of the plate and causes interference colors *higher* in the scale than that of the plate alone, the colors are said to *rise*. When the colors caused are *lower* in the scale than that of the plate alone, the colors are said to *fall*.[3]

Practice

To identify a vibration direction as D_h or D_l, one need only know the high-index direction of the accessory plate and observe the effect on the interference color of the plate produced by a crystal not at extinction when the polars are crossed. The object of doing this is ultimately to orient the desired D north–south so that phenomena caused by light vibrating in that D alone can be observed.

Most microscopes and accessories are so arranged that the high index direction of the plates is NE, which is the arrangement assumed in this book. If one inserts such a plate over a nonextinct crystal and the colors *rise*, D_h of the crystal is also NE; if they *fall*, D_l is NE.[4] If the polarizer is N–S, one may then study the direction of interest by turning counterclockwise to extinction and removing the analyzer to observe color, refractive index relations, etc. If the polarizer is E–W, then the desired direction can be studied if the stage is turned clockwise to the next extinction before removing the analyzer.

Gypsum and Mica Plates. The gypsum (1°) plate (now usually made of quartz) enhances retardation by a whole order; it is best used with grains showing first- or second-order interference colors, as illustrated

[3] Equivalent terms used by various authors include compensation (colors fall), in or out of phase (rise, fall), and path difference increase/decrease (rise/fall). Biot used the terms attractive and repulsive.

[4] The opposite effects may occur with some microscopes, so the actual effect needs to be verified for the particular make and model of microscope used.

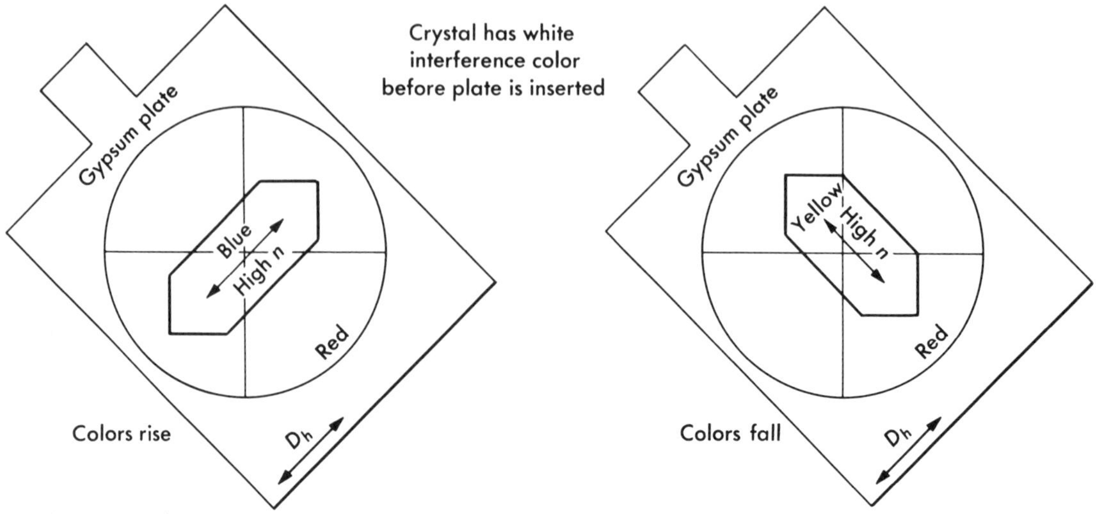

Fig. 6-2. Effect of gypsum plate on a white first order color.

in Fig. 6-2. Grains of quartz in thin section or of apatite in immersion or a mica plate placed on the stage illustrate this effectively. The gypsum plate is less useful in higher orders because colors higher than third order are difficult to tell apart.

The mica plate may be used to distinguish D_h from D_l, but is confusing with first-order colors. It is occasionally useful with colors of higher orders than first. Since it has no outstanding advantages, it is not an essential accessory.

Quartz Wedge. With grains that have thinner edges than centers, the quartz wedge[5] is most effective in determining the rise and fall of colors. The colors on the edges are the lowest colors because the grain is thinnest here. Bands of color parallel to the edge of the grain move either up or down the slope as the wedge is inserted thin edge first. If, as the wedge is introduced, they move up-slope, they replace higher colors nearer the center. Compensation is being effected. Colors in the center of the grain are falling. If the colors move down-slope, high colors are taking the place of low ones. Colors are rising. To avoid confusing the motion of colors of the crystal and the motion of the wedge, one should observe margins of crystals trending parallel to the

[5] See frontispiece. The quartz wedge, gypsum plate, and mica plate were first used by Biot around 1812. The quartz wedge was later used by Senarmont but then forgotten until reinvented by Sorby in 1877 (Johannsen, 1918).

wedge motion so that the motion of the crystal edge is at right angles to the direction of the wedge motion.

Introducing the quartz wedge, thin edge first, the abbreviated rule is as follows.

When the interference colors on the thin edge,

Move up-slope, colors fall

Move down-slope, colors rise

Good examples are provided by grains of quartz in an immersion mount.

When accessory plates are used to determine the high and low index directions by observing falling and rising colors, it is recommended (as a precaution and to make sure of the decision) that, after observing rise or fall of colors, the grain be rotated 90° so that the opposite relationship between vibration directions in plate and crystal occurs. In this new position the effect on the interference colors should be opposite to that in the original position.

MEASUREMENT OF RETARDATION

Retardation ($\Delta = tB$, equation 5-3) is a function of the thickness t and birefringence B of a crystal section. For this reason, measurement of any two of the properties (Δ, t, B) will define the third. For most crystal identification purposes the determination of B is of prime interest, and since direct measurement of the two refractive indices on the same grain in random immersion mounts is a difficult matter, the determination is usually approached through measurement of Δ and t.

Visual Inspection Methods

The retardation of a section is related directly to the maximum interference color observed. Colors in the first three or four orders can be identified with ordinary equipment (see frontispiece), but it is rarely possible to identify even the order of higher-order colors. Several visual methods are possible.

Identification of a Unique Interference Color. The gray interference colors of the first order may be identified by inspection, for gray is not repeated in higher orders. First-order white can be confused only with very high-order colors. With some experience, higher first-order and lower second-order colors may be determined by inspection. Colors higher than second-order blue are easily confused, and thus not easily identified unless their position in a series of colors can be ascertained.

Interference Colors on Thin Edges: Identification of Colors in a Series. The interference colors on thin edges of crystal fragments include much of the series of interference colors from the first order at the thin edge to the order represented by the thickest part of the fragment. Within the order each color appears but once. The red in each order is of low intensity and frequently appears with the adjacent blue as a dark band. Each successive band represents an additional order. The order of a given band may be determined by counting bands.

Compensation with Accessory Plates. If thin edges are not present, accessory plates may be used to discover the position of the interference color in the scale if it is fourth order or lower. Because of the limited range of the quartz wedge or other compensators usually available, identification of the orders above fourth is not usually practical. A quartz wedge may be inserted in the optical system when the crystal is in such a position that compensation will take place (i.e., the colors will fall). The order of color seen in an isotropic area of the field outside the grain when the color in the crystal has been reduced to gray by the quartz wedge indicates the order of color of the crystal. To determine that the crystal being investigated has a first-order gray color, use is made of the fact that when compensation is complete, colors in the crystal will have fallen to gray, and on further increase in compensation the colors will rise. Once colors start to rise again one has just passed the gray color. For accurate work, a wedge with an engraved scale is used. The use of a Wright (slotted) ocular to receive the wedge permits the scale to be read while the color is observed.

Frederick Eugene Wright was an early appointee to the Geophysical Laboratory of the Carnegie Institution of Washington, which published his influential treatise on the petrographic microscope (Wright, 1911). Wright's measurements of polarized light from the moon first showed, in 1935, the presence there of light-colored rocks (Adams, 1954).

Measurement with Compensators. It is sometimes desired to measure retardation quite accurately, and it commonly happens that the retardation to be measured is small. Accurate measurement can be made with a variety of compensators, each having its particular advantages for the kind of accuracy desired. All compensators operate on the fundamental principle of reducing retardation to zero by means of a physical motion of some kind, and by linking the motion to a scale that can be accurately read. The scale readings are usually converted to retardation by means of a calibrating equation or chart. The more widely used compensators are described as follows.

The *Wright combination quartz wedge* is a laminated wedge with an engraved scale for direct reading of retardation to about ±5 nm. The *Berek compensator* is the most familiar device for routine retardation measurement with minerals. It consists of a calcite plate cut normal to the optic axis, which can be rotated about a horizontal axis by means of a calibrated drum. It can compensate colors up to the fourth order, and is especially suited to measurement of small or moderate retardations. This is a Leitz product. The *Ehringhaus compensator* is about equivalent in scope and function to the Berek compensator, but uses two quartz plates instead of a calcite plate, and is made by Zeiss.

The *Sénarmont compensator* and the *elliptic compensator* both make use of elliptic polarization, the first by use of a rotating analyzer and the second by means of a plate that rotates about a vertical axis. They are both suited to measurements of small retardation, and they require the use of monochromatic light. All of the compensators described are more accurate with monochromatic light.

More detailed treatment of the theory and manipulation of compensators will not be attempted here. Much useful information can be supplied by the microscope manufacturers who make the various compensators. Perhaps the best general summary of the compensators mentioned above is that given by Burri (1950). Good coverage of the last two types is given by Hartshorne and Stuart (1970). Of historical interest is the treatment of many types of compensating devices given by Johannsen (1918, 1968).

MEASUREMENT OF THICKNESS

The thickness of a single grain can be measured if B is accurately known and Δ measured (equation 5-3). The thickness of grains in immersion mounts is ordinarily of little interest, since the thickness will vary from grain to grain. In thin sections, however, thickness is of considerable interest because it is approximately uniform for all grains. The maximum interference color of quartz is brought to a straw-yellow in the grinding of thin sections to a standard thickness of 30 microns. The relationship between interference color, thickness, and birefringence is discussed below.

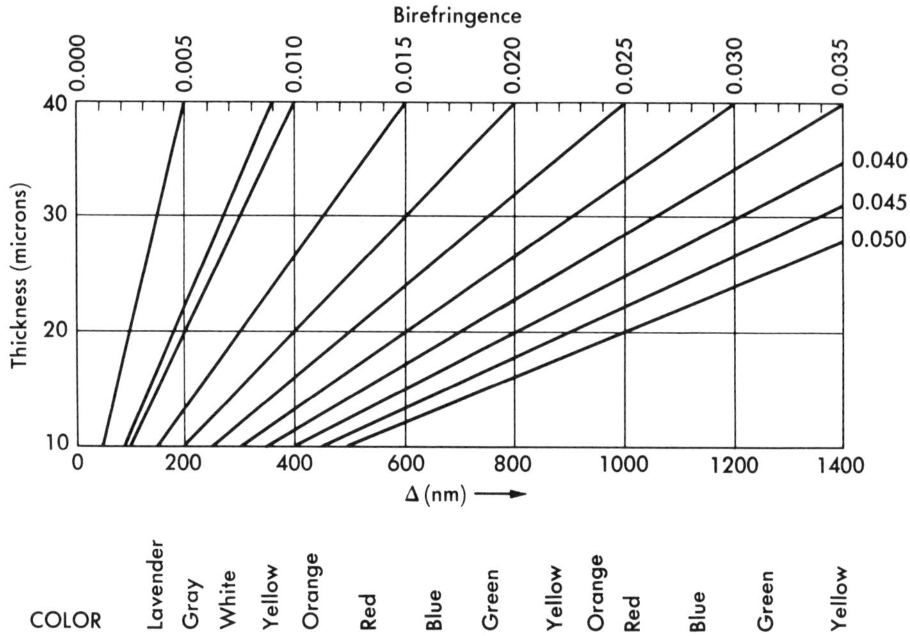

Fig. 6-3. Interference color chart (Michel-Lévy chart). For reference to colors, see photo of quartz wedge in frontispiece.

BIREFRINGENCE OF SECTIONS

The birefringence of the section of an anisotropic grain in the plane of the slide may be estimated if the interference color is of low enough order for its position in the color scale to be recognized and if the thickness of the grain is known. The relative birefringence of two grains showing color of low order can be roughly compared if the thickness is approximately equal though unknown. The relationship between interference color and retardation is graphically illustrated in Fig. 5-10 and the relationship among retardation, thickness, and birefringence is that of equation 5-3. These relationships are also shown by a chart like that of Michel-Lévy (1888) on which are plotted retardation as abscissa, thickness of section as ordinate, and a family of lines radial from the origin, each a different selected value of birefringence.

A simple chart of this type, on which the sequence of interference colors is superposed, is illustrated by Fig. 6-3. The chart is used by following a vertical line along the observed interference color to the horizontal line representing known thickness, thence diagonally upward to the stated birefringence. The problem of determination of birefringence resolves itself into the identification of the position of the interference color in the sequence and the determination of the thickness of the grain.

THE USE OF VIBRATION DIRECTIONS

The remainder of this book is concerned almost solely with how to identify vibration directions and how to make use of them. There are certain principal vibration directions that are of greatest interest. Biaxial crystals have three of these, one associated with the greatest refractive index, one associated with the least refractive index, and an intermediate one. The intermediate one is the vibration direction whose associated refractive index is that of light passing along either optic axis. Uniaxial crystals have only two principal vibration directions associated with the greatest and least refractive indices, respectively. One of these is analogous to the intermediate biaxial direction in that it lies normal to the single optic axis, but in uniaxial crystals this vibration direction can be associated with either the least or greatest refractive index. Uniaxial crystals can therefore be considered a special kind of biaxial crystal in which the intermediate refractive index equals either the lowest or highest refractive index.

7

Uniaxial Crystal Optics

Detailed knowledge of the optics of uniaxial crystals is required for their identification. One must be able to identify the vibration directions in a uniaxial grain and to relate the optical directions to the crystallographic directions. The uniaxial indicatrix, a device that summarizes the values of the refractive indices and the orientation of the associated vibration directions for all directions in a uniaxial crystal, is described in this chapter.

VIBRATION DIRECTIONS AND REFRACTIVE INDICES

Several statements have been made in preceding chapters regarding light in uniaxial crystals:

- Light propagated along the one optic axis has a single velocity.
- Light, except along the one optic axis direction, is resolved into two plane polarized components having different velocities and vibrating in planes normal to each other so that every grain contains a high index vibration direction D_h and a low index vibration direction D_l.
- The vibration directions are normal to the wave propagation direction.

To these generalities may be added several details regarding the plane polarized light components in a uniaxial crystal.

One component travels with the same velocity in every direction through the crystal. This is known as the O ray or ordinary ray. The refractive index of the O ray is ω. The other component is the E ray or extraordinary ray. It travels with a velocity that depends upon the propagation direction within the crystal. The E ray velocity is equal to that of the O ray if the light is propagated along the optic axis direction. The E ray velocity is most different from the O ray velocity when the propagation direction is perpendicular to the optic axis. The refractive index associated with this extreme value of E ray velocity is ϵ. ω and ϵ are the two principal refractive indices of uniaxial crystals.[1]

The refractive index associated with the E vibration direction when the E ray velocity is neither maximum nor minimum is designated as ϵ'. Since there are infinitely many values of ϵ', a specific value may be indicated by citing the orientation of the section perpendicular to which light is traveling. For example, ϵ' on $(10\bar{1}1)$ of calcite refers to the refractive index associated with the E vibration direction of light traveling perpendicular to the $(10\bar{1}1)$ section of this mineral.

INDICATRIX

As defined by Fletcher (1892) the indicatrix of a uniaxial crystal is a geometric figure based on the indices of refraction. It is drawn to scale, and from it the magnitude of the refractive indices and the directions of vibration in any plane in the crystal may be ascertained. The uniaxial indicatrix is an ellipsoid of revolution, as illustrated in Fig. 7-1. The axis of revolution has length 2ϵ and is the direction of vibration of the E ray when ϵ is maximum. The diameter of the ellipsoid has length 2ω and is the direction of vibration of the O ray. Any vector from the origin to the surface of the ellipsoid has the magnitude of a refractive index. Light for which the crystal has that refractive index vibrates parallel to the vector and is propagated normal to the vector. It is important to note that the magnitude of a refractive index is plotted along the *vibration* direction of the polarized component having this refractive index, *not* the propagation direction.

Lazarus Fletcher was Keeper of Minerals in the British Museum after an earlier career as a teacher and Fellow of University College, Oxford. His 1892 book on the indicatrix is reprinted from an article of the previous year in the *Mineralogical Magazine*. It contains a historical introduction in which it is made clear that Fletcher's work extends the published work of Maskelyne, for whom the rare natural glass maskelynite (fused plagioclase) is named.

[1] Other symbols for refractive indices of uniaxial crystals are used by some authors. These include the following:

For ω: N_o, n_g, n_p

For ϵ: N_e, n_g, n_p

The subscripts *g* and *p* stand for *grand* and *petit*, and n_g is assigned to whichever index is greater, ω or ϵ. When this is done, the vibration direction associated with n_g is called Z and the other vibration direction X.

Examination of the indicatrix indicates that all but one of the sections through its center are ellipses, as shown in Fig. 7-2. One axis of every elliptical section has the same length, 2ω, and therefore the refractive index associated with one of the vibration directions in every uniaxial crystal section is ω.

There is one circular section of radius ω. The direction normal to this circular section is the *optic axis*. Because the velocities of the O and the E rays along the optic axis are equal, both rays are called O and the associated refractive index is ω. Light propagated along the optic axis is not resolved into two unlike components; it is not plane polarized. Vibration normal to the optic axis occurs in all directions.

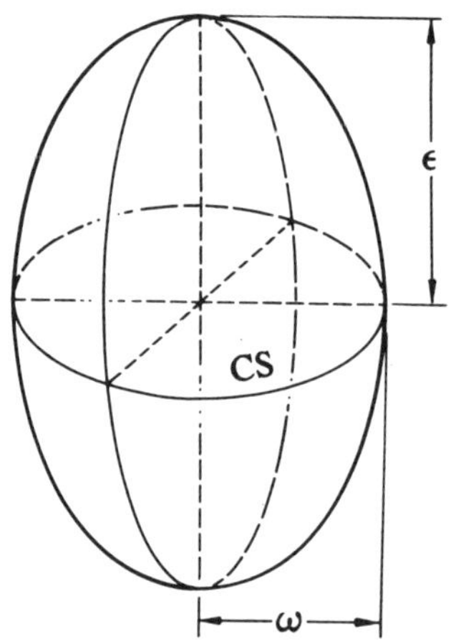

Fig. 7-1. Uniaxial indicatrix for a crystal having $\varepsilon = 3.0$, $\omega = 2.0$. CS = circular section.

Besides the indicatrix there are at least seven other geometrical surfaces used to elucidate (or confuse) the optical properties of crystals. Equations for all of these may be found conveniently grouped in one place by Shubnikov (1960).

Symmetry Planes

The circular section is a unique symmetry plane in the uniaxial indicatrix. An infinite number of symmetry planes identical to each other and containing the rotation axis exist normal to the circular section. These trivial observations are made here to introduce the concept of indicatrix symmetry, which is useful in discussions of interference figures and biaxial crystals (Chapters 10 and 11).

Relationship of Indicatrix to Crystallographic Directions

Uniaxial crystals are those of the tetragonal and hexagonal crystal systems. It is convenient to remember that the orientation of optical or

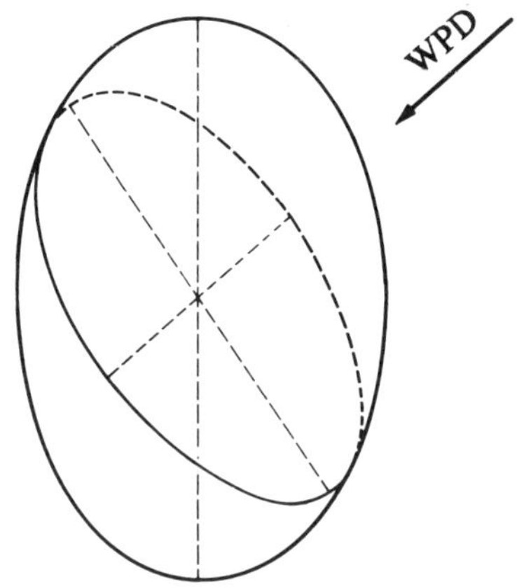

Fig. 7-2. Indicatrix showing an elliptical section normal to a wave propagation direction (WPD).

indicatrix directions in a crystal must always be consistent with the morphological crystal symmetry of the holosymmetric class of a given crystal system, since both optical and morphological features are manifestations of the structural symmetry. The optic axis of a uniaxial crystal, which is the rotation axis of the indicatrix ellipsoid, must therefore always be the *c* crystallographic axis. The *c* axis therefore coincides with the E vibration direction for which ϵ is a limiting value. Light propagated normal to the *c* axis therefore has refractive index ϵ if it vibrates parallel to *c*, or refractive index ω if it vibrates normal to *c*.

The indicatrix is considered to be oriented at the center of a crystal as in Fig 7-3. Any random slice through the center of the crystal will cut an elliptical section through the indicatrix, permitting the two directions of vibration and their associated refractive indices to be ascertained. The propagation direction is considered to be normal to the slice and the elliptical section. The major and minor axes of the ellipse are D_h and D_l for that section. The refractive indices for that section are given by the lengths of the major and minor semiaxes. One of these will be ω, the other ϵ'. Omega is the one perpendicular to a plane containing the *c* axis and the normal to the section. Such a section through the indicatrix alone (without the enclosing crystal) is shown in Fig. 7-2.

Optic Sign

The value of ϵ may be greater or less than ω. This is the basis for subdivision of uniaxial crystals into two groups characterized as being of different optic sign:

$$\epsilon > \omega, \text{ crystal positive}$$
$$\epsilon < \omega, \text{ crystal negative}$$

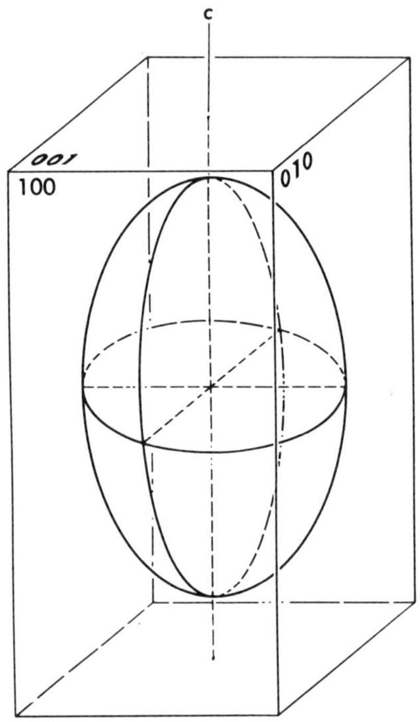

Fig. 7-3. Skeleton drawing of an indicatrix appropriately oriented in a tetragonal crystal.

For positive crystals the axis of rotation is the long axis; the indicatrix is a prolate ellipsoid. For negative crystals the short axis is the axis of rotation; the ellipsoid is oblate. Examples are illustrated in Figs. 7-4 and 7-5. The rotation axis of the ellipsoid is the E vibration direction in the positive as well as the negative indicatrix.

If the birefringence of a uniaxial crystal is defined as

$$B = \epsilon - \omega$$

rather than as in equation 1-13, the plus or minus value of B is the *optic sign* or the *sign of the birefringence*, as some writers call it.

Structural Basis for the Optic Sign. The kind and arrangement of atoms in a uniaxial crystal determines whether the high or low index vibration is parallel to the optic axis (the *c* axis). Some generalities can be stated. The carbonate minerals of the rhombohedral division of the hexagonal system are negative. The relation of atomic structure to ease of vibration and hence the refractive index in different directions in these carbonate minerals is discussed by Bragg (1937, pp. 120–126). The carbonate atomic groups, consisting of three oxygen atoms in a single plane surrounding one carbon atom, are oriented so that the plane of the groups is normal to *c*. The O ray vibrates in this plane.

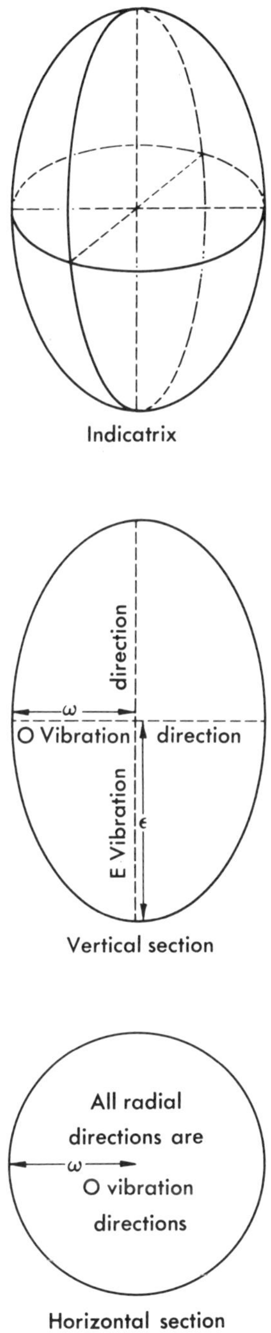

Fig. 7-4. Sections through a uniaxial positive indicatrix.

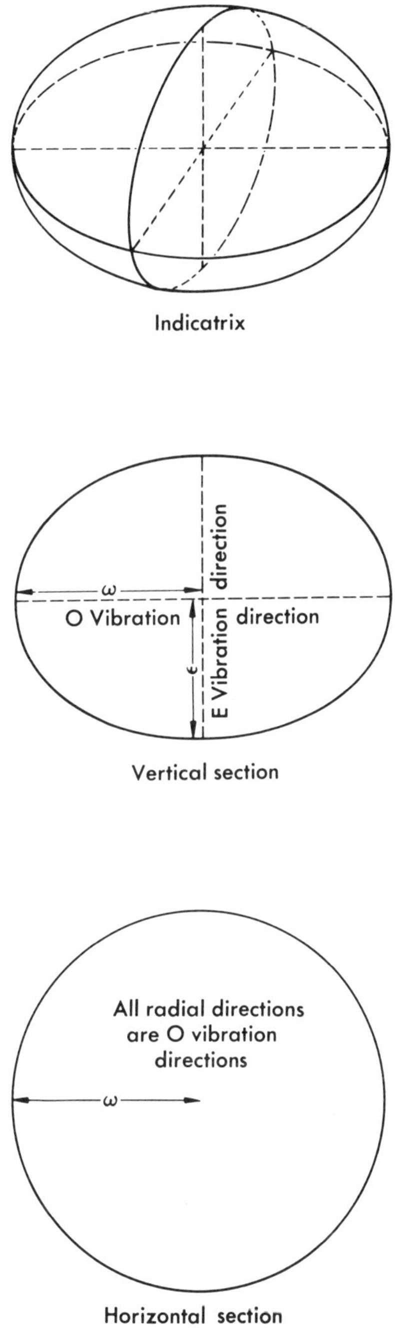

Fig. 7-5. Sections through a uniaxial negative indicatrix.

There is a much larger force opposing vibrations parallel to a carbonate group than opposing the vibrations normal to such a group. The O ray is retarded more (has a higher refractive index) than the E ray, which vibrates normal to the plane of the group, so that the crystal is negative.

The refractive index is also affected by the other atoms present but the influence of the carbonate groups of atoms is dominant. Nitrate groups present similar opposing forces to light. In rhombohedral crystals represented by $NaNO_3$ (the mineral soda niter), nitrate groups are oriented with the plane of the group normal to the *c* axis, and the crystals are negative. The orientation of the planar anion group (for example carbonate or nitrate) determines the optic sign.

Silicates in which the silica tetrahedra are arranged in a ring structure (represented by the minerals beryl and tourmaline) are negative. The individual silica tetrahedra are essentially isotropic in their effect on electrical disturbances, but the combination of tetrahedra into a planar ring structure oriented normal to the *c* axis accounts for the negative sign of the crystal. Many sheet silicates such as mica are also negative. Although these minerals are biaxial, they tend to have a small optic angle and they are negative for the same reason as the strictly uniaxial examples cited above—cause the light vibrating parallel to the silica tetrahedral sheets is slowed down by its interaction with the electric field of the anion array. Further examples are cited in Chapter 10, p. 167.

8 Uniaxial Interference Figures

INTRODUCTION

The first part of this chapter concerns the general properties of interference figures common to both uniaxial and biaxial crystals. The second part deals solely with uniaxial interference figures. Biaxial figures are treated separately in Chapter 11. Photographs of interference figures can be seen in Figs. 8-1 and 11-2, for example.

An interference figure, also called a "directions image," is obtained with conoscopic illumination. This utilizes the upper (swing-out) converging lens of the condenser system, and either the Bertrand lens or a direct view of the back focal plane of the objective. The interference figure is the image seen at the back focal plane when the strongly convergent light passes through a birefringent crystal between crossed polars. The vibration directions at the center of the figure are due to light traveling normal to the plane of the object slide, and therefore correspond to the vibration directions in the crystal under orthoscopic observation. All other parts of the interference figure are illuminated by the strongly convergent light passing obliquely through the crystal, and vibration directions in these parts of the figure are therefore associated with oblique wave normals. Specifically, these vibration directions are projections of the vibration directions encountered by each wave normal as it passes through the crystal.

In obtaining an interference figure, an objective of large numerical aperture (equation 2-1) is used, so that a wide-angle cone of oblique rays can enter. This is commonly a high-power objective of 40× or 50× magnification with N.A. = 0.85. The image of the back focal plane is observed either unmagnified, by removing the eyepiece, or magnified, by inserting the Bertrand lens. A pinhole ocular may be used to center the observer's eye in the former case; a pinhole diaphragm may be used to restrict the image to a small grain in the latter case. The conoscopic field may need to be filled by slightly raising the focus from that of the orthoscopic image.

Emile Bertrand in 1878 was the first to use a lens in the tube of the polarizing microscope to see and measure interference figures, although such a lens had been advocated for the independent conoscope by Amici as early as 1844. The trick of seeing an interference figure by merely removing the eyepiece is due to Lasaulx (also in 1878) and is often called the Lasaulx method. Historical references to the study of interference figures may be found in Johannsen (1918).

GENERAL PROPERTIES OF INTERFERENCE FIGURES

Isogyres

All interference figures consist of one or more black bars, which are known as *isogyres*. The remaining area of the field is colored by interference colors, very often gray or white. The isogyres are bands of extinction in the interference figure. They are formed where the vibration directions in the figure are parallel to those of the polars, which are usually set N–S and E–W. The form and position of the isogyres in the interference figure depend upon the orientation and axial character of the grain. Rotation of the grain (with the stage) usually changes the form of the isogyres, because in almost all instances such rotation changes the orientation of the vibration directions in the grain relative to those of the polars.

Color Curves

With crystals of sufficiently large thickness or birefringence, the interference colors seen in the interference figure are higher than white or gray, and these may form *color curves* (isochromatic curves, color bands, isochromes). The color at each point in a color curve is the result of the retardation (phase difference) of the two plane polarized components that are propagated along the same oblique path. Any one curve is the locus of all points, in the figure, of equal retardation. The form and position of the color curves usually change on stage rotation, just as do those of the isogyres.

The origin of isogyres and color curves is discussed more fully below, in connection with the specific types of interference figures to be encountered.

Effect of Crystal Orientation

Each orientation of an anisotropic crystal fragment produces an interference figure of somewhat different form. Some of these forms are specifically recognizable, and can instantly be used to determine the orientation of the indicatrix in space. In uniaxial crystals the optic axis and E vibration direction are always parallel to *c*, so a knowledge of the orientation of the indicatrix in space carries with it some knowledge of the crystallographic orientation of the crystal in space.

Many interference figures have forms that cannot be uniquely related to some specific indicatrix orientation. Such figures can be assigned to one of several categories, however, and many of them can be used for information about the crystal. In uniaxial crystals, these figures are produced by all sections of grains not parallel or perpendicular to *c*.

Classification of Interference Figures

All interference figures may be classified with respect to the symmetry planes of the indicatrix, and within this classification certain figures have specific names. The systematic classification is introduced here for simplicity of discussion, even though its chief use is in connection with biaxial interference figures. Certain very useful biaxial interference figures have no specific names other than those conferred by the systematic classification.

Table 8-1
Classification of Interference Figures

General Name	Specific Name
Uniaxial figures	
1-symmetry plane	Uncentered optic axis or uncentered flash
2-symmetry plane	Flash figure
n-symmetry plane	Optic axis (centered)
Biaxial figures	
0-symmetry plane	No specific names
1-symmetry plane	
Counter-rotating (CR)	Includes optic axis figure
Same-rotating (SR)	No specific names
2-symmetry plane	Acute bisectrix (Bxa), obtuse bisectrix (Bxo), optic normal (ON)

Interference figures may be named both generally and specifically. A summary of all possible kinds of interference figures is given in Table 8-1. The general classification is based on the number of symmetry planes of the indicatrix that are normal to the section under observation; for example, a 1-symmetry plane figure is obtained from a section cut or broken perpendicular to one symmetry plane, so that the symmetry plane in question contains the microscope axis and is perpendicular to the stage. The specific classification identifies the exact symmetry plane or special indicatrix direction normal to the section. The special indicatrix direction is often the axis of intersection of two or more symmetry planes.

The various types of interference figures are recognized by the behavior of isogyres upon stage rotation. Certain general properties of each type can be summarized briefly.

The *n-symmetry plane figure* (uniaxial only) is obtained from sections normal to an infinite number of symmetry planes, hence normal to the rotation axis of the uniaxial indicatrix. Two isogyres form a cross that remains coincident with the crosshairs throughout stage rotation. This is the uniaxial centered optic axis figure.

2-symmetry plane figures, obtained from sections normal simultaneously to two symmetry planes, contain two isogyres that form a cross coincident with the crosshairs only four times in a complete rotation of the stage. At intermediate positions of stage rotation, the isogyres are curved, and in the uniaxial figure called a flash figure (and in most biaxial figures), they leave the field altogether. The uniaxial flash figure is formed by sections parallel to the *c* axis.

1-symmetry plane figures result from sections perpendicular to only one symmetry plane of the indicatrix. They have one or two isogyres, and one isogyre is coincident with a crosshair four times on complete stage rotation. The chief uniaxial figure of this type is the uncentered optic axis figure. All sections of uniaxial crystals except those normal or parallel to the *c* axis are 1-symmetry plane sections; any section through the center of an ellipsoid of revolution is normal to at least one symmetry plane.

0-symmetry plane figures (biaxial only) have one or two isogyres, at no time coincident with a crosshair. The biaxial indicatrix contains only three planes of symmetry, being a triaxial ellipsoid; hence it is possible for a section to be normal to no plane of symmetry.

UNIAXIAL INTERFERENCE FIGURES

The Centered Optic Axis Figure

The centered optic axis figure (*n*-symmetry plane figure) consists of a black cross formed by isogyres coincident with the crosshairs. In the four quadrants, color curves concentric with the center of the cross may be present. The interference figure does not change shape or position when the stage is rotated. This figure, illustrated in Fig. 8-1, is produced from a grain in which the circular section of the indicatrix is parallel to the stage. The grain is a section normal to the *c* axis, a basal section. Sections that will produce this figure may be recognized orthoscopically because they are black or a constant gray at all positions of rotation under crossed polars.[1]

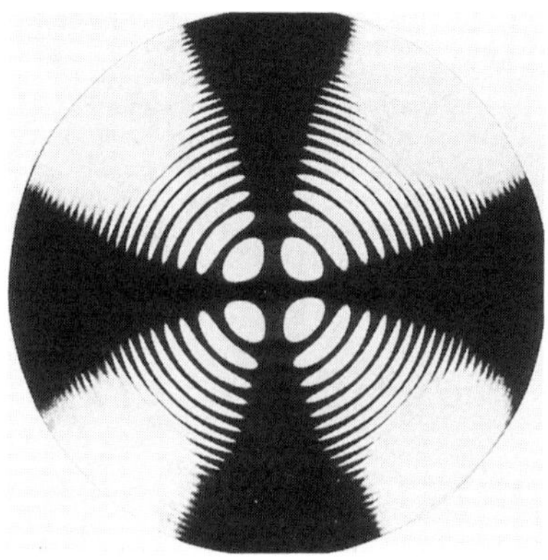

Fig. 8-1. Centered optic axis figure with color curves, which may be absent in crystals of low birefringence. The isogyres form a black cross. Hauswaldt 2-1, calcite, 1/2-mm thick, sodium light.

Isogyres. The form of the isogyres and color curves in a centered optic axis figure may be explained by reference to an appropriately oriented indicatrix, representing the crystal and the wave propagation directions of three representative convergent rays, as shown in Fig. 8-2. The sections of the indicatrix normal to each of the three directions of wave propagation may be plotted in correct relative position in a circle representing the field of the interference figure. The sections plotted are part of a plan of the optical properties in all parts of the field. Since the indicatrix is symmetrical about the *c* axis, any other convergent wave propagation directions making similar angles with the optic axis

[1] The photo in Fig. 8-1 is from Hauswaldt (1902), courtesy of Dartmouth College Library. This rare collection contains a great number of striking photographs of interference figures, mostly of very thick sections as compared to the normal thickness of 30 micrometers. Although the Hauswaldt photos included in this book show more color curves than would be seen in most immersion mounts or thin sections, they are useful for illustrating the shapes of color curves. Isogyres in the Hauswaldt photos are much broader than will normally be seen (compare Fig. 11-2). Numbers in captions of Hauswaldt photos are the original plate-figure designations.

will be normal to analogous sections of the indicatrix. Appropriate sections of the indicatrix may therefore be inserted along other radii in the plan of the microscope field as in Fig. 8-3. The directions of the vibrations in the field of the interference figure with respect to the planes of vibrations of the polars determine the position of the isogyres. Light with east–west or north–south vibration directions will not be transmitted because these are the vibration directions of the polars. Black isogyres envelop these directions in the interference figure, where the sections of the indicatrix in the plan show north–south or east–west vibration directions.

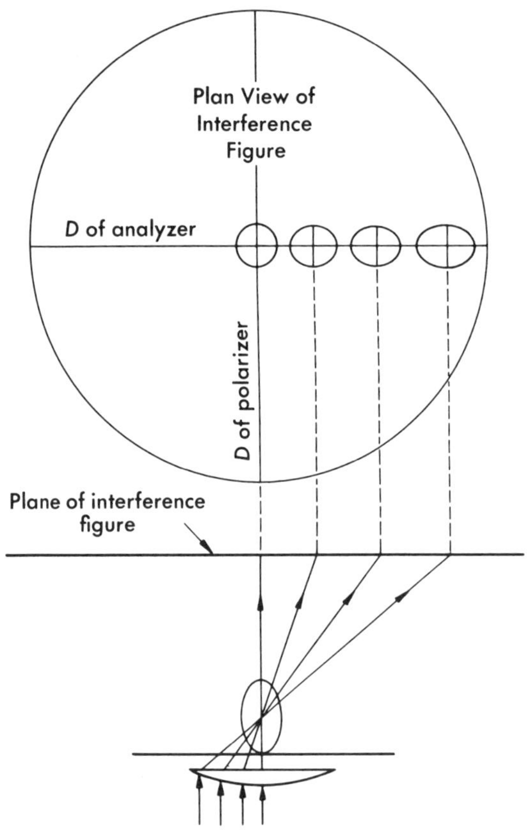

Fig. 8-2. Four wave normals passing through a uniaxial positive indicatrix. Vertical section below, plan view above.

Color Curves. Where vibration directions are not north–south or east–west, the interference figure will be illuminated by interference colors. The color nearest the center of the figure is first-order gray, and, farther from the center, colors appear that are progressively higher in the interference color scale. If more than one complete order of colors is present, the red curve at the upper end of an order and the blue curve

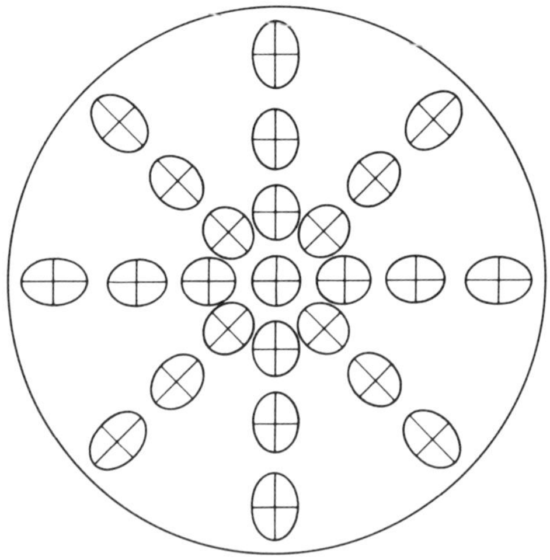

Fig. 8-3. Plan of vibration directions in a uniaxial optic axis figure. Derived from the upper part of Fig. 8-2 by rotation of the field about an axis normal to the page.

of the order just above appear as a marked ring, in contrast to the adjacent paler colors of the interference color scale.

The colors are higher toward the margin of the figure because retardation increases outward from the center of the figure. The retardation increases because both the birefringence and the effective thickness of the crystal (see equations 5-1 and 5-2) increase with more oblique wave normals. The increase of birefringence is by far the more important of the two effects. The manner in which birefringence increases with more oblique wave normals is shown quantitatively in Fig. 8-4.

Figure 8-4(a) is a scale drawing of a section through a uniaxial positive indicatrix. Two wave normals, WN_1 and WN_2, pass through the section in the plane of the page. Both have one vibration vector in common; this is the O direction, with magnitude ω, normal to the page. The other vibration vector lies in the plane of the page and has a magnitude dependent on the obliquity—ϵ_1' for wave normal 1 and ϵ_2' for wave normal 2. The plan views of Fig. 8-4(b) and (c), also to scale, illustrate the difference in the birefringence resulting from the different obliquity. Recalling that a vector in the indicatrix has the magnitude of refractive index, the birefringence is the difference in length between the ϵ' and ω vectors. This difference is not noticeable for wave normal 1, which is close to the normal to the circular section (the optic axis). It is very noticeable for wave normal 2, which has much higher obliquity. Since the birefringence becomes vanishingly small for wave normals close to the optic axis, the increase in retardation due to birefringence for more oblique wave normals may be many orders of magnitude. The contribution from the greater thickness t (traversed by the more oblique wave normal) is not likely to exceed a factor of 5.

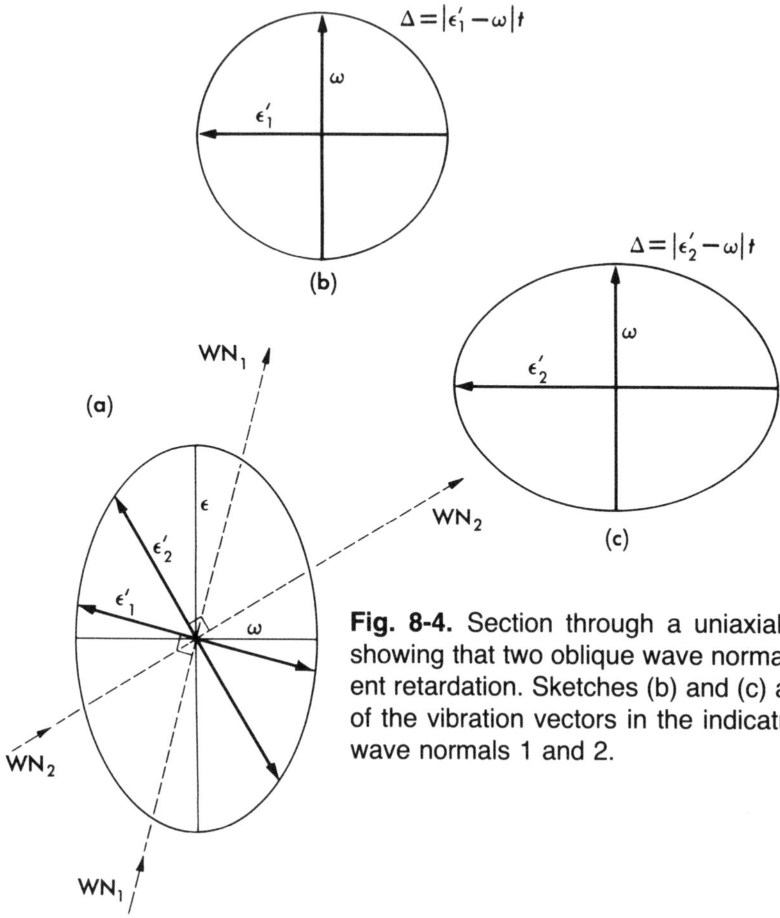

Fig. 8-4. Section through a uniaxial indicatrix (a) showing that two oblique wave normals have different retardation. Sketches (b) and (c) are plan views of the vibration vectors in the indicatrix sections of wave normals 1 and 2.

It is apparent from Fig. 8-3 that the loci of points of equal birefringence and retardation are circles or isopleths, comparable to contours. For each value of retardation there is a circle of an interference color concentric about the center, and interrupted by black isogyres where extinction occurs. For grains of equal thickness, the number of color curves or kind of interference color relates to the birefringence of the crystal species being examined; the higher the colors, the greater the birefringence.

Vibration Directions in the Figure and the Grain. The vibration direction in each of the radial directions in the centered optic axis figure is the vibration direction related to E. All vibration directions normal to radial directions are O vibration directions, related to the refractive index ω as shown in Fig. 8-3. In the exact center of the figure, directions normal to the radial directions are also radial; hence all are O directions related to the refractive index ω. The vibration directions in the center of the field result from light propagated normal to the section, the same light that illuminates the orthoscopic image. The vibration directions in the *center* of any interference figure are the same as the vibration directions in *all parts* of the grain viewed orthoscopically. Hence grains forming centered interference figures show the ω index at all positions of stage rotation.

Uncentered Optic Axis Figure

A section of a uniaxial crystal cut neither parallel nor perpendicular to the optic axis exhibits an uncentered optic axis figure (1-symmetry plane figure). Because the normal to the section is not the axis of rotation of the indicatrix, the figure moves as the section is rotated. The center of the cross is not in the center of the field. In a section almost perpendicular to the *c* axis the center of the black cross describes a small circle on rotation of the stage. The center of the figure rotates in the same direction as the stage is rotated. The color curves are not quite circular. The larger the angle between the optic axis and the plane of the section, the greater the radius of the circle described by the center on rotation. If the angle is sufficiently large the center of the cross never appears in the field, but the almost straight bars sweep across the field, lie parallel to the crosshair as the section is rotated, and coincide with the crosshair at the orthoscopic extinction positions (Fig. 8-5). The point of emergence of the optic axis, marked by the

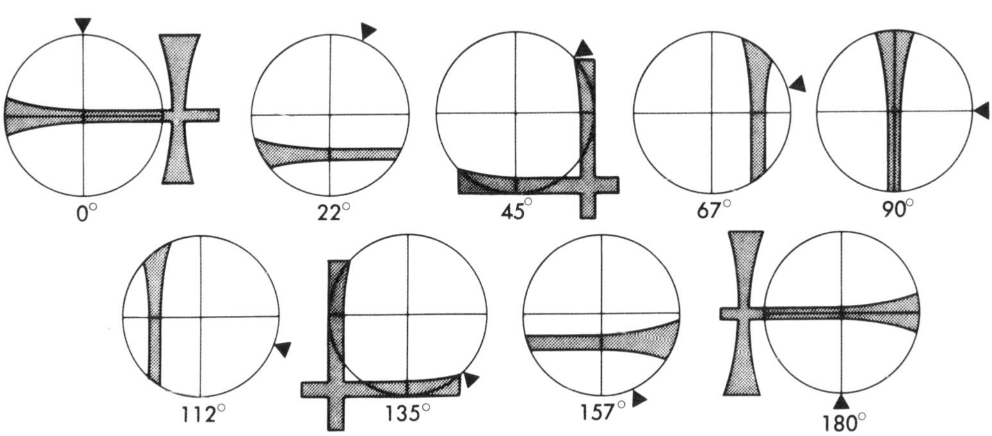

Fig. 8-5. Uncentered optic axis figure at several positions of stage rotation. Stage and center of cross both rotate clockwise.

center of the cross, can be visualized if it is outside the field by noting the sequence of positions of the bars on rotation. The location of the center of the cross can also be determined by observing color curves if they are present. These are always concave toward the center of the cross.

Optic axis figures farther offcenter than the one pictured in Fig. 8-5 show a bar that is not quite parallel to a crosshair except when coincident with it.

Fragments appropriately oriented to give uncentered optic axis figures are characterized, when observed orthoscopically, by interference colors intermediate between the lowest and highest colors observed in a group of fragments of diverse orientation and comparable thickness.

Location of Vibration Directions. Vibration directions in uncentered optic axis figures have the same relationship to the center of the cross as in centered optic axis figures: E′ associated with ϵ', radial, and O, associated with ω, normal to E′. At a position of stage rotation such that an isogyre coincides with a crosshair, E′ lies along this crosshair and O along the other crosshair. In the grain viewed orthoscopically the refractive indices at the two extinction positions are ϵ' and ω. Cleavage fragments of calcite and soda niter are suitably oriented to form uncentered optic axis figures.

Flash Figure

The flash figure (2-symmetry plane figure) is formed by all sections parallel to the optic axis, hence all sections in the vertical zone, i.e., parallel to the *c* axis. Grains are parallel to the section of the indicatrix of maximum ellipticity, so they may be recognized orthoscopically because they exhibit higher interference colors than other grains of the same crystal of equal thickness.

The isogyres form a broad gray or black cross at positions in which the grain, viewed orthoscopically, is at extinction. On rotation, the cross breaks up into two branches of a hyperbola, which leave the field in opposite quadrants after the stage has been rotated only a few degrees. On further rotation the isogyres return again, and after 90°, are at the crossed position, as illustrated in Fig. 8-6. During the next 90° of rotation the isogyres leave, and subsequently return, in the other two quadrants. The isogyres and the direction in which they leave the field are less easily recognized in figures formed from crystals of strong birefringence.

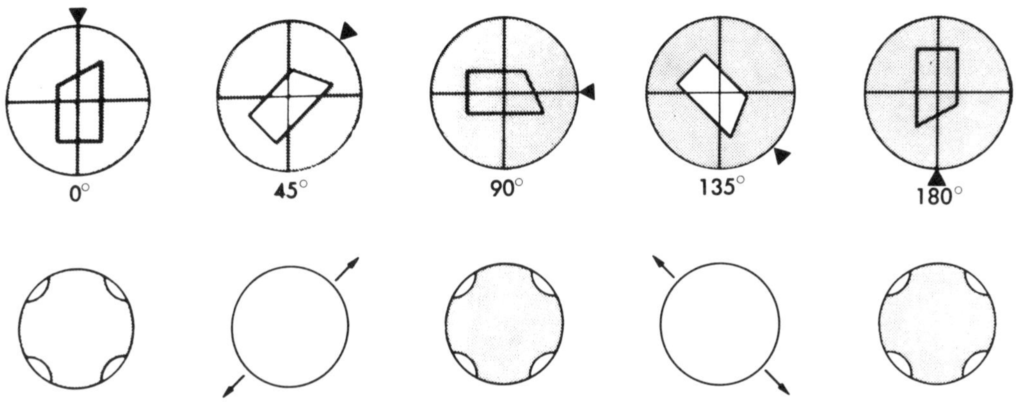

Fig. 8-6. Orthoscopic and conoscopic images of grain showing flash figure. The grain lies with *c* axis parallel to the stage. Arrows show direction in which isogyres have left the field.

Location of Vibration Directions. The vibration directions in the center of the field are E and O, associated with ϵ and ω. These are also the vibration directions of the orthoscopic image. This is the one uniaxial figure in which the vibration direction associated with the principal index of refraction ϵ appears in the center. It follows that grains showing this type of figure are the only ones furnishing the principal refractive index ϵ. The E and O directions are not readily distinguished in the figure at crossed position, but at 45° from the crossed position, in the flash figure and the associated orthoscopic image, E lies 45° from the crosshairs in quadrants from which the isogyres have left the field. This is also the *c* axis direction. Prismatic cleavage fragments of cancrinite, which are therefore elongated parallel to *c*, will serve to test this important generality.

Color Curves. Color curves in uniaxial flash figures are of little importance. They are hyperbolic, symmetrical with the center as illustrated in Fig. 8-7, and rotate about the center on rotation of the stage. They are obscured, in part, by the isogyres at the crossed position. At 45° from the crossed position one may observe, in some favorable instances, as in the cancrinite cleavages, that curves of lower color are in the quadrants in which the isogyres have left the field, in the E direction. This observation is occasionally useful in some crystals to check one's determination of which quadrants are those in which the isogyres have left. In crystals with strong birefringence, all colors are of such high order that they are not readily distinguished.

Fig. 8-7. Flash figure of a grain rotated 45° from extinction. Color curves reach higher orders in NW and SE quadrants, so isogyres left NE-SW, so optic axis lies NE-SW. Hauswaldt 6-4, calcite, sodium light.

SUMMARY

Table 8-2 summarizes information regarding the three types of uniaxial interference figures. All uniaxial figures provide a means of identifying O in the figure or orthoscopic image. The flash figure provides a means of identifying both E and O, both in the figure and in the orthoscopic image.

The identification of vibration directions, first in the figure and then

Table 8-2
Uniaxial Interference Figures

Name	Isogyres	Crystal Orientation	Section Viewed Orthoscopically	Vibration Direction in the Figure	Vibration Direction in Orthoscopic View
Centered optic axis (*n*-symmetry plane [SP])	Stationary cross	$\perp c$	B = 0	O and E′	O
Uncentered optic axis (1-SP)	Cross moving in circle; center may not appear	Neither $\parallel$ nor $\perp c$	B = low	O and E′	O and E′
Flash (2-SP)	Two hyperbolic branches which leave field	$\parallel c$	B = maximum	O and E	O and E

in the orthoscopic image in most cases, is the key to the determination of several optical properties of uniaxial substances. Absorption and pleochroism associated with each of the principal vibration directions may be determined. The optic sign is found by determining which is the higher index, ω or either ϵ or ϵ'. The details of the determination of these and other properties of uniaxial crystals are discussed in Chapter 9.

9 Identification of Uniaxial Crystals

PROPERTIES USED IN IDENTIFICATION

The properties of uniaxial crystals discussed in this chapter are of two kinds, optical and crystallographic. Chief among the optical properties used in identification in immersion mounts are those related to refractive index, optic sign, and color. In thin section, the most useful optical properties are relief, birefringence, and optic sign. Crystallographic properties that are most useful include those related to crystal habit, cleavage, and twinning.

AXIAL CHARACTER

Uniaxial crystals have one optic axis, one direction along which light travels with the same velocity. Crystals of the tetragonal and hexagonal systems are uniaxial. In Winchell and Winchell's tables of artificial inorganic crystals (1964) about 25 percent of the entries are uniaxial. Larsen and Berman (1934) found that about 21 percent of the 1,700 entries in their tables on nonopaque minerals were uniaxial. In the newer tables of Fleischer et al. (1984), about 23 percent of the 3,300 entries are uniaxial, about 15 percent negative and 8 percent positive. Important rock-forming minerals included in the uniaxial class are quartz, the rhombohedral carbonates, some of the feldspathoids, and many of the accessory minerals in rocks such as zircon, apatite, rutile, corundum, and tourmaline. Some members of the chlorite group are uniaxial and some biotite is uniaxial.[1]

[1] Uniaxial crystals may be considered as a group of biaxial crystals in which the angle between the two optic axes is zero for all wavelengths. Chlorite and biotite are normally biaxial, but at some compositions they degenerate to uniaxial character for certain wavelengths.

The uniaxial character of a grain is indicated unequivocally by certain interference figures. The only uniaxial figures that could not be mistaken for biaxial figures are those in which the isogyres form a cross, the center of which remains within the field throughout a complete rotation of the stage. This is a centered or nearly centered optic axis figure. The grains that should be chosen in the search for these typically uniaxial figures can be recognized in an assemblage of diversely oriented grains viewed orthoscopically between crossed polars. They remain dark or almost evenly lighted at all positions of rotation, or have very low interference colors compared with other grains of the same substance of similar thickness.

COLOR, ABSORPTION, AND PLEOCHROISM

Most colored uniaxial crystals show differential absorption of transmitted light related to vibration directions associated with ϵ and ω. If a different amount of some wavelengths of light is absorbed in different directions of transmission and vibration then the intensity of light in these directions is different. The difference in intensity of light in the two principal directions of vibration is *absorption*.

Absorption is reported by a formula that indicates which vibration direction, O or E, has the greater absorption (i.e., transmits the least light). The absorption formula is either $O > E$ or $O < E$. For example, in dravite, the brown magnesian tourmaline, the absorption formula is $O > E$.

If different wavelengths are absorbed in the O vibration direction than in E, the color of the crystal will be different when each of these vibration directions is observed separately. The difference in wavelength and intensity of color related to the different vibration directions is *pleochroism*. Along vibration directions corresponding to values of ϵ' intermediate between ϵ and ω the color is intermediate because the color changes continuously from O to E. In uniaxial crystals there are two principal vibration directions and hence in pleochroic uniaxial crystals there are two principal associated colors, so that instead of the general term pleochroism the more restrictive term *dichroism* is sometimes employed. It is customary to express pleochroism as formulas relating color to the vibration directions associated with the two refractive indices. In dravite, for example, the pleochroism is O = brown, E = light brown.

Recognition of absorption and pleochroic colors depends upon the intensity of the color, which in turn depends upon the thickness of the section and the strength of illumination. Data for minerals usually refer to sections 30 microns thick unless otherwise noted. Neither absorption nor pleochroism is determined quantitatively with the ordinary petrographic microscope, but they may be determined using methods discussed by Burns (1966; 1970) and Mandarino (1959).

Determination

The determination of pleochroism and absorption requires that the vibration directions O and E associated with the principal indices ϵ and ω be identified and the color associated with each observed separately in polarized light without crossed polars. Both O and E lie in the plane of the slide in a grain that forms a flash figure. E is identified by noting the quadrant from which the isogyres leave the field. E and O are then successively oriented parallel to the vibration direction of the lower polar. The grain is observed without crossed polars, and the color is noted. The pleochroism and absorption formulas are derived from these observations. Any grain forming an interference figure that allows identification of the O direction can be used to identify the color and intensity associated with the vibration direction O. Only the grain forming a flash figure can be used to identify the color associated with E.

Determination Without Interference Figures. A simple method of determining pleochroism and absorption of uniaxial crystals without the use of interference figures is illustrated in Fig. 9-1. It is applicable if many diversely oriented grains of a pleochroic crystal are present in a slide. The grains are observed while rotating the stage. Polars are not crossed. In such an assemblage the grains that do not change color on stage rotation (without crossed polars) must be basal sections and their color must be that of O. This is true because basal sections are the only sections that show only O vibration directions and only one color. In the sections that show the most marked change of color on stage rotation, the color which is not that of O must be the color of E.

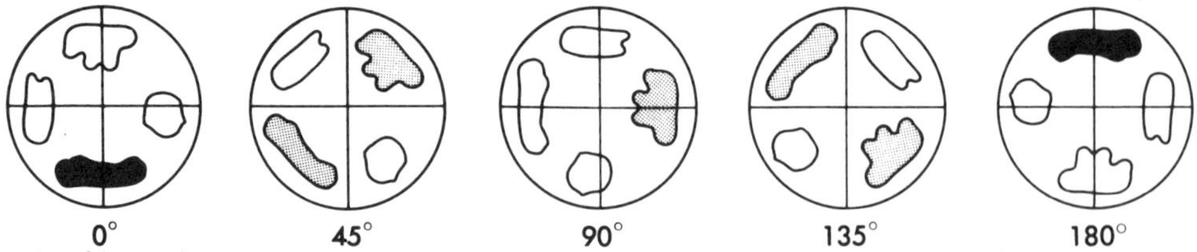

Fig. 9-1. Dichroic grains at several positions without crossed polars. The dichroic formula is O = white, E = black. The absorption formula is E > O.

Fragments of brown tourmaline provide effective examples of pleochroism and absorption. Pleochroism can be determined with or without the aid of interference figures, because this mineral has no good cleavage that would cause preferred orientation of grains in an immersion mount.

BIREFRINGENCE

In uniaxial crystals birefringence may be defined more specifically than for anisotropic substances in general (equation 1-13):

$$B = |\epsilon - \omega| \tag{9-1}$$

Birefringence is described as weak, moderate, strong, very strong, or extreme. Numerical equivalents (Larsen and Berman, 1934) of these adjectives are given in Table 9-1. Examples are cited of representative minerals not infrequently present in thin sections of rocks. Birefringence is in general stronger in uniaxial substances containing planar atomic groups such as carbonates than in those containing atomic groups of almost equidimensional configuration such as sulfates.

Estimation in Immersion Mounts

The birefringence of a crystal species is that of a grain lying parallel to the optic axis and containing both the O and E vibration directions. The methods of determining the birefringence of a particular species rather than that of a grain of unknown orientation are discussed in this section. The birefringence of sections of crystals without regard to optical orientation has been considered earlier, in Chapter 6.

The birefringence of a crystal may be calculated from the refractive indices. This is the method employed for crystals of very strong or extreme birefringence and for grains whose thickness is not known or not easily estimated. In immersion mounts, since the thickness of the grains is not known, this method is the practical one. The error in determination will include the error in both refractive index determinations, and the percentage error may be very large if the birefringence is small.

There are several very general measures of birefringence. An approximate measure can be obtained by noting the maximum interference color and the prevalence or infrequency of low-order colors of grains in a diversely oriented group despite the effect of variation in thickness of each grain. It can be easily calculated from equation 5-1 that crystals with weak birefringence will not exhibit colors higher than second order until they are about 100 microns thick or thicker, more than triple the thickness of a standard thin section. On the other hand, fragments of crystals with very strong or extreme birefringence will show colors of the lower three orders only in the few sections that are oriented normal or nearly normal to the optic axis. Examination of crushed grains, not specially sized, of each of the minerals apatite, zircon, and rutile, cited as examples of different strengths of birefringence in Table 9-1, will illustrate the effect of birefringence on interference color.

Table 9-1
Categories of Birefringence Strength with Examples[a]

Birefringence	Interference Color in Sections 0.03 mm Thick	Examples, and Birefringence of Example
Weak: 0–0.010	First-order gray, white, or yellow	Apatite: 0.003
Moderate: 0.010–0.025	First-order red to second-order green	Cancrinite: 0.023–0.029
Strong: 0.025–0.100	Upper second-order into fifth order	Zircon: 0.062
Very strong: 0.100–0.200	High order: sixth and higher	Calcite: 0.172
Extreme: 0.200 and higher	Very high order	Rutile: 0.285

[a]The numerical designations are those of Larsen and Berman (1934).

Birefringence may also be estimated from the number of color curves present in centered and uncentered optic axis interference figures. The birefringence, the thickness of the grain, and the angular aperture of the objective affect the number of rings that may be observed in an optic axis figure. In grains of comparable thickness, viewed with the same optical equipment, the number of rings increases, though not directly, with birefringence. A number of rings characterize figures from crystals of strong birefringence, and color curves are absent from figures from crystals of weak birefringence. Some of the following may be used as illustrative material if sections cut parallel to the base are available, or if, as in the instances noted, there is basal cleavage:

Quartz	B = 0.009
Alunite	B = 0.020 (cleavage)
Cancrinite	B = 0.026 (cleavage)
Meionite	B = 0.039
Zircon	B = 0.058
Calcite	B = 0.172

It is necessary to size the cleavage fragments by screening to make them of comparable thickness.

Estimation in Thin Section

Thin sections of standard thickness (30 microns; 0.03 mm) permit determination of birefringence by measurement of retardation. They also permit easy location of grains giving flash figures, since these grains have the highest interference colors of all grains in the section. Birefringence may be estimated by comparing this highest color with an interference color chart like that of Fig. 6-3.

The number of color curves in an optic axis figure may also be used to estimate birefringence in thin section, as in immersion.

Dispersion of the Birefringence; Abnormal Interference Colors. Dispersion of the birefringence occurs if birefringence varies with the wavelength of light used. Data on dispersion of the birefringence of crystals are scarce, and measurements must be made in monochromatic light, hence this property is seldom used in crystal identification.

When dispersion of the birefringence is sufficiently strong, *abnormal (anomalous) interference colors* result. These are so distinctive that they are helpful in crystal identification. Abnormal interference colors are those that do not appear in the regular color sequence as seen on an interference color chart or in a quartz wedge placed between two crossed sheets of Polaroid. They are most commonly blue or violet. They arise when the birefringence for some wavelengths is near zero, so that the crystal is effectively isotropic for those wavelengths. For example, if a crystal is isotropic for green and yellow wavelengths, these hues will be absent from the interference color, which will then be predominantly a combination of red and blue. This can be seen with the aid of Fig. 5-10 by imagining the intensity of all colors but red and blue to be practically zero. Abnormal interference colors can be seen in thin sections of certain kinds of chlorite that are nearly uniaxial. They are also not uncommon in members of the biaxial epidote group of minerals. Vivid green colors may be seen from near-optic axis sections of allanite.

OPTIC SIGN

Crystals for which ϵ is greater than ω have, by definition, a positive optic sign, and those for which ϵ is less than ω are negative. The high index vibration direction, D_h, is parallel to the optic axis and *c* in positive crystals (Fig. 7-4) and the low index vibration direction, D_l, is parallel to the optic axis and *c* in negative crystals (Fig. 7-5). These facts follow from the definition of the optic sign and the fact that E is parallel to the optic axis and to the *c* crystallographic axis of uniaxial crystals.

Far more uniaxial minerals are negative than positive. Of the tabulated entries in Fleischer et al. (1984), only 8 percent are uniaxial positive and 15 percent are uniaxial negative.

Determination

The optic sign may be determined by inspecting the values of the refractive indices ω and ϵ or ϵ' to determine which is larger, or it may be determined directly. Direct determination usually involves two steps: the identification of O and E directions (usually from interference figures) and the determination (using an accessory plate) of whether E is a higher or lower index vibration direction than O. Whether the E vibration direction chosen is related to ϵ or ϵ' makes no difference, for if one of these is a higher index direction than O, so is the other. Similarly, if ϵ' is lower than ω, so is ϵ. Accessory plates are used either to alter the colors in the interference figure or to alter the colors in the grain viewed orthoscopically in which the directions of vibration have been identified from the interference figure.

Use of Flash Figure. One recalls that a grain likely to produce a flash figure is one with high interference colors relative to others of the same crystal. When the isogyres of a flash figure leave the field, they leave in the direction of E (the *c* axis). E may therefore be positioned NE–SW (for example) by rotating 45° from extinction if the direction in which the isogyres leave is NE–SW. Once this is done, an accessory plate can be used to determine whether E is D_h or D_l relative to O. To do this, *the grain is viewed orthoscopically with crossed polars.* Assuming that D_h of the accessory plate is parallel to E of the grain:

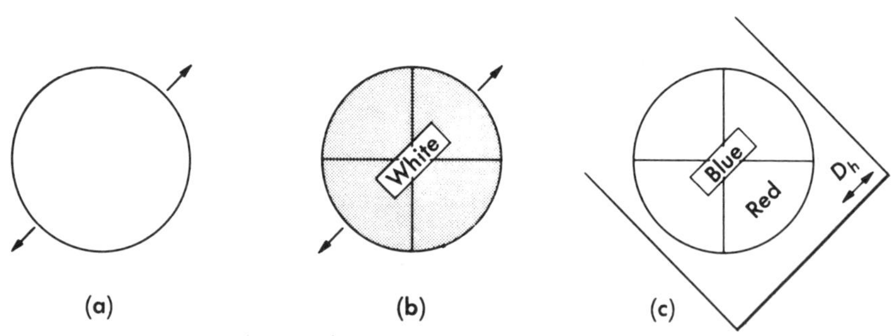

Fig. 9-2. Determination of optic sign from grain showing flash figure. (a) Figure 45° from cross. (b) Same but *without* Bertrand lens. (c) Same but with gypsum plate. Colors rise, E is D_h, $\varepsilon > \omega$, crystal (+).

If colors rise, E is D_h, crystal (+), ($\epsilon > \omega$)

If colors fall, E is D_l, crystal (−), ($\epsilon < \omega$)

The sequence of operations described here is illustrated in Fig. 9-2.

Use of Centered and Uncentered Optic Axis Figures. The flash figure and E are not necessary for optic sign determination. If a flash figure is not obtained, the sign may be determined from E′ in an optic axis figure, recalling that E′ is the vibration direction in the figure that is radial with respect to the center of the cross. The other vibration direction is O. That the radial direction is E′ and associated with a refractive index ϵ' is demonstrated in Fig. 8-4.

A centered or uncentered optic axis figure is likely to be formed by a grain showing relatively low interference colors between crossed polars. The figure can be oriented as in Fig. 9-3 to make E′ NE–SW, or parallel to D_h of the accessory plate. The accessory plate is inserted and the results noted *in the figure* rather than the orthoscopic image, since only the interference figure of a basal section has E′. E′ is not

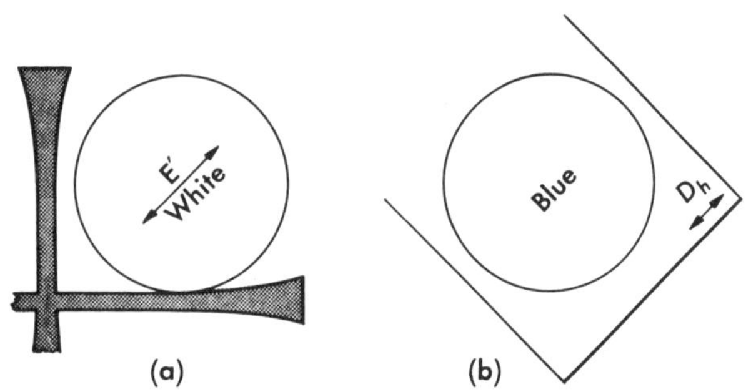

Fig. 9-3. Determination of optic sign with uncentered optic axis figure. (a) Figure rotated until E′ is NE. (b) Same, accessory plate inserted. Colors rise, E′ is high *n*, crystal (+).

visible in the orthoscopic image. The interference colors noted are either those of the whole field, as in Fig. 9-3, or those of whichever quadrant (usually NE or SW) has E′ parallel to D_h of the accessory plate, as in Fig. 9-4. Using these conventions:

If colors rise, crystal (+)

If colors fall, crystal (−)

There is some advantage in adopting a standard procedure so that in all determinations of optic sign with accessory plates the signal "rising colors" means "crystal (+)." This is the scheme adopted in this book. Other schemes can be used, but they may lead to confusion unless one consistent pattern is followed. After some experience is gained, the interpretation of signals from the accessory plate becomes a reflex action, as do most operations with the petrographic microscope. This fact permits concentration on the variables and problems of crystal identification rather than on procedure.

Gypsum Plate. The use of accessory plates with the interference figure rather than with the grain viewed orthoscopically offers some possibility of confusion. The part of the optic axis figure adjacent to the center of the cross is always gray and white of the first order. If the birefringence is low, the entire area of each quadrant is gray or white. The vibration directions are as in Fig. 9-4(a); E′ is radial. Insertion of a gypsum plate causes the area formerly black to become first-order red, the interference color of Gips alone. Two of the quadrants show rising colors (blue), and two show falling colors (yellow) because the gray interference color of the field is either added to or subtracted from the red of the gypsum plate. Which two quadrants are yellow, and which blue, depends upon the sign of the crystal, as shown in Fig. 9-4(b).

Quartz Wedge. When birefringence is high enough to cause one or two color curves in the optic axis figure, Gips may still be used, either by observing the gray area adjacent to the cross or by observing the curves themselves. When curves become numerous, however, the quartz wedge is used. Confusion is possible, but easy to avoid by use of a standard procedure. When the quartz wedge is introduced *thin edge first* [from the NW in Fig. 9-4(c)] and kept moving, color curves either *contract* toward the cross or expand away from it depending on whether the colors are rising or falling. Contracting color curves indicate rising colors because the higher-order colors near the margin of the field are displacing the lower-order colors adjacent to the cross; the reverse is true of expanding color curves. The rule below applies also to biaxial optic axis figures.

The quartz wedge causes color curves in interference figures to move relative to the point of emergence of the optic axis:

If they contract, colors rise
If they expand, colors fall

If attention is limited to the NE or SW quadrant, or to whichever quadrant contains E′ parallel to D_h of the quartz wedge, contraction implies crystal (+) and expansion implies crystal (−). There is no analogy between the direction of movement of color bands on edges of crystal fragments and color curves in interference figures, and the rules for the two effects should not be associated.

Color-blind people should use the quartz wedge whenever possible, since the motion it causes eliminates the need to identify colors.

Illustrations of Sign Determination. To illustrate the determination of the optic sign using centered or uncentered optic axis figures it is convenient to observe cleavage flakes of the strongly birefringent mineral calcite. The quartz wedge will be the appropriate accessory plate to use because of the strong birefringence. With interference figures from cleavage plates about 0.03–0.05 mm thick of the mineral alunite the gypsum plate is a suitable accessory. The weak birefringence of this mineral results in there being few or no color curves and therefore large areas of first-order colors in the interference figure.

Examples that may conveniently be used for finding the optic sign employing the flash figure are a quartz crystal mounted on the prism or cleavage fragments of the mineral cancrinite.

REFRACTIVE INDEX

Determination in Immersion Mounts

Determining refractive index in immersion mounts involves the same method as for isotropic substances except for preliminary steps. Before comparison of the index of solid and liquid, a vibration direction associated with a principal refractive index must be identified (often from an interference figure). The orthoscopic image must be observed and rotated to the extinction position at which the vibration direction will be parallel to the polarizer vibration direction. Only then can the associated refractive index be compared with the refractive index of the immersion medium using conventional tests.[2]

Since ω is to be found in any grain it is useful to determine how ω compares with the refractive index of the immersion medium and also find the sign. Thereafter ω can be identified in any grain with an accessory plate superimposed on the orthoscopic image rotated 45° from extinction. One of the two vibration directions is surely O, the other E′ or E. If the crystal has a negative sign the high index direction is O. It is to be noted again that the other refractive index in a randomly

[2] A discussion indicating that this fact is not always appreciated appears in Emmons and Gates (1957). For an incorrect practice that is approximately valid with minerals of low birefringence see p. 13 of Hess (1960).

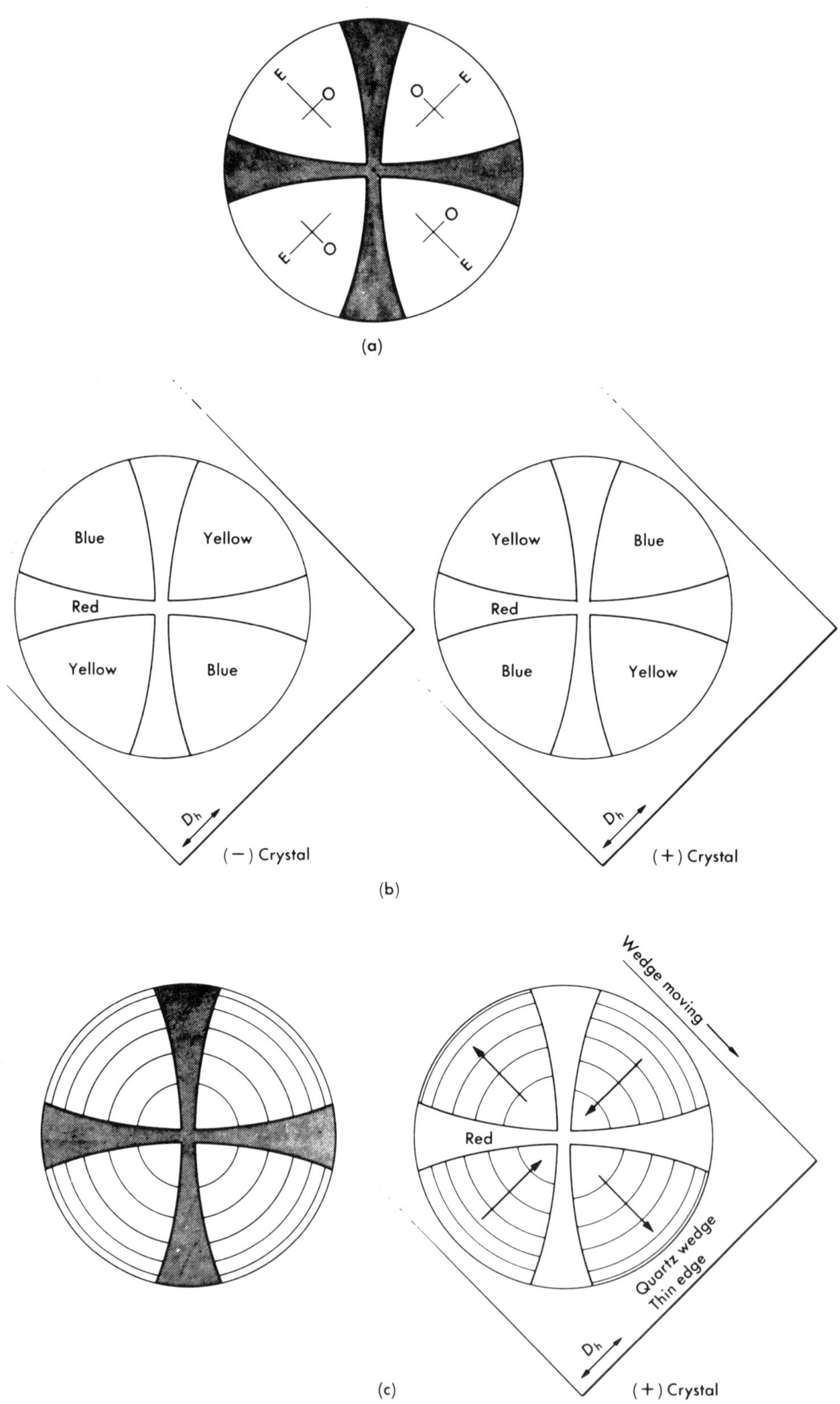
E
O
E
O
E
O
O
E
(a)
Blue
Yellow
Red
Yellow
Blue
Dh
(−) Crystal
Yellow
Blue
Red
Blue
Yellow
Dh
(+) Crystal
(b)
Wedge moving
Red
Quartz wedge
Thin edge
Dh
(c)
(+) Crystal

oriented section is ϵ', *not E,* unless the grain exhibits a flash figure. To find a grain oriented so that the principal index ϵ can be measured, a centered flash figure must be obtained. This should be the second object of search.

Random Search Methods. It is efficient to seek, by refractive index tests, uniaxial grains that have E, before seeking flash figures to confirm that *c* really lies parallel to the stage. Knowledge that *c* lies parallel to the stage, and that the full value of E can be tested, is useless unless the immersion liquid has a good chance of being close to ϵ. A random search[3] is made by observing many grains throughout the slide under oblique illumination and noting from color fringes or shadows the relative index of liquid and crystals. Suppose ϵ is the higher refractive index [crystal (+)]. Liquids of successively higher index are used in the immersion mounts, usually in a leapfrog manner so that the value of E is approached alternately from above and below. A quick glance with oblique illumination may show the presence of yellow color fringes on some grains, indicating that some crystal index is higher than the liquid. Since ϵ is the higher index, this must be ϵ or ϵ'. There is no point in examining this slide further; a new mount of higher index should be made at once. As long as yellow fringes appear, ϵ of the crystal is sure to be high. Finally a mount is made in which almost all grains show brown color fringes, but a few, in certain positions of rotation, show orange-red color fringes indicating a match with the liquid. These few grains are prime candidates for having *c* parallel to the stage, and their interference figures should be examined. If they show centered flash figures, the search for ϵ is over.

If the crystal is negative, the search proceeds in a reverse fashion; as long as brown or red color fringes appear, a new mount with lower liquid is needed.

The initial search for ϵ may be aided by removing the lower polar, so that the color fringe depends only on the inclination of *c* to the stage, without respect to stage rotation. Such a search may turn up grains having the best chance of showing the ϵ value, which can then be confirmed in the usual way with polarized light, setting the grain to extinction after observing that a flash figure is obtained.

[3] Statistical method of some authors.

Fig. 9-4. Determination of optic sign with centered optic axis figures. (a) Vibration directions. (b) Gypsum plate inserted; negative and positive crystals. (c) Color curves in a figure from a grain of high birefringence.

Caution is needed in the use of random search methods, for if a prominent cleavage prevents *c* from lying parallel to the stage, only ϵ' will be observed. The random search method can still be used to set limits on the value of ϵ.

Random search methods allow the determination of the optic sign and both ω and ϵ without use of accessory plates or interference figures, a fact of some interest to beginning students. Immersion mounts may be made until every grain, at one extinction position or another, matches the liquid. This liquid must give the value *of* ω, which is the only index of constant value from grain to grain. A grain is then turned to the other extinction position to examine the relation of ϵ or ϵ' to the liquid. If this index is high, the crystal is positive; if low, negative. The search for ϵ may then proceed as before.

Confirmation of epsilon. Once a grain is found that is considered likely to give ϵ, either from its high interference colors or from the fact that one of its refractive indices matches the liquid suspected of having a value like ϵ, confirmation of the crystal orientation should be sought from the flash figure. If this is centered, *c* lies parallel to the stage, and all is well. E and *c* lie in the direction in which the isogyres leave the field. If the flash figure is not centered, or if an uncentered optic axis figure results, another grain should be sought.

Use of Pleochroism. Knowledge of the pleochroism formula may offer a shortcut in refractive index determination. Every grain will, at one or the other extinction position, show the pleochroic color of O. If this color is known, the O direction may be recognized in every grain without having to use an interference figure or accessory plate. The use of pleochroic colors to distinguish between ϵ' and the principal value of ϵ is not effective, for the color change is gradational between O and E. The pleochroic formula may be determined without interference figures or an accessory plate as shown in Fig. 9-1.

Effects of Prominent Cleavage Not Parallel to *c*. Crystal fragments with prominent cleavage parallel to a rhombohedron frequently lie in immersion mounts so oriented that no fragment will allow determination of the principal value of ϵ. The constant value of ϵ' on the cleavage is associated with E′, the radial direction in the uncentered optic axis figure formed by the cleavage fragment. The value of ϵ' on $(10\bar{1}1)$ of calcite and some few other substances is recorded in collections of optical data. If it is not recorded it may sometimes be calculated. Knowing the values of ϵ, ω, and the angle between the normals to the cleavage and the basal pinacoid (the interfacial angle $0001 \wedge 10\bar{1}1$), ϵ'

may be calculated from equation 9-2. The value of any radius ϵ' of an ellipsoid of revolution with axis of revolution 2ϵ and radius ω normal to 2ϵ is:

$$\epsilon' = \frac{\omega\,\epsilon}{\sqrt{\omega^2 \sin^2\theta + \epsilon^2 \cos^2\theta}} \tag{9-2}$$

in which θ is the angle between the normal to the radius ϵ' and the axis of revolution. θ is therefore the angle $0001 \wedge 10\bar{1}1$. This equation has been used by Loupekine (1947) to prepare graphs relating ϵ', ω, and ϵ in carbonate minerals. In these graphs the average value of the angle between *c* and the normal to the cleavage for the carbonate group is used as θ. Accuracy is ± 0.004, but is increased if one has some knowledge of the chemical composition of the carbonate so that a more nearly correct value of θ may be used.

Equation 9-2 may of course also be used to compute ϵ if ϵ', ω, and θ are known.

Use of the spindle stage (Chapter 15) allows rapid and unequivocal determination of all principal refractive indices despite the presence of prominent cleavages.

Determination in Thin Section

Determination of refractive index in thin section is devoted to comparisons of indices of the specimen with the index of the adjacent material using Becke line or oblique illumination methods. Frequently, both indices are higher or both are lower; one need not then consider the orientation of the grains. Occasionally (in minerals for which one index is greater than the mounting medium and one less), it may be necessary to observe each vibration direction separately and make an index comparison. Comparison of indices in thin section is seldom conclusive evidence as to the identity of a mineral, but in some special cases is very effective in distinguishing minerals with otherwise similar properties. The carbonates are an example: calcite, magnesite, and dolomite containing up to about 40 percent $(Mn,Fe)CO_3$ have ϵ less than 1.54, the index of common mounting media. In rhombohedral carbonates with greater than about 40 percent $(Mn,Fe)CO_3$, ϵ is higher than 1.54.

Relief

Relief in uniaxial crystals is due to the difference in index of refraction between crystal and surrounding medium just as in isotropic crystals. In uniaxial crystals, however, relief varies with orientation of the grain. Each vibration direction has characteristic relief when observed separately if the vibration direction is first oriented parallel to the vibration direction of the lower polar.

In immersion mounts relief is a rough measure of the nearness of match between the index of refraction related to the vibration direction being observed and the index of refraction of the immersion liquid. In thin section, relief relative to the adjacent mineral or the mounting medium of the slide is useful in estimating the difference in refractive index.

The amount of change of relief on rotation of the stage is a function of the birefringence of the section of the grain that lies in the plane of the slide. This in turn is related to the crystallographic orientation of the grain and the birefringence of the crystal. Basal sections show no change of relief, and a maximum change is shown by grains parallel to *c*. In quartz, for example, because the birefringence is small, the change of relief even in prismatic sections is scarcely noticeable. The rhombohedral carbonate minerals such as calcite, which have a high birefringence, show marked change in relief on rotation, unless the grain is normal or almost normal to *c*. (The change of relief on rotation of the stage when very marked is called "twinkling" by English writers on optical crystallography.) The above remarks have little validity if both refractive indices of the crystal are very different from the refractive index of the immersion medium, except insofar as cracks in the crystal may show change in relief.

Dispersion of the Refractive Indices. Dispersion affects both refractive indices of uniaxial crystals; ω and ϵ both change with the wavelength of light used. Dispersion may be measured by determining both refractive indices with monochromatic light of several different wavelengths. It is rarely a diagnostic quantity by itself, but is often determined in the course of crystal identification by the dispersion method (Chapter 16).

SIZE AND CRYSTAL FORM

Crystal size, while not important to identification, is often of interest for petrogenetic reasons. Crystal form and habit are often valuable in identification and almost always of some concern insofar as they affect measurement of optical properties.

The size, shape, and crystal habit of uniaxial crystals are conveniently studied with crossed polars, which may reveal, within aggregates, crystal boundaries that were unsuspected in plain light. The area of a single crystal, or rarely two adjacent crystals of similar orientation, is optically similar as indicated by simultaneous extinction of all parts.

Single crystals in which extinction is not simultaneous over the entire area include strained crystals that exhibit undulatory extinction and most twinned crystals in which extinction positions in the two parts are not parallel because of the difference in orientation of two parts of the twin.

Grains of tetragonal and hexagonal crystals are commonly elongate, acicular, or prismatic. This is unlike the more common equidimensional habit of isometric crystals. In thin section there are usually several differently oriented cross-sections of the same mineral available for study. In immersion mounts the external form of a crystal or the cleavage planes of a fragment often control the grain orientation in the mount. Both tend to produce elongate development of fragments of tetragonal and hexagonal crystals. Mounted in nonviscous media, cleavage flakes and euhedral grains lie in preferred orientations related to the largest flat surface developed on each grain. All possible cross-sections are not, therefore, equally available for study.

Crystals grown on slides often lie in a preferred orientation. The shape of the crystals coupled with the orientation determined from an interference figure may be diagnostic if the conditions of growth and method of preparation of the sample are taken into consideration.

The angle between the crystal faces in grains is the silhouette angle. Measurement of this angle in grains of known orientation would seem to offer possibilities of identification. Measurements are useful as confirmatory information of a general sort, but are subject to large errors because the measured angle varies as the plane of the section, in which the angle is measured, departs from being perpendicular to the faces. Measurement of silhouette angles is of little or no value in differentiating between members of a solid solution series because the variation in interfacial angle with compositional change is usually smaller than the error in measurement.

Elongation

The elongation of a euhedral crystal of specific habit may be coincident with either the high or low index vibration direction in the crystal. This is the basis for defining the *elongation of euhedral* crystals as *positive or negative:* positive if the high index vibration direction is parallel to the length, and negative if the low index vibration direction is parallel to the length.

To determine the sign of elongation, the elongate grain, oriented with the long axis approximately 45° from extinction, is viewed between crossed polars. An appropriate accessory plate is inserted and if the

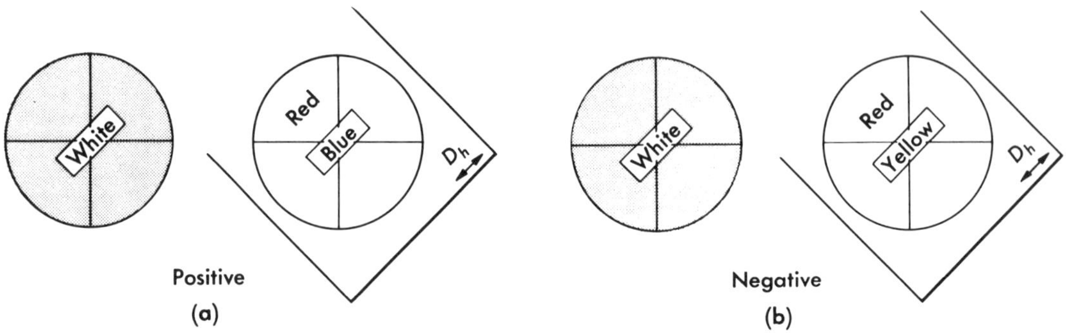

Fig. 9-5. Positive and negative elongation determined with the gypsum plate.

high index vibration direction in the plate and the long axis of the crystal are both NE–SW as illustrated in Fig. 9-5:

1. Colors rise [Fig. 9-5(a)]
   High index direction in plate and crystal are parallel
   Crystal has positive elongation

or

2. Colors fall [Fig. 9-5(b)]
   High index direction in plate is parallel to the low index direction in the crystal
   Crystal has negative elongation.

Some authors use the terms *length-slow* for positive elongation and *length-fast* for negative elongation.

Euhedral crystals of quartz and apatite serve as examples of crystals with positive and negative elongation, respectively.

The orientation of E and O in elongate grains not bounded by crystal faces is likely to be due to the cleavage or fracture that results during the preparation of the sample, so the vibration directions are not necessarily related to the length of the crystal. The sign of elongation applies only to grains whose elongation is a consequence of crystal form.

CLEAVAGE

General

Cleavage in uniaxial crystals is most commonly basal, prismatic, or rhombohedral. Pyramidal cleavage is rare. Minerals of the alunite group and certain uniaxial chlorite and micas, as well as some other micaceous minerals such as brucite, exhibit basal cleavage. Rhombohedral cleavage is characteristic of the common rhombohedral carbonate minerals. Prismatic cleavage is characteristic of several of the uniaxial zeolites. Pyramidal cleavage is characteristic of scheelite and a few other minerals.

In Thin Section

Cleavage is identified in thin section by determining the relation between the traces of cleavage planes (cracks) and vibration directions. Vibration directions are located by extinction positions. Sections showing the trace of basal cleavage show parallel extinction. Prismatic cleavage, parallel to the *c* axis, also shows parallel extinction in any section parallel to *c* and therefore in the vertical zone. Sections normal to prismatic cleavage are also normal to the optic axis, and *c*, and hence equally dark at all positions of stage rotation.

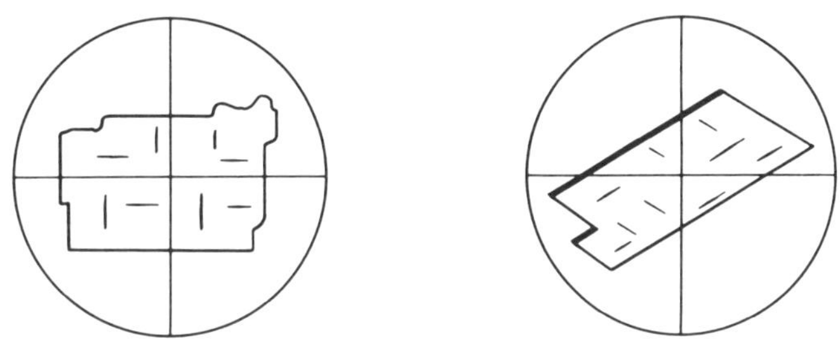

(a) (b)

Fig. 9-6. Cleavage fragments. (a) Basal and prismatic. (b) Rhombohedral cleavage. If polars were crossed, both fragments would be at extinction.

Prismatic and basal cleavages, illustrated in Fig. 9-6(a), can be differentiated in sections showing parallel extinction by determining which cleavage is parallel to the E vibration direction (and therefore *c*) and which is parallel to O (and therefore in the plane normal to *c*). Vibration directions are identified as O and E from an interference figure, or if the optic sign is known, simply by use of an accessory plate.

Rhombohedral cleavage is illustrated in Fig. 9-6(b). Grains showing traces of rhombohedral cleavage are characterized by symmetrical extinction in all sections parallel to a crystallographic axis. In other sections extinction is not symmetrical. The carbonate minerals in a thin section of marble illustrate rhombohedral cleavage.

In Immersion Mounts

The trace of a cleavage plane appears in immersion mounts as a straight edge, or the cleavage plane forms the flat upper surface of a fragment lying on a cleavage just as in isometric crystal fragments. Parallelism of the top and bottom surfaces of the grain is indicated by the persistence of the same interference color over the whole central area of the grain. The two parallel surfaces thus indicated may be presumed to be cleavages lying in the plane of the slide. This use of interference colors is of course impossible in isotropic crystals. The crystallographic orientation of cleavage is identified from the type of extinction or interference figure shown by these flat-topped cleavage fragments. If a large number of the fragments produce centered optic axis interference figures and are extinct or gray at all positions of rotation between crossed polars, basal cleavage is indicated. If many flat-topped fragments show flash figures, prismatic cleavage is indicated. Symmetrical extinction and an uncentered optic axis figure identify fragments lying on a rhombohedral cleavage.

Crushed fragments of the following minerals will serve as examples of cleavage as indicated:

Cancrinite	Prismatic
Calcite	Rhombohedral
Melilite	Basal
Nepheline Willemite from Franklin, N.J. }	Prismatic distinct Basal imperfect

TWINNING

To determine that a crystal is twinned, the composition plane must be recognized, unless external forms can be observed that are characteristic of a twinned crystal. The composition plane may be recognized between crossed polars whenever the vibration directions and hence the extinction positions are not parallel in two parts of the twin. If twinning is polysynthetic, several composition planes are present and alternate segments or twin lamellae extinguish simultaneously. Polysynthetic twin lamellae may be sharp or diffuse. As a rule the lamellae are thinnest and sharpest when normal to the plane of the stage.

Twinning is rarely observed in minerals of the hexagonal system, hexagonal division. Minerals of the hexagonal system, rhombohedral division, exhibit several types of polysynthetic twinning. In calcite the composition plane of polysynthetic twins may be the base (0001), the positive rhombohedron $(10\bar{1}1)$, or the negative rhombohedron $(01\bar{1}2)$. Simple twins of calcite also follow other laws. In dolomite that has been deformed, polysynthetic twinning on a steep rhombohedron $(02\bar{2}1)$ is characteristically developed, but this twin is not usually present in calcite. Twinning is common in quartz, but it cannot be observed with the petrographic microscope because the composition plane is parallel to the *c* axis and hence the vibration directions in both parts of the twin are parallel. Other less common laws in minerals of the tetragonal and hexagonal systems are not commonly encountered.

Identification of the twin law is suggested by identification of the crystallographic orientation of the composition plane. This is determined by the relationship of the composition plane to the cleavage or to an optical direction determined from an interference figure. The relationship of the composition plane to the extinction position may be pertinent. If the crystal does not have rhombohedral cleavage and extinction is not parallel to the composition plane of the twin, the supposition is that the twin law is one of the pyramidal laws.

DETERMINATION OF CRYSTAL SYSTEM

There are no optical properties that distinguish tetragonal from hexagonal crystals, but several crystallographic properties may be useful. The shape of optic axis sections of euhedral grains may be suggestive: they are frequently hexagonal in hexagonal crystals and square or oblong in tetragonal crystals. Cleavage in optic axis sections is not commonly encountered, but is occasionally diagnostic: if cleavage directions are at approximately 60° the hexagonal system is indicated; if 90°, the crystal is tetragonal. The angle between the traces of cleavages or crystal faces may only be measured approximately. Unless basal cleavage is also present, optic axis sections that lie normal to prismatic cleavages will be found only in thin section.

WORKED EXAMPLE OF UNIAXIAL CRYSTAL IDENTIFICATION

We first review the observations in white light, as before, and then discuss the application of the dispersion method.

White Light

The crushed grains were immersed in a 1.630 oil after some trials at higher and lower indices. The observations were:

1.630 oil:
Crystal clear—nonopaque, not colored
Straight edges with parallel extinction—high symmetry implied

Inclusions like rods—characteristic?
Low gray interference colors on elongate grains, to dark on some rounder grains—low birefringence probable.

The following observations apply only after the grain was turned to extinction and the analyzer then removed, to make sure that a vibration direction lay parallel to the polarizer.

Omega (every grain) color fringe orange-yellow, xl hi by about 0.005 (Fig. 3-15)

Epsilon (rare grains) color fringe red-orange, xl hi by about 0.002. Flash figure seen.

Provisional birefringence -0.003

Optic axis figure confirms uniaxial $(-)$

The data indicate $\omega = 1.635$, $\epsilon = 1.632$, $B = -0.003$, uniaxial negative. The low birefringence is distinctive, and allows identification from the tables as fluorapatite. The nearest entries in USGS 1627 with birefringence < 0.017 are both fluorapatite: 1.633, 1.630, $B = -0.003$; 1.638, 1.634, $B = -0.004$. Note that the optic sign was deduced from the relative abundances of the matching refractive indices before it was confirmed with an interference figure.

Dispersion Method

A worked example of the identification of a uniaxial mineral is shown in Fig. 9-7 using the same worksheet (Fig. 16-4) as used in the isotropic example in Chapter 4. In this case, the birefringence was small and both refractive indices were obtained in the same oil by the dispersion method.

The basic observations are the shuttle readings and temperatures recorded in the first two columns of the worksheet. The sheet is devised so that it can be used for biaxial and uniaxial crystals. The center column is used for uniaxial omega, and whichever other column is then used as appropriate, depending on whether epsilon is low (as here) or high. The calculations are done as required by the lines in the worksheet, and as described for an isotropic example in Chapter 4. The wavelength of match is obtained from the monochromator manufacturer's tables. The temperature is obtained by thermometer. Because the same oil matches both indices (at different wavelengths), the calculation is the same for both columns except in matters related to wavelength.

For the crystal, the dispersion is obtained from the formula as 0.011.

REFRACTIVE INDEX WORKSHEET

SAMPLE NO. AP-1

	Epsilon ~~ALPHA α~~ (NLOW)	~~BETA β~~ (OMEGA)	GAMMA γ (NHI)
Oil:	1.630	1.630	____
Shuttle reading	105 mm	94 mm	____ mm
Lambda match = λ_0	568 nm	541 nm	____ nm
k_D, from table	0.1140	0.2791	____
t match	23.8 °C	23.8 °C	____ °C
dn/dt (bottle)	− 4.1 E-4	− 4.1 E-4	− ____ E-4
$n_D{}^L = n_{25} + [(t-25)(dn/dt)]$	1.63049	1.63049	____
Δ^L(F-C) (bottle or graph)	0.0255	0.0255	____

Crystal:

Calculate Δ^S(F-C) = 10↑(5.9 log n - 3.2) or = exp (5.9 ln n -7.368):

Provisional Dispersion, Δ^S(F-C)	0.011	0.011	____
$n_D{}^S = n_D{}^L + k_D(\Delta^L - \Delta^S)$	1.6321	1.6345	____

(If warranted, recycle with refined values of the crystal dispersion)

$\gamma - \beta$ = 0 B = 0.0024 $Vx \approx \arcsin\sqrt{(\gamma - \beta)/B}$ = 0

$2V_{calc}$ — $2V_{obs}$ — from figure —

Sign (−) Cleavages — — — Color, etc. colorless

Optic orientation E = c Crystal system (Hex.)

Distinguishing features Inclusions, parallel ext.

Mineral Name Apatite Composition ~$Ca_5(PO_4)_3F$ (prob.)

Observer Morse Date Oct. '87, '89

Fig. 9-7. Worked example of uniaxial crystal identification, using the worksheet of Fig. 16-4.

(The accepted value for apatite is 0.010, so recycling of the calculation is not warranted.) The final results of the calculation are, as shown, $\omega = 1.6345$, $\epsilon = 1.6321$, $B = -0.0024$. These are within 0.0005 and 0.0001 of the values determined in white light, and can be regarded as refinements of the earlier values. The accuracy of the dispersion determination on this easy material is estimated at ± 0.0005 or better.

10 Biaxial Crystal Optics

VIBRATION DIRECTIONS AND REFRACTIVE INDICES

Biaxial crystals, as their name indicates, have two optic axes along which light is propagated with the same velocity. In every other wave propagation direction there are two plane polarized components at right angles to each other and to the propagation direction; these have unequal velocities. The angle between the optic axes is a useful variable related to the optic sign.

There are three principal indices of refraction: alpha (α), beta (β), and gamma (γ). β is the index associated with light propagated along an optic axis and is thus analogous to ω of uniaxial crystals. By convention, α, β, and γ are respectively *always* the least, intermediate, and greatest refractive indices of the crystal. This is unlike the uniaxial case, where either ϵ or ω may be greater, depending on the optic sign. The principal refractive indices α, β, γ and are associated respectively with vibration directions called X, Y, and Z. Alternative designations of refractive index and vibration direction are summarized in Table 10-1.

Table 10-1
Nomenclature of Principal Biaxial Refractive Indices and Associated Vibration Directions

	Refractive Index		Vibration Direction	
Relative Size of Refractive Index	In Most Common Use	Less Often Used[a]	In Most Common Use	Less Often Used
Lowest	α	n_α, n_p, N_p, N_X, n_X, X	X	α
Medium	β	n_β, n_m, N_m, N_Y, n_Y, Y	Y	β
Highest	γ	n_γ, n_g, N_g, N_Z, n_Z, Z	Z	γ

[a]The symbols n_p, n_m, and n_g represent French terminology. Subscripts *p, m,* and *g* signify, respectively, *petite, moyen, grand* (small, intermediate, large). *X, Y,* and *Z* were adopted for *Dana's System of Mineralogy* (Palache et al., 1944, p. 47). Boldface Greek symbols have also been used, but it is probably desirable to reserve boldface for vector quantities.

Indicatrix

The biaxial indicatrix (Fig. 10-1) is a geometrical figure summarizing the refractive indices of all vibration directions in a crystal. The form is a triaxial ellipsoid. Any vector drawn from the origin to the surface of the indicatrix is a vibration direction for light propagated normal to it, and the length of the vector is a refractive index of the crystal for such light.

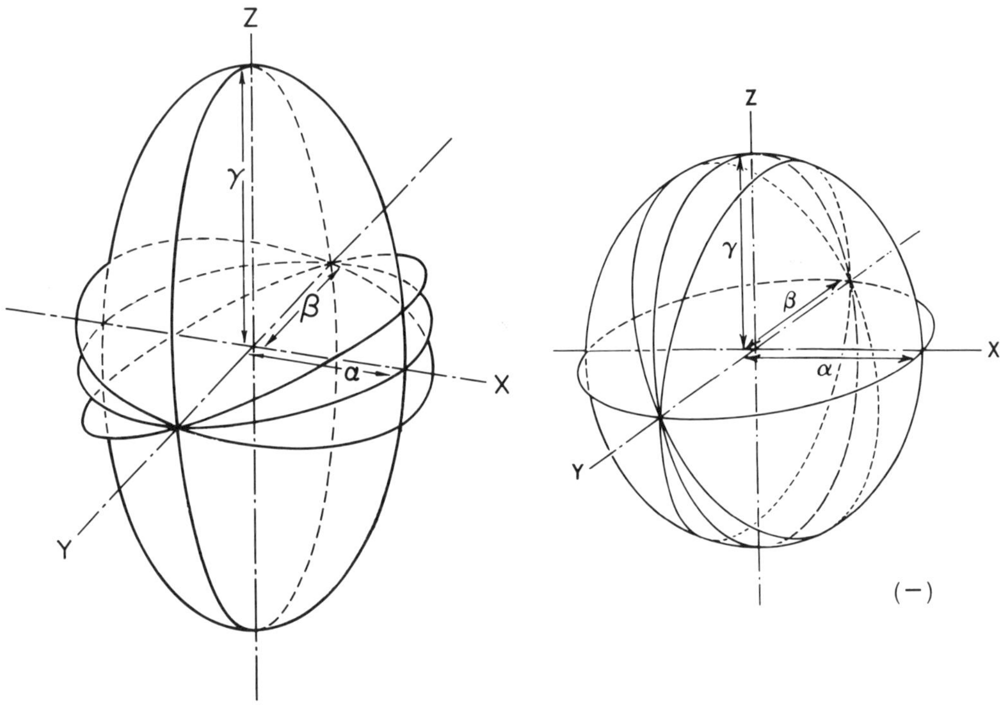

Fig. 10-1. Important planes in positive and negative biaxial indicatrices. External surfaces not shown. Circular sections nearer X in positive indicatrix and nearer Z in negative indicatrix.

The biaxial indicatrix has a long axis of length 2γ (which is the Z vibration direction), an intermediate axis at right angles of length 2β (which is the Y vibration direction), and a short axis normal to the other two of length 2α (which is the X vibration direction). There are three symmetry planes: XY, XZ, and YZ. There are two circular sections. The directions normal to these are the optic axes. Models to aid in visualizing the indicatrix may be made from transparent plastic following the suggestions published by Smith (1938).[1]

[1] It is recommended that markings on the model, not properly part of the indicatrix, be omitted. For "radius" [pp. 632 and 634 of Smith (1938)] read "diameter."

In using the indicatrix, one assumes it to be located and properly oriented at the center of a crystal or crystal fragment. Any slice through the crystal or fragment is assumed to cut through the center of the indicatrix. Any section of the indicatrix is an ellipse whose axes show the directions of vibration of light propagated normal to that section. The half-lengths of the axes give the refractive indices of the two plane polarized components having these vibration directions. Two special sections intersecting in the Y axis are circular sections. If a section of the indicatrix is not perpendicular to any one of the three symmetry planes, both refractive indices will be partial indices not equal to any of the three principal indices. If there is no coincidence with Y (the common situation), the higher refractive index of the section is designated γ' and its vibration direction Z′. Similarly, the lower refractive index is designated α' and its vibration direction X′.

Symmetry Planes. A section of a crystal may be perpendicular to one or two planes of symmetry, or to none. A 1-symmetry plane section is sketched in Fig. 10-2. Such a section may have any orientation normal to the symmetry plane, other than those in which it is normal to two planes. Any 1-symmetry plane section contains one of the principal indicatrix axes (vibration directions) X, Y, or Z. If the symmetry plane perpendicular to the section is

XZ, the contained axis is Y
XY, the contained axis is Z
YZ, the contained axis is X

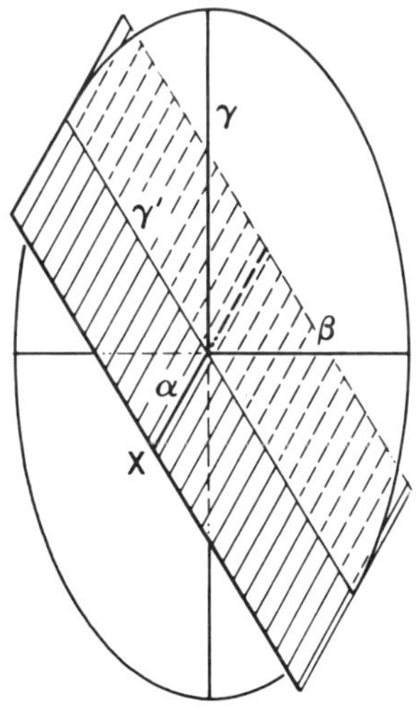

Fig. 10-2. Perspective view of a 1-symmetry plane section of the biaxial indicatrix. The section (ruled) is normal to the YZ plane, and therefore contains X.

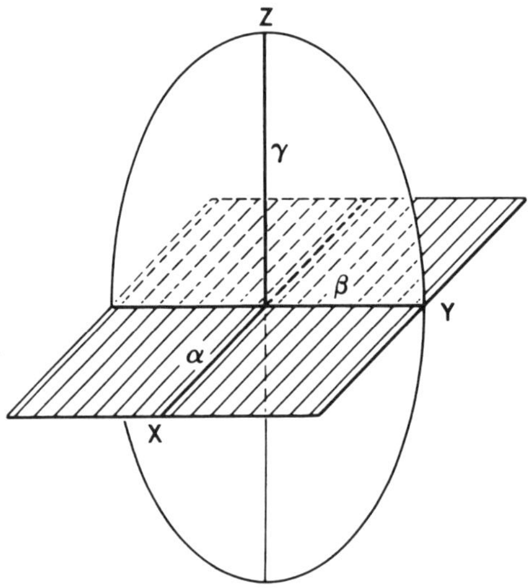

Fig. 10-3. Perspective view of a 2-symmetry plane section of the biaxial indicatrix. The section (ruled) is normal to both the YZ and the XZ planes, and is therefore the XY plane.

Therefore, any 1-symmetry plane section furnishes one of the principal indices β, γ, or α (in the order given to the contained vibration directions above), and another index which is a partial index, α′, rarely β′ or γ′.

A 2-symmetry plane section (Fig. 10-3) contains two of the principal vibration directions and is normal to a third; in other words, a 2-symmetry plane section is *itself* a plane of symmetry. Such a section furnishes two of the three principal refractive indices α, β, and γ.

Circular Sections and Optic Axes. The existence of the circular sections can be elucidated as follows. A triaxial ellipsoid contains a plane XZ having semiaxes of length α and γ. The axis normal to this plane, Y, has a semiaxis of length β intermediate between α and γ. In any quadrant of the ellipse XZ there must be some vector from the origin to the perimeter having magnitude β, since β is a number between α and γ. A plane containing this vector and the axis Y (whose vector length is β) must therefore be a section with semiaxis β in at least two directions. Since the ellipsoid is a smooth form without reentrants, the section must be part of a circle of radius β.

In reality there are four vectors of length β in the full XZ ellipse, one in each quadrant. The vectors in opposite quadrants are paired against each other, and are thus contained in the same circular section. There are therefore two circular sections having Y as the common axis of intersection. The vectors in two adjacent quadrants may be at any angle to each other between 0° and 180° inclusive. Hence the circular sections may lie anywhere symmetrically to one another from near X (positive indicatrix, Fig. 10-1) to near Z (negative indicatrix, Fig. 10-1).

The angular departure from X or Z of the vector which is the radius β depends on how near the magnitude β is to α or γ.

As in the case of uniaxial crystals, light propagated normal to a circular section is not plane polarized; it vibrates along any vector of magnitude β, and has a single velocity. The normals to circular sections are therefore designated optic axes, hence the name "biaxial." The angle between the optic axes is the supplement of the obtuse angle between the circular sections themselves, and is one of the measurable parameters used to describe biaxial crystals.

Note that the optic axes are directions of wave propagation, not vibration directions related to a principal refractive index.

Terminology of the Indicatrix. Some terms used in the description of biaxial optics are conveniently listed with explanatory comments:

Optic axes: Normals to the circular sections of the indicatrix. Sometimes designated separately as A and B, for although they are indistinguishable in the indicatrix, they may be differentiated by their relation to crystallographic directions.

Optic plane: Plane containing the optic axes, always the XZ plane; normal to Y, normal to the circular sections.

Optic normal or ON: The vibration direction (indicatrix axis) normal to the optic plane XZ; it is the axis Y.

Acute bisectrix or Bxa: One of the vibration directions (indicatrix axes) in the optic plane, X or Z, whichever bisects the acute angle between the optic axes.

Obtuse bisectrix or Bxo: The vibration direction X or Z in the optic plane which bisects the obtuse angle between the optic axes.

Optic angle (axial angle, optic axial angle) or 2V: The angle between the optic axes, i.e., the angle between the normals to the circular sections. The symbol for this angle is 2V, V being the angle between an optic axis and a bisectrix. There are two optic angles that sum to 180°, one measured around X and the other measured around Z (Fig. 10-4). If no specification is made, the acute angle is understood to be implied by "2V." Other notations are sometimes used:

$2V_X$ or $2V_\alpha$ indicates the optic angle measured over X, ranging anywhere between 0° and 180°.

$2V_Z$ or $2V_\gamma$ is the optic angle measured over Z, ranging anywhere between 0° and 180°.

$2V_{ac}$ is the acute optic angle.

$2V_{ob}$ is the obtuse optic angle.

Optic sign: By convention, positive (+) if Z lies in the acute optic angle (Bxa = Z), and negative (−) if X lies in the acute optic angle (Bxa = X). Hence $2V_x$ ranges between 0° and 180°, and the optic sign changes at 2V = 90°.

The Optic Angle in the Crystal and in Air or Oil. The Bxa, Bxo, and optic angle are illustrated in diagrams of the optic plane (XZ plane) in Fig. 10-4.

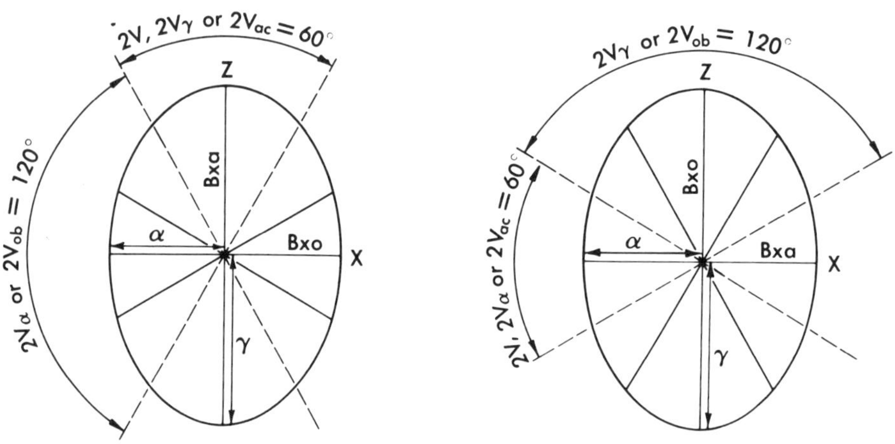

Fig. 10-4. Features of the optic plane in positive and negative crystals. Dashed lines (the optic axes) are perpendicular to traces of the circular sections.

Oblique light may strike a Bxa section (section normal to the acute bisectrix) at such an angle that it travels along the optic axes. Such light propagated along the optic axes is refracted upon entering air, as shown in Fig. 10-5. The angle between the paths of the refracted wave propagation directions is the *apparent* optic angle in air, 2E. This angle is always greater than 2V; from equation 1-11 (Snell's law, page 18),

$$\sin E = \beta \sin V \tag{10-1}$$

If an oil immersion objective is used, light leaving the microscopic preparation enters oil rather than air. The apparent optic angle in immersion oil (often of conventional refractive index 1.515) is designated 2H. If n_o is the index of the oil,

$$n_o \sin H = \beta \sin V. \tag{10-2}$$

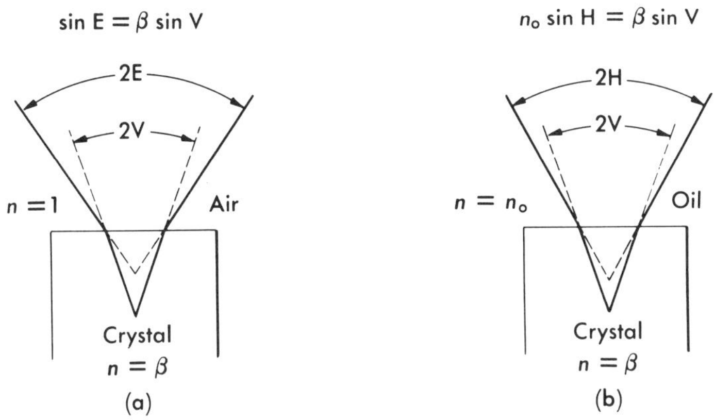

Fig. 10-5. Apparent optic angle in air or oil, related to true optic angle 2V in the crystal. (a) In air, apparent angle 2E; (b) in oil, apparent angle 2H.

Vibration Directions: Biot–Fresnel Law. The directions of vibration of the polarized components in any section of a biaxial crystal depend upon the orientation of the section with respect to the axes of its indicatrix. These directions may be found for any crystal section by application of the Biot–Fresnel law (Biot, 1820; Fresnel, 1827). This law states that one of the vibration directions is the bisector of the angle made by the two construction planes, each containing an optic axis and the normal to the section. These planes are shown in Fig. 10–6 in which *ON* is the wave propagation direction.

Jean-Baptiste Biot demonstrated the relationship between the vibration directions and the optic axes experimentally, and Augustin Jean Fresnel produced the theoretical proof. It is ironic that these two names should be joined in the law attributed to them, for Biot was a champion of the corpuscular theory of light, while Fresnel was the chief architect of its discreditation and supplantation by the wave theory.

Mathematical Relationships. Any point on the surface of the indicatrix is located, in terms of the conventional rectangular coordinates, x, y, and z by the indicatrix equation

$$\frac{x^2}{\alpha^2} + \frac{y^2}{\beta^2} + \frac{z^2}{\gamma^2} = 1 , \qquad (10\text{-}3)$$

which is the equation of a triaxial ellipsoid with semiaxes α, β, and γ.

The optic angle and the refractive indices are related by several expressions (Johannsen, 1918, 1968, pp. 103–4):

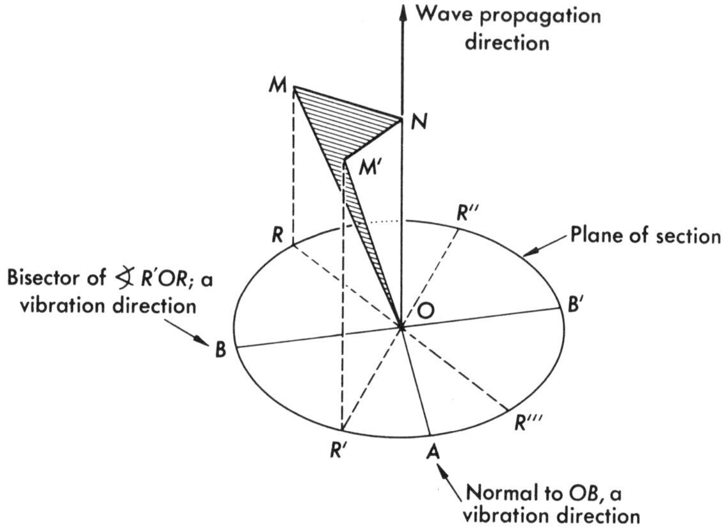

Fig. 10-6. Illustration of the Biot-Fresnel law, showing the vibration directions *OB* and *OA* as bisectors of planes formed by the optic axes (*OM, OM'*) and the wave propagation direction (*ON*).

$$\cos V_x = \sqrt{\frac{\gamma^2(\beta^2 - \alpha^2)}{\beta^2(\gamma^2 - \alpha^2)}} \qquad (10\text{-}4)$$

$$\tan V_x = \sqrt{\frac{\alpha^2(\gamma^2 - \beta^2)}{\gamma^2(\beta^2 - \alpha^2)}} \qquad (10\text{-}5)$$

$$\sin V_x = \sqrt{\frac{\alpha^2(\gamma^2 - \beta^2)}{\beta^2(\gamma^2 - \alpha^2)}} = \sqrt{\frac{\beta^{-2} - \gamma^{-2}}{\alpha^{-2} - \gamma^{-2}}} \qquad (10\text{-}6)$$

If the birefringence is weak it may be shown (Wright, 1951) from equation 10-4 that

$$\cos V_x \approx \sqrt{\frac{(\beta - \alpha)}{(\gamma - \alpha)}} \qquad (10\text{-}7)$$

or

$$\cos 2V_z \approx \frac{(\gamma - \beta) - (\beta - \alpha)}{(\gamma - \alpha)} . \qquad (10\text{-}8)$$

From equation 10-8 it can readily be seen that when $2V = 90°$ ($\cos 2V = 0$), β is about halfway between α and γ. Similarly, from equation 10-6,

$$\sin V_x \approx \sqrt{\frac{(\gamma - \beta)}{(\gamma - \alpha)}}\,; \quad V_x \approx \arc\sin \sqrt{(\gamma - \beta)/B}\,. \qquad (10\text{-}9)$$

Nomograms for the determination of the optic angle from partial birefringences are given on page 17 of Fleischer et al. (1984). However, it will often be found preferable and simpler to make the calculation with a hand calculator, using equation 10-8 or 10-9, or, if the birefringence is too strong, one of the other equations. The limitations of Wright's approximation can readily be evaluated for the purpose at hand. If an error of 3 percent in the calculated 2V can be tolerated, equations 10-8 or 10-9 can be used in cases where B is as large as 0.100, i.e. bordering on very strong.

Optic Sign. By convention, the optic sign is:

$$(+) \text{ if Bxa} = Z$$

$$(-) \text{ if Bxa} = X$$

That the optic angle may be acute around either X or Z can be seen with the aid of Fig. 10-1; the critical factor is the relative size of the indicatrix axes α, β, and γ. More specifically, when the right-hand side of equation 10-8 is positive (β closer to α than γ), $\cos 2V_z$ is positive and the optic angle is less than 90° (acute) around Z; Z is the Bxa. When β becomes closer to γ than α, $\cos 2V_z < 0$ and $2V_z > 90°$, i.e., X is the Bxa. These rules will not apply accurately when $(\gamma - \beta) \approx (\beta - \alpha)$ and $2V \approx 90°$, as mentioned above.

A simple way to remember the generalities of the relation between the optic sign and the refractive indices is to recall the manner in which the biaxial indicatrix degenerates to a uniaxial indicatrix. When β is nearer γ, the circular sections tend toward radius γ, and the indicatrix tends toward a uniaxial negative ellipsoid with $\epsilon = \alpha$ and $\omega = \gamma$. The biaxial sign is then (−), with optic angle acute around X. If β is near α, the circular sections tend toward radius α, and the indicatrix tends toward a uniaxial positive ellipsoid with $\epsilon = \gamma$ and $\omega = \alpha$. The biaxial sign is then (+), with optic angle acute around Z. The uniaxial indicatrix is a special case of the more general biaxial indicatrix in which 2V = 0° for all wavelengths.

Structural Basis for the Optic Sign

The size of $2V_x$ or $2V_z$ (hence the sign) is a function of the principal refractive indices (equation 10-2). The refractive indices in turn are functions of the crystal structure. In general, as discussed in Chapter

7, pronounced planar orientation of anions or groups such as CO_3 or SiO_4 has the effect of retarding waves vibrating parallel to them more than waves vibrating normal to them. The more retarded waves have the higher indices of refraction. Sheet structures with Si_4O_{10} layers such as micas are usually negative, although when the stacking and cation content of tetrahedral and octahedral layers is more complex, as with chlorites, a number of positive varieties are found. Silicate minerals with double chain structures (amphiboles) tend to be negative. Most silicates with single chains (pyroxenes except hypersthene) are positive. Of the independent tetrahedral silicates (nesosilicates), more are positive than negative, but the exceptions are numerous.

OPTIC ORIENTATION

The relationship between optical directions and crystallographic directions is termed *optic orientation.* Whereas in uniaxial crystals this is always fixed by symmetry with E = *c*, in biaxial crystals there may be no coincidence of indicatrix axes and crystal axes. The optic orientation is described completely by the angular relationship among X, Y, and Z, and *a, b, c* for all wavelengths.[2]

Optic orientation and crystallographic symmetry are fundamentally related by crystal structure, hence, optic orientation must be consistent with the symmetry of the crystal, and is therefore useful in deducing the crystal system. Specifically, orthorhombic crystals have all three indicatrix axes X, Y, and Z, coincident with crystal axes in any of six combinations. In monoclinic crystals, the unique diad axis (*b*) coincides with an indicatrix axis; this also guarantees that a symmetry plane of the indicatrix lies normal to *b*, parallel to the morphological mirror plane of the holosymmetric class. In triclinic crystals, the indicatrix can have any orientation. In monoclinic and triclinic crystals, there may be chance coincidence of indicatrix and crystal axes not demanded by symmetry.

Diagrammatic Representation of Optical and Crystallographic Properties

Many optical measurements in biaxial crystals are made in sections perpendicular to two of the three planes of symmetry of the ellipsoid,

[2] Optic orientation may vary slightly with temperature, but this is not of importance in crystal identification.

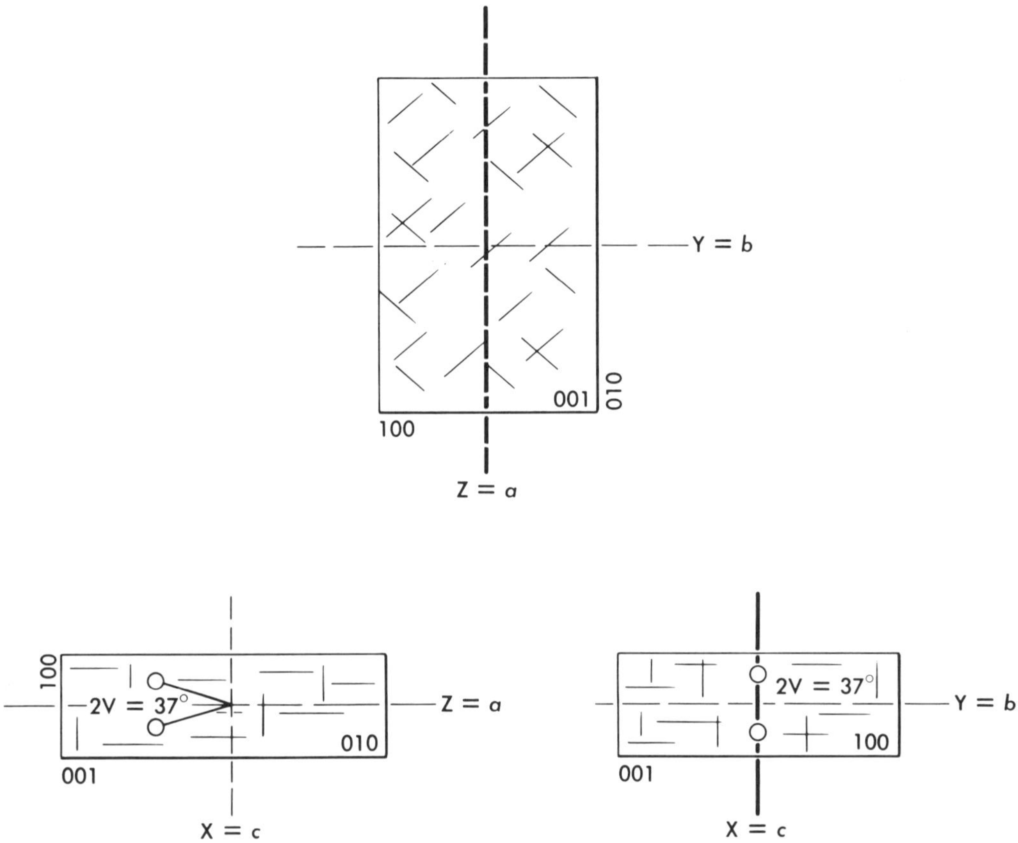

Fig. 10-7. Orientation diagrams of three planes of barite.

XY, XZ, or YZ. In these sections the major and minor axes of the elliptical section of the indicatrix are the axes of the indicatrix, the refractive indices are the principal indices, and the interference figures are centered 2-symmetry plane figures. The planes XY, YZ, and XZ are respectively perpendicular to the Bxa, Bxo, and optic normal directions in positive crystals and the Bxo, Bxa, and optic normal directions in negative crystals. It is often convenient to diagram, without drawing the indicatrix, the symmetry planes XY, XZ, or YZ individually, or chosen crystallographic planes. The diagrams can show optical features such as the optic axes and vibration directions and crystallographic information such as the traces of the cleavage planes or composition planes of twins. These diagrams, frequently referred to as orientation diagrams, appear in one form or another in almost all collections of optical information to illustrate the optic orientation. Examples are

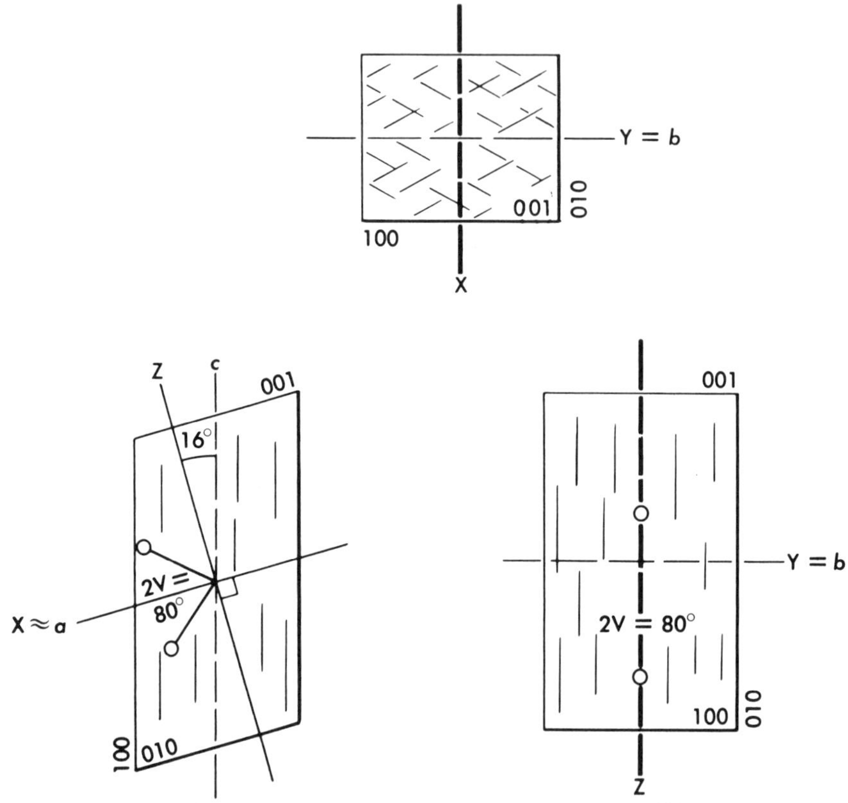

Fig. 10-8. Orientation diagrams of three planes of hornblende.

sketched in Figs. 10-7 and 10-8. Various conventions are used in such diagrams. Further discussion of this important topic is reserved to Chapters 13 and 14, in which perspective drawings are included.

Dispersion of the Vibration Directions

The orientation of the indicatrix in a crystal may or may not be the same for all colors of light. When it is not, the crystal is said to show dispersion of the vibration directions, which are the symmetry axes of the indicatrix. When vibration directions coincide with crystal axes for reasons of symmetry, they must do so for all colors, and therefore cannot be dispersed. This restriction therefore applies to the three orthorhombic symmetry axes and to the unique monoclinic *b* axis. There is a rare exception (crossed axial plane dispersion) in which vibration directions coincide with entirely different crystal axes for different colors. Dispersion of the vibration directions is common for the three triclinic and two monoclinic vibration directions where it is permitted; see, for example, Fig. 12-11.

Dispersion of the vibration directions may affect extinction, as mentioned in Chapter 6. Further discussion of dispersion and its utility is presented in Chapter 12.

11

Biaxial Interference Figures

Biaxial interference figures are similar to uniaxial figures but show more variety in the shape and movement of isogyres. Biaxial figures often contain only a single isogyre, which is not always parallel to a crosshair. One new term must be introduced: a *melatope* is the point of emergence of an optic axis. It is always a point on an isogyre.

CLASSIFICATION

An effective classification of interference figures useful in describing both uniaxial and biaxial figures has already been discussed in Chapter 8, and summarized in Table 8-1. In this classification the figures are divided on the basis of how many (if any) symmetry planes of the indicatrix are perpendicular to the plane of the microscope stage. The three biaxial categories are 0-, 1-, and 2-symmetry plane figures; *n*-symmetry plane figures can only result from uniaxial crystals, as discussed in Chapter 8. An interference figure may be assigned to the appropriate category by observing the shape and movement of isogyres on stage rotation. If one symmetry plane of the indicatrix is normal to the stage, one isogyre is coincident with a crosshair four times in a complete rotation of the stage. The reasons for this behavior can be deduced by an extension of the argument used in connection with the uniaxial centered optic axis figure (Fig. 8-2). For the 1-symmetry plane (1-SP) figure, consider only the converging wave propagation directions contained in the north–south plane of the microscope, i.e., those that impinge on the N–S crosshair. Each propagation direction has associated with it two mutually perpendicular vibration directions. These vibration directions are the major and minor axes of that elliptical section of the indicatrix which is normal to the wave propagation direction in question. If a symmetry plane of the indicatrix is normal to the stage and north–south, one of these ellipse axes lies in the symmetry plane and the other lies normal to it. This is true for any of the convergent wave propagation directions impinging along the N–S crosshair. Hence all of these propagation directions have vibration directions that are N–S (lying in the symmetry plane) and E–W (normal to the symmetry plane). If the vibration directions of the polars are N–S and E–W, as usual, the region along the N–S crosshair will be at extinction; an isogyre will be formed parallel to the crosshair.

If the symmetry plane is now made E–W, the isogyre lies along the E–W crosshair, since light convergent in the E–W plane must have vibration directions contained in and normal to the symmetry plane, as before. With continued rotation, each of the above circumstances is repeated once, and the isogyre is therefore coincident with a crosshair four times on complete rotation. It can be shown that an isogyre is coincident with a crosshair only if a symmetry plane is vertical and N–S or E–W. Note that the behavior of the isogyre while the symmetry plane is *not* N–S or E–W is unspecified in the above discussion. The isogyre may do anything from rotating on a point at the center of the field to leaving the field altogether, and it may rotate counter to stage rotation or in the same direction, depending on whether it is the optic plane or some other symmetry plane that is normal to the stage.

If two symmetry planes of the indicatrix are so oriented in a grain that they are perpendicular to the stage, the isogyres of the 2-SP interference figure coincide with the crosshairs four times during every complete rotation of the stage. They form a cross each time they coincide with the crosshairs. The origin of the cross can be deduced by analogy with the above discussion of 1-symmetry plane figures. The cross forms at the four extinction positions of the orthoscopic image. At other positions of stage rotation the isogyres break into a hyperbola; they may or may not remain in the field. Note that in either 1- or 2-SP figures an isogyre when straight marks the position of a symmetry plane.

Two-symmetry plane figures are centered figures; examination of the biaxial indicatrix (Fig. 11-1) will reveal that there are three sections that give such figures. These sections are perpendicular to the three axes (X, Y, Z) of the indicatrix.

Biaxial interference figures are named for the significant indicatrix direction that emerges in the center of the field (i.e., perpendicular to which the section is cut). These directions are the Bxa, Bxo, optic normal (ON), and optic axis. Figures named for these directions are shown diagrammatically in Fig. 11-1. There are also figures in which one of these four directions emerges within the field but not at its center. These figures are called uncentered Bxa, uncentered Bxo, uncentered

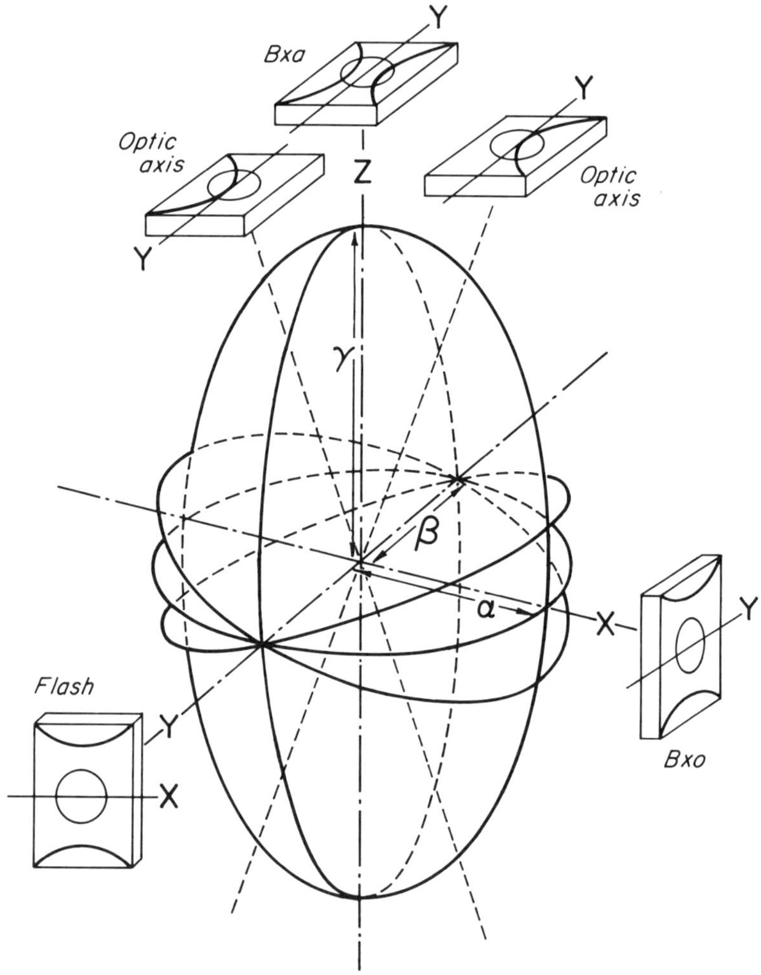

Fig. 11-1. Relationship of interference figures to a positive indicatrix at the 45-degree position.

optic normal, and uncentered optic axis figures. The centered and uncentered optic normal figures are also known as centered and uncentered flash figures.

There are two types of very useful uncentered 1-SP figures in which the Bxa, Bxo, ON, or optic axis emerges outside the field. In this book these are named the "same-rotating 1-symmetry plane figure" (SR) and the "counter-rotating 1-symmetry plane figure" (CR). Table 11-1 summarizes the characteristics of biaxial interference figures in more detail than does Table 8-1.

2-SYMMETRY PLANE FIGURES

General Description

There are three types of 2-symmetry plane figures, each from a section normal to an indicatrix axis. All of them consist of two isogyres forming a cross coincident with the crosshairs at each of the four extinction positions. The cross breaks into a hyperbola at other positions of rotation. The branches of the hyperbola leave the field except in the case of the Bxa figure when 2V is small or moderate. The branches always leave the field in the direction of the optic plane. Two-symmetry plane figures are named for the vibration direction (indicatrix axis) normal to the section: Bxa, Bxo, or ON. This is the vibration direction coincident with the optical path of the microscope; the vibration directions in the center of the figure are those in the plane of the stage and hence those of the orthoscopic image.

Table 11-1
Characteristics of Biaxial Interference Figures

0-Symmetry Plane Figures
No isogyre is coincident with a crosshair at any position of stage rotation.
No special names.

1-Symmetry Plane Figures
One isogyre is coincident with a crosshair at four positions of stage rotation.

The symmetry plane is the optic plane—counter-rotating (CR) figures: One end of the isogyre that coincides swings counter to the direction of stage rotation.
Names: Uncentered Bxa or Bxo if two isogyres are present.
Centered optic axis if one isogyre rotates on a point in the center of the field.
Uncentered optic axis if one isogyre rotates on a point that describes a circular path within the field.
CR 1-SP figure if none of the above.

The symmetry plane is the XY or YZ plane—same-rotating (SR) figures: One end of the isogyre that coincides swings in the direction of stage rotation.
Names: Uncentered flash, Bxa, or Bxo if two isogyres are present.
SR 1-SP figure if only one isogyre is present.

2-Symmetry Plane Figures
Two isogyres are both coincident with the crosshairs at four positions of stage rotation.
Names: Bxa if perpendicular to XZ plane and the Bxa. If 2V is small, isogyres will never leave the field.
Bxo if perpendicular to XZ plane and the Bxo. Isogyres always leave the field.
Flash or optic normal if perpendicular to both XY and YZ planes. Isogyres always leave the field.

The Bxa Figure

Since the three 2-symmetry plane figures are closely related, a detailed description of the elements of one will suffice in principle for the others. The Bxa figure of small 2V is chosen.

The interference figure formed by a section normal to the Bxa direction consists of two isogyres of hyperbolic form. The two branches of the hyperbola join to form a cross when the crystal is at extinction. As the stage is rotated through 45° from the crossed position the branches of the hyperbola move out from the center in opposite quadrants. They leave the field entirely only if the angle subtended by the field is less than 2E of the crystal. As the stage is rotated through the next 45° the branches of the hyperbola return in the same quadrants and join to form a cross at the other extinction position. During the rotation between 90° and 180° the isogyres leave and return in the other two quadrants. The sequence is repeated on further rotation. The movement of the isogyres is illustrated in Fig. 11-2 for a crystal in which the isogyres do not leave the field.

The vertices (nearest points of the two branches) of the hyperbola at the 45° position mark the points of emergence of the optic axes, the melatopes. The transverse axis (line connecting the vertices) is therefore the trace of the optic plane. This is a useful fact, since it follows that if the isogyres leave the field, they leave *in the direction of the optic plane*. In the case of the Bxa figure they leave in the direction of the Bxo, which is the other bisectrix in the optic plane. There are many occasions when it is useful to know where the Bxo or the trace of the optic plane lies. For example, the interference figure of Fig. 11-2, rotation position 45°, shows that the trace of the optic plane lies NE–SW when the trace of the cleavage in the orthoscopic image lies NE–SW. It may be concluded from this that the optic plane lies parallel to the cleavage, a fact relating to optic orientation, which is discussed further in Chapter 13.

Vibration Directions. The isogyres can be taken as the loci of all points in the figure where the vibration directions are N–S and E–W. The form of the isogyres and color curves can be explained in a non-rigorous fashion by the use of simple diagrams. It should be emphasized that a complete theory takes into account rotation of the plane of polarization by the refracting surfaces of the lenses and the crystal, as noted by Kamb (1958).

To determine the orientation of the vibration directions associated with each oblique wave propagation direction passing through the crystal one may apply the Biot–Fresnel law (page 165) and project the

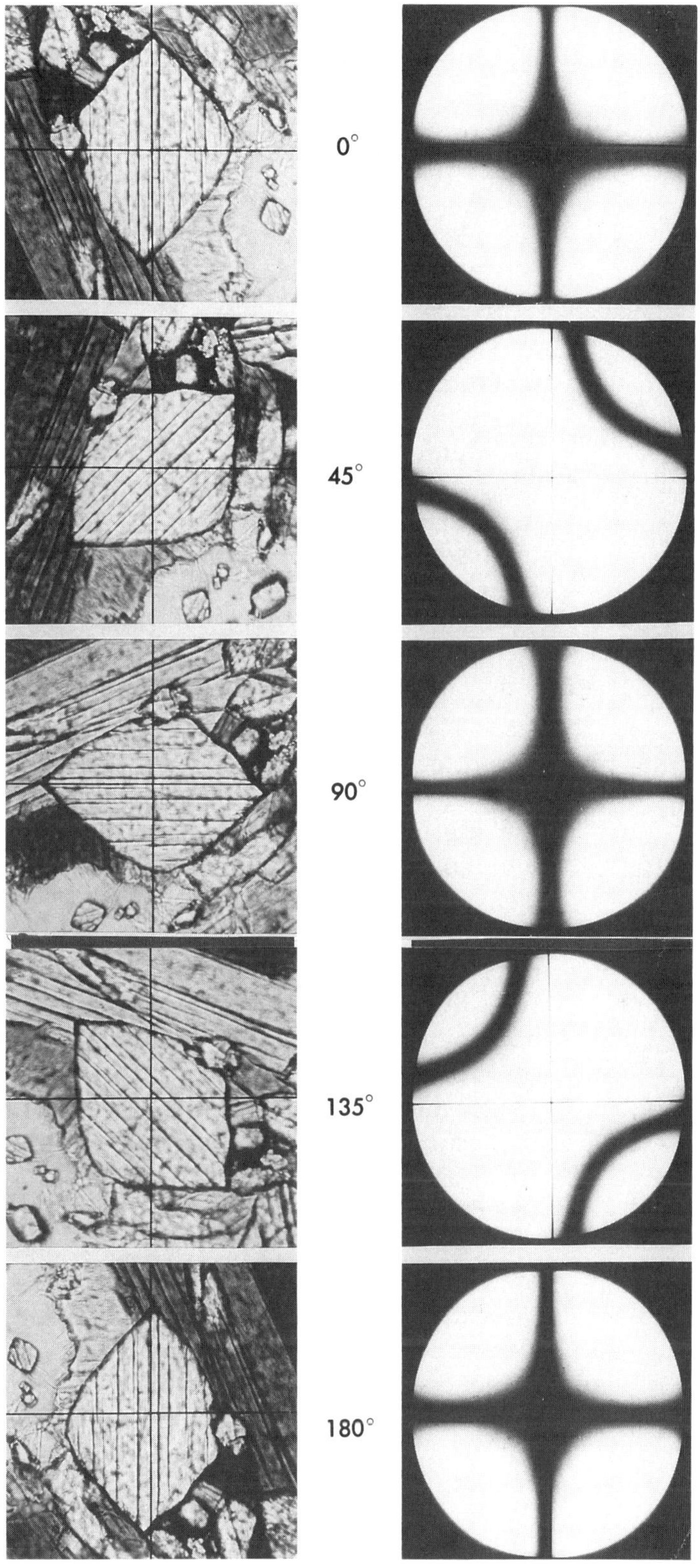

Fig. 11-2. Centered Bxa figure (right column) and orthoscopic view of the crystal (left column). The crystal is a thin section of sillimanite, Al_2SiO_5, orthorhombic, viewed down the *c* axis. The crystal (surrounded by pleochroic biotite) is bounded by prism faces, and has one excellent cleavage parallel to (010), shown by the fine cracks. The optic plane is parallel to this cleavage, as demonstrated by the interference figure. Note that the optic plane isogyre is thinner than the other isogyre at the crossed position. (Thin section of sillimanite-biotite schist from the Santa Lucia Mountains, CA, courtesy of Prof. R.A. Wiebe.)

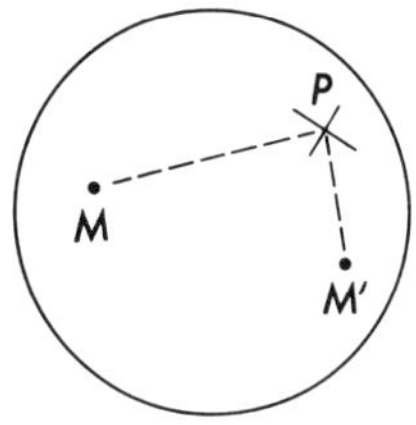

Fig. 11-3. Approximation of the Biot-Fresnel law in a Bxa figure. One of the vibration directions at point *P* is the bisector of the angle *MPM*′. *M* and *M*′ are melatopes.

directions thus ascertained onto the plane of the interference figure. There is, however, a graphical simplification of the Biot-Fresnel law that determines the approximate orientation of the vibration directions at any point in an interference figure. Connect the point in question to each melatope by two straight lines. It is an approximate rule that the bisector of the angle formed by these two lines is a vibration direction; the other vibration direction is normal to this, as shown in Fig. 11-3. This approximate rule approaches exactness when the wave propagation direction makes a very small angle with the optic axes.

The relationship between the approximate rule and the exact formulation is illustrated in Fig. 11-4. If the angle between the wave propagation direction *ON* and the optic axes *OM* and *OM*′ is small, *VV*′, the bisector of angle *MPM*′, is very nearly the projection in the plane of the interference figure of *BB*′ as shown by Chapman (1940). *BB*′ is the vibration direction in the plane *RBR*′*A* determined by the Biot–Fresnel law (Fig. 10-6), and *VV*′ is its analogue in the interference figure. To explain the isogyres, one needs to know the orientation of the vibration directions throughout the field. Using the approximate Biot–Fresnel rule, the vibration directions at selected points in a Bxa figure have been plotted in Fig. 11-5. The crystal is oriented with the trace of the optic plane E–W. The construction is further developed in Fig. 11-6, which shows for three positions of stage rotation that the isogyres are the loci of points at which the vibration directions are N–S and E–W. At a position of rotation in which the trace of the optic plane is parallel to a polar, the isogyres form a cross. As the stage (and simultaneously the trace of the optic plane) is rotated, the cross breaks into a hyperbola, the foci of which are farthest apart at the 45° position. Note that the three drawings of Fig. 11-6, except for the isogyres, are identical except for rotation about an axis normal to the page. The vibration directions do not change their orientations relative to each other, but only relative to the polars.[1]

[1] A model that illustrates the vibration directions and the form of the isogyres at all positions of rotation is described by Chapman (1940). A vector theory and a FORTRAN program that synthesizes interference figures for crystals in any orientation are described by Bethke and Birnie (1980).

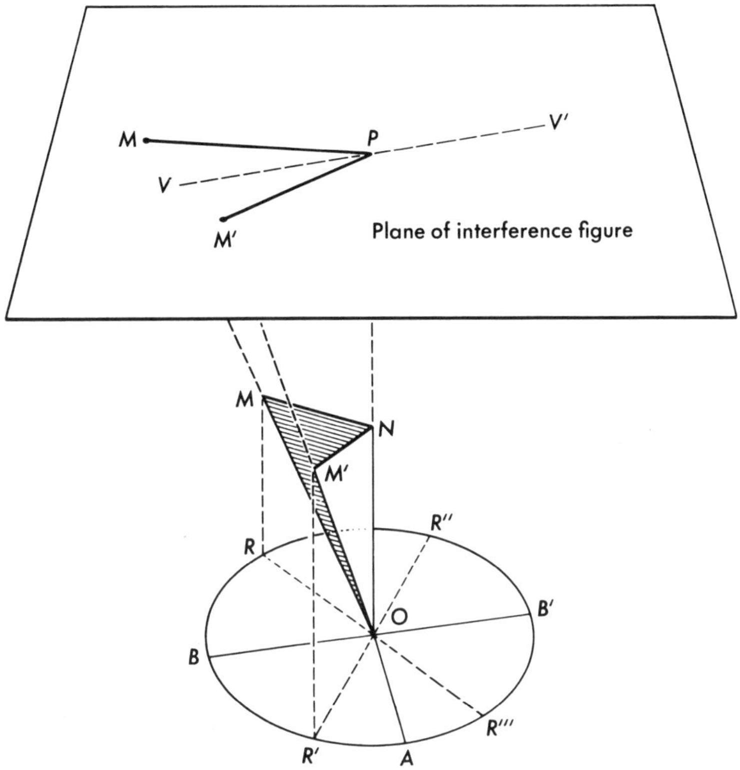

Fig. 11-4. Relationship of the Biot–Fresnel law (lower part, from Fig. 10-6) to its approximation in interference figures. Vibration direction *VV′*, analogous to BB′, is found as the bisector of angle MPM′.

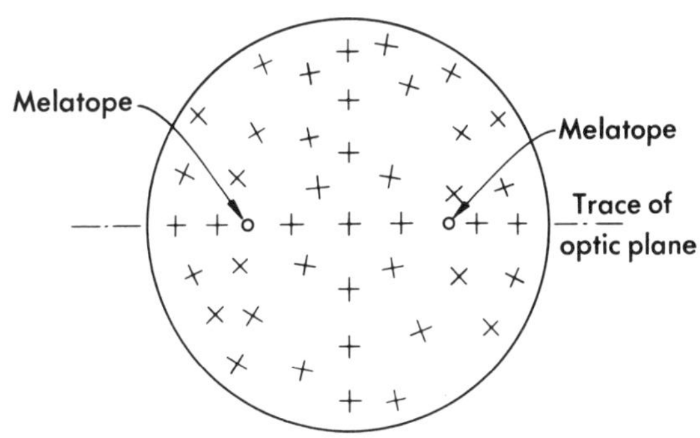

Fig. 11-5. Vibration directions at arbitrarily selected points in a Bxa figure, found by the approximation of the Biot–Fresnel law.

Vibration Vectors. In addition to knowing the orientation of vibration directions in the Bxa figure, it is also necessary to know the relative magnitudes of the associated refractive indices. An understanding of these relative magnitudes provides the basis for optic sign determination. As in the case of uniaxial crystals (see Fig. 8-4), the vibration vectors projected in the Bxa figure can be analyzed in terms of the vectors in the indicatrix normal to the wave propagation directions

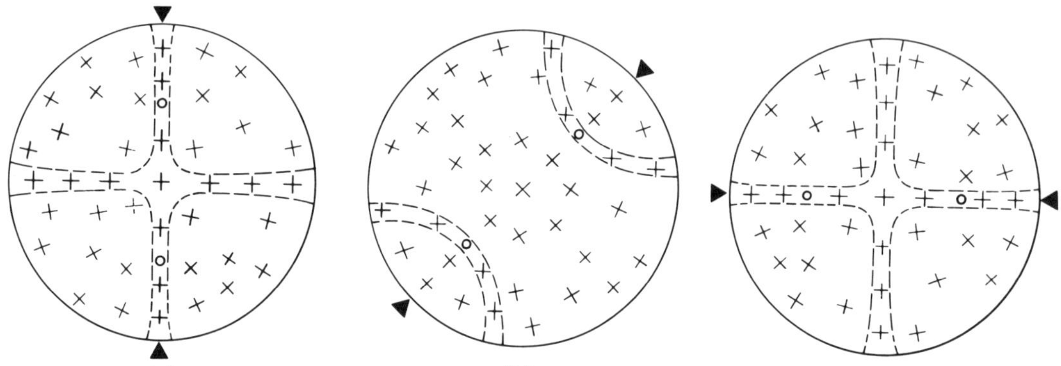

Fig. 11-6. Melatopes, vibration directions, and isogyres at three positions of rotation in a Bxa figure, a further development of Fig. 11-5. Ticks mark the trace of the optic plane.

that form the figure. It turns out that the nearly radial direction on the concave side of an isogyre is a projection of the bisectrix (Bxa in this case) that is normal to the stage. This is illustrated in Fig. 11-7, which is a quantitative development of a portion of Fig. 11-6(c).

In Fig. 11-7(a), the optic plane is E–W in the plan of the interference figure above line AA'; below this is a vertical XZ section of a positive indicatrix. In the vertical section, five wave normals lie in the optic plane and pass through the indicatrix with increasing obliquity. As is generally true in anisotropic crystals, light is resolved into two mutually perpendicular vibration vectors. One of these is the optic normal Y for all the five wave normals. This is the vector perpendicular to the page in the vertical section and N–S above line AA'; it has magnitude β for all wave normals. The other vibration vector lies in the optic plane and has a different magnitude for each wave normal, depending on the obliquity. For wave normal 1, the vector is the X axis and the magnitude is α. For wave normal 2, the magnitude is α', between α and β. For wave normal 3, which coincides with an optic axis, the magnitude is β (the diameter of a circular section). For wave normal 4, the magnitude is γ', and for wave normal 5 another, yet larger γ'. Note that γ' is associated with Bxa$'$, a partial value of the vibration vector of the acute bisectrix, which means a value between β and, in this case, γ. The vector of magnitude γ is of course unobtainable in this case since light vibrating with that magnitude is propagated at right angles to the microscope axis.

In the upper part of Fig. 11-7(a), elliptical sections are located at the point of emergence of each wave normal in the interference figure. It

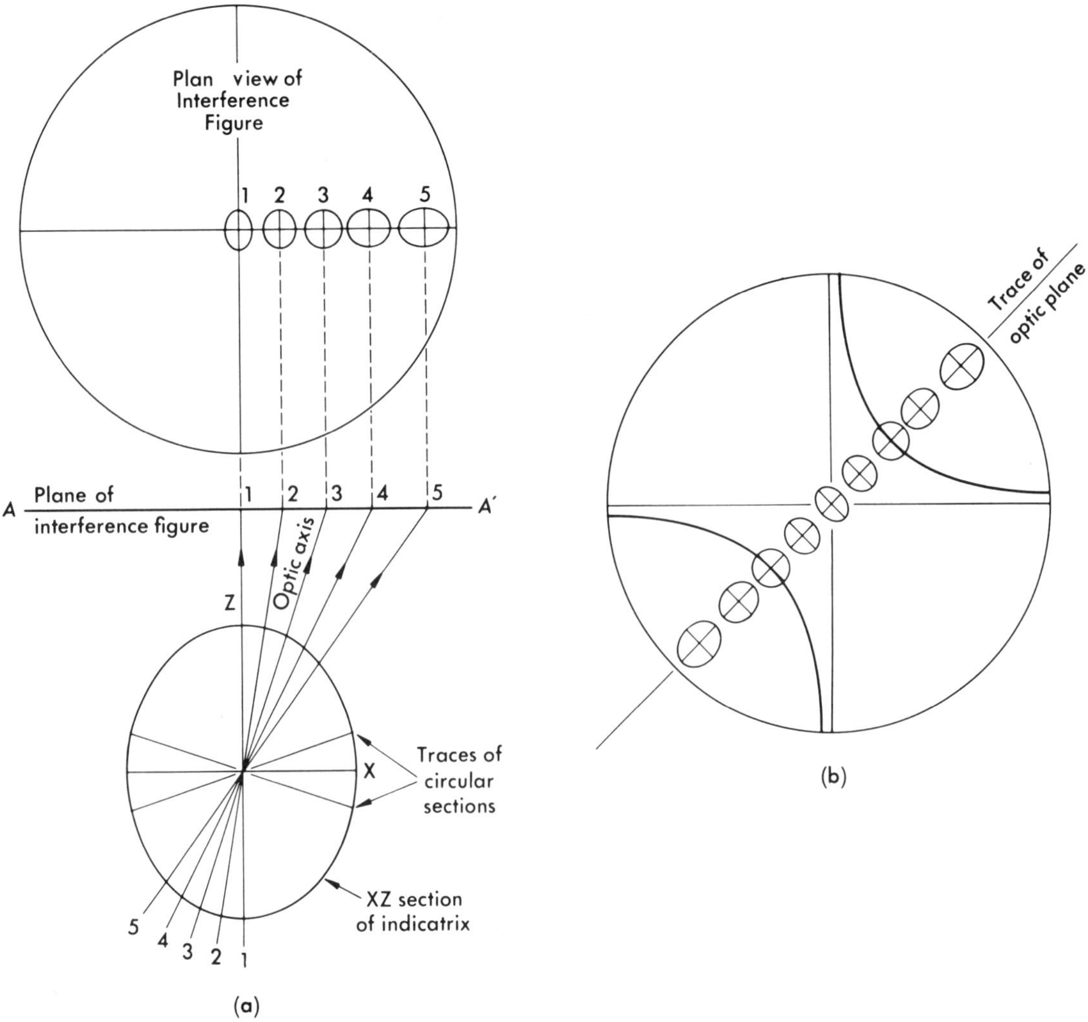

Fig. 11-7. Vibration vectors in a Bxa figure in relation to propagation directions through the indicatrix. (a) Plan view above, section below; optic plane E–W. (b) Interference figure only, optic plane NE–SW.

is apparent that for all possible wave normals from the center of the field to the margin, there is a continuum of E–W vector magnitudes from α through β to some maximum γ'. All the wave normals emerging nearer the center of the interference figure than the melatope have vibration vectors of magnitude α' and β, whereas all those emerging beyond the melatope have vectors of magnitude γ' and β.

If the optic plane is turned NE–SW instead of E–W, the vibration vectors in the figure are as shown in Fig. 11-7(b), unchanged except for orientation and the addition of more ellipses to complete the symmetry, as actually seen. Isogyres pass through the melatopes, where both vibration vectors have magnitude β. Figure 11-7(b) represents a rotation position commonly used for optic sign determination, since the radial or nearly radial direction on the concave side of an isogyre is Bxa′, and the insertion of an accessory plate allows instant determination of whether Bxa′ is Z′ or X′ and whether the crystal is therefore positive or negative.

Color Curves. The ellipticity of the sections of the indicatrix projected in Fig. 11-7 is a measure of the birefringence at the selected point in the plane of the interference figure. Knowledge of the birefringence is a guide to the approximate positions of the color curves. Each color curve in a Bxa figure is the locus of all points of equal retardation, and this locus is approximately the same as the locus of points of equal birefringence. (See the discussion of color curves in the uniaxial optic axis figure, page 128.)

A sheet of muscovite mica provides an effective example of a Bxa interference figure. Sheets of several appropriately selected thicknesses illustrate the position of the color curves for several values of retardation, as shown in Fig. 11-8. The color curves are marked as in uniaxial figures by the position of the red colors of the first, second, and higher orders, or by bands of red and blue together.

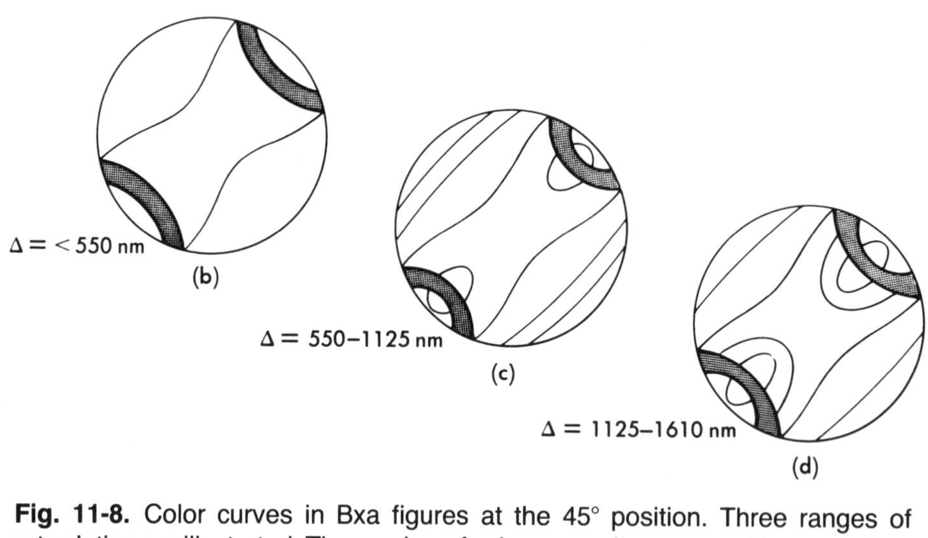

Fig. 11-8. Color curves in Bxa figures at the 45° position. Three ranges of retardation are illustrated. The number of color curves increases with retardation, hence birefringence.

Along the trace of the optic plane the retardation in the figure increases away from the melatopes toward both the center and the perimeter of the figure. Along the optic normal direction (which makes a 90° angle with the trace of the optic plane), retardation increases from the point of emergence of the Bxa, the center, toward the circumference. In the Bxa figure, the melatopes, points of zero retardation, are surrounded by oval curves enveloped by lemniscates of which the melatopes are foci. Each oval or lemniscate is marked by a dark band composed of the red color at the top of one order and the blue at the bottom of the succeeding order. The value of retardation at the point of emergence of the Bxa determines the number of complete ovals that will surround a melatope. The pattern of color curves for three ranges of retardation values is shown in Fig. 11-8. If the retardation of the two polarized components propagated along the Bxa direction is more than 550 nm, the upper limit of first-order interference colors, each melatope is surrounded by one or more ovals, as illustrated. If the retardation is less, there are no ovals.[2]

Summary of 2-Symmetry Plane Figures

Bxa Figure. At the center of the figure, the vibration vectors are Bxo, parallel to the trace of the optic plane, and Y, normal to it. On the concave side of the isogyre at the 45° position the vectors are Bxa′, parallel to the trace of the optic plane, and Y, normal to it. Bxa′ is Z′ for (+) crystals and X′ for (−) crystals, as illustrated in Fig. 11-9. The isogyres leave (if at all) in the direction of the optic plane, specifically in the direction of the Bxo.

The vibration vectors in the orthoscopic image of a grain normal to the Bxa do not include partial values of the Bxa itself, but only Bxo and Y, as in the center of the interference figure.

[2] The longer path in the crystal traveled by the more oblique wave normals affects the retardation and hence the interference color. Furthermore, all wave normals are refracted at the upper surface of the grain. The refraction of wave normals traveling along the optic axes depends upon the β index of the crystal, and therefore a line connecting the melatopes in the figure subtends the angle 2E and not 2V. In all other directions the two wave normals within the crystal are unequally refracted at the upper surface of any but a spherical grain. Therefore, the two wave normals emerging at a single point in the figure are not both normal to the same section of the indicatrix. These and other relatively slight effects cause the color curves to be different from curves of equal birefringence. Hence, rigorous demonstration of the position of color curves in interference figures cannot be achieved by plotting elliptical sections of the indicatrix at all points as in Fig. 11-7.

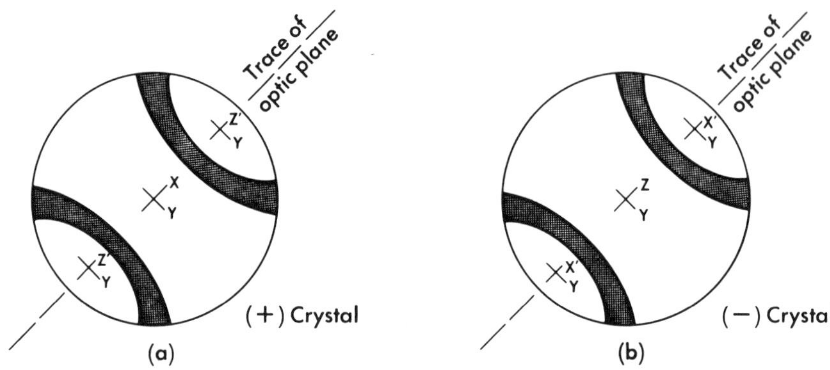

Fig. 11-9. Vibration directions along the trace of the optic plane in Bxa figures at the 45° position. (a) Positive crystal; (b) negative crystal.

Bxo Figure. The Bxo figure is formed from a section normal to X or Z, and is analogous to the Bxa figure except that the isogyres always leave the field because $2E_{Obtuse}$ is always greater than the angle subtended by the field. *Isogyres leave the field in the direction of the optic plane, specifically in the direction of the Bxa.* The color curves are the same as those in the Bxa figure, except that one sees only portions of curves enveloping the melatopes at the 45° position, since the melatopes lie outside the field. For this reason, the color curves in the Bxo figure (Fig. 11-10) resemble those near the center of the Bxa figure [Fig. 11-8 (c)].

The vibration vectors in the center of a Bxo figure (and in the orthoscopic image of the grain) are Y and Bxa, as illustrated in Fig. 11-11. The vector parallel to the trace of the optic plane is Z for positive crystals and X for negative crystals. The concave side of an isogyre is of little practical interest since it is observed only fleetingly upon rotation of the stage.

Cleavage fragments of the mineral barite or a piece of clear cellulose tape form a Bxo figure and are useful as illustrations.

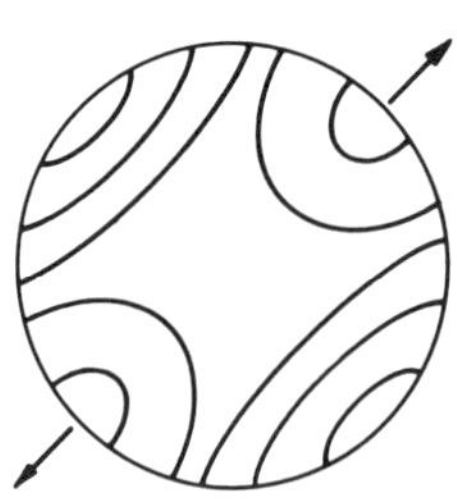

Fig. 11-10. Color curves in a Bxo figure at the 45° position. Arrows indicate trace of optic plane and directions in which isogyres lie outside the field.

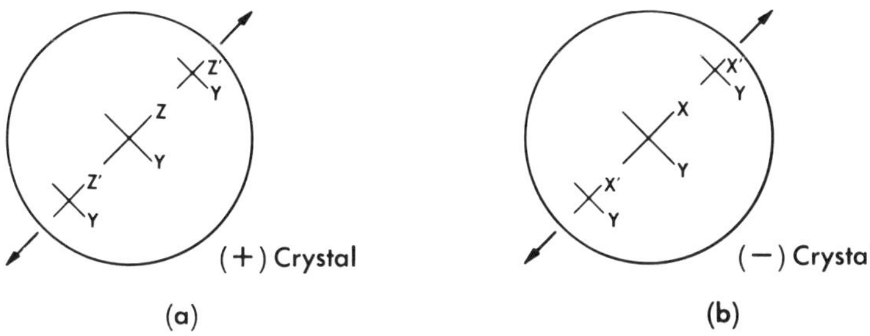

Fig. 11-11. Vibration directions along the trace of the optic plane in Bxo figures at the 45° position. Isogyres lie outside the field in the direction of the arrows.

Optic Normal or Flash Figure. The optic normal figure (flash figure) is formed from a section cut normal to Y. The isogyres are similar to those of a uniaxial flash figure, broader and fuzzier than those of the Bxa or Bxo. *Isogyres leave the field in the direction of the Bxa,* but this direction is difficult to identify in crystals having 2V near 90°. Color curves are hyperbolic, as illustrated by the very thick section shown in Fig. 11-12; their number increases with birefringence and grain thickness and decreases with increasing 2V. They are less closely spaced in the direction of the Bxa, as shown by Becke (1896).

The vibration vectors in the center of the figure and in the orthoscopic image of the grain are X and Z (Fig. 11-13).

Methods of Differentiating 2-Symmetry Plane Figures

In many cases the identity of a 2-symmetry plane figure need not be directly determined, since the three figures all furnish two of the vibra-

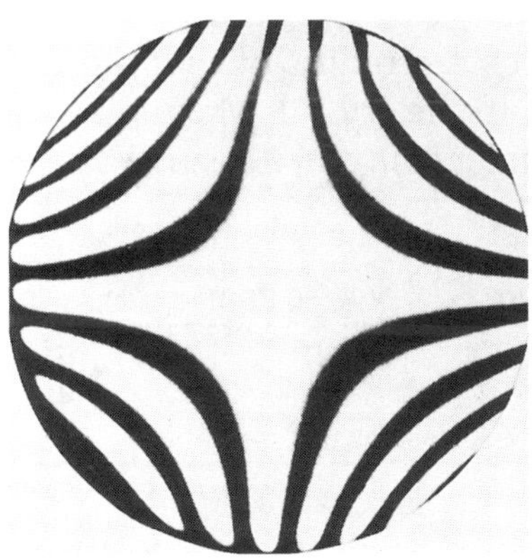

Fig. 11-12. Color curves in a biaxial flash figure at the 45° position. The color curves are less closely spaced in the direction of the Bxa. Isogyres lie outside the field in the NE-SW direction. Hauswaldt 28-3, euclase, sodium light.

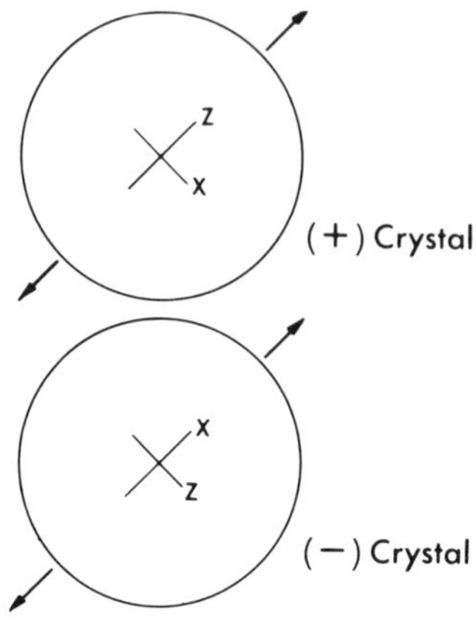

Fig. 11-13. Vibration directions in flash figures at the 45° position. Isogyres lie outside the field in the direction of the arrows. (a) Positive crystal; (b) negative crystal.

tion directions X, Y, or Z. The identity of the two vibration directions is often apparent from refractive indices in the immersion mounts. The isogyres always leave in the Bxa or Bxo direction (X or Z, never Y). If the figures must be differentiated by direct methods, some difficulty may be encountered except in the case of a Bxa figure in which the isogyres do not leave the field. This condition is met when 2E does not exceed the diameter subtended by the field. This diameter is governed by the numerical aperture of the objective. Setting E for μ in equation 2-1, we have for air sin E = N.A., and from Fig. 10-5 we have sin E = β sin V, hence

$$\sin V = \frac{\text{N.A.}}{\beta} \,. \tag{11-1}$$

This relation describes the radius of the interference field in terms of the path of light within the crystal. Maximum values of 2V that do not exceed the field diameter using an objective having the following values of N.A. are:

β	N.A.=0.85	N.A.=0.65
1.5	69°	51°
1.6	64°	48°
1.7	60°	45°

To differentiate among Bxo, flash, and Bxa figures in which the isogyres do leave the field, any of several procedures may be used.

One may identify the direction in which the isogyres leave the field, as shown in Fig. 11-14. Observe the orthoscopic image between crossed polars after turning the interference figure so that the isogyres have left in two opposing quadrants. Use an accessory plate to test the rise or fall of interference colors. If the sign of the mineral is known then the Bxa is known to be Z or X. If the isogyres leave in the Bxa direction, the figure is a Bxo or flash figure. If the isogyres leave in the Bxo direction the figure is a Bxa. The several steps are illustrated in Fig. 11-14. Cleavages of oligoclase (−) that show no twinning and are therefore parallel (010) will serve as an example.

Another method originates with the rotation angle. This is the angle through which the stage must be rotated to bring an isogyre from the crossed position to a position at which the median line of the isogyre is tangent to the field. The size of the rotation angle depends not only on

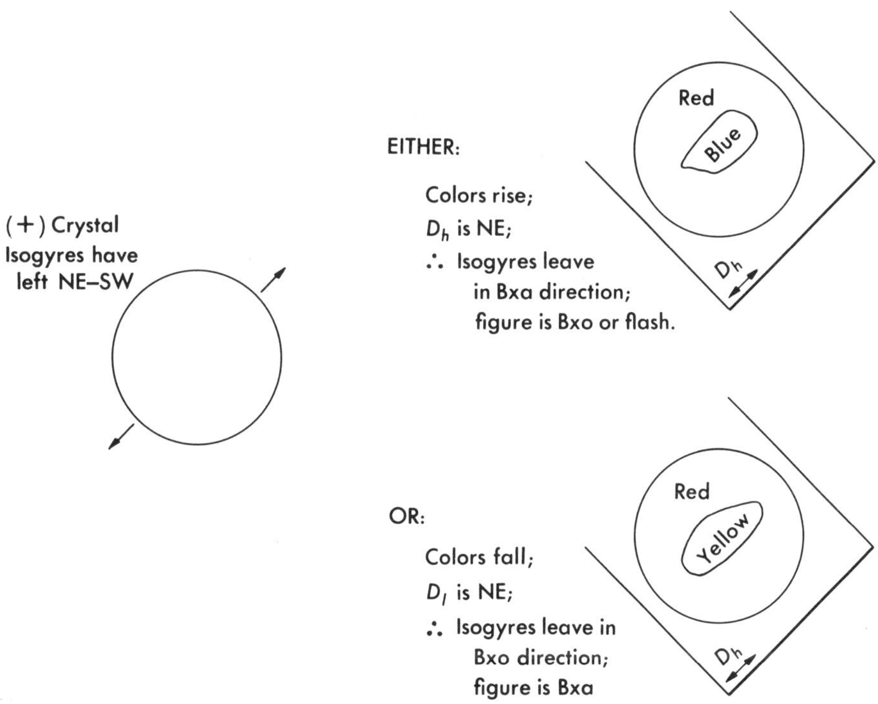

Fig. 11-14. A Bxa figure is distinguished from a Bxo or flash figure when the optic sign is known to be positive. The empty circle depicts the interference figure. The other sketches depict the orthoscopic view after the gypsum plate is inserted.

the type of figure but also, according to Kamb (1958) on the numerical aperture of the objective, 2V, and the average refractive index of the crystal. If μ is the rotation angle,

$$\sin 2\mu = \frac{\left(\frac{\text{N.A.}}{n}\right)^2 [1 - 1/2(\sin V)^2]}{(\sin V)^2\left[1 - \frac{1}{2}\left(\frac{\text{N.A.}}{n}\right)^2\right]}, \tag{11-2}$$

where n = the average refractive index of the crystal, $(\alpha+\beta+\gamma)/3$,
V = one half the optic angle, and
N.A. = numerical aperture.

Values of the rotation angle for specific values of N.A. and n are plotted against 2V in Fig. 11-15.

It can be difficult to measure the rotation angle because of the lack of clarity of the field boundary, the isogyre, or both. This is particularly true in flash figures, which leave the field quickly. In slightly uncentered figures, the angles necessary to bring each isogyre to tangency must be averaged. Rotation angles for 2V between 0° and 90° and average refractive index in the range 1.45–1.80 are as follows, using curves of Kamb (1958):

	N.A.=0.85	N.A.=0.65
Bxa	≥ 11°	≥ 5°
Bxo	3–25°	2–10°
ON	≤ 6°	≤ 3°

The angle can be diagnostic of the type of interference figure if either 2V or the β refractive index of the mineral is known. When one of these is known, the rotation angle can be helpful in identifying the figure. It is also true that when the figure and β are known, an estimate for 2V can be obtained, and in some cases this may be the best available basis for estimating 2V. The tabulated data make it clear that the objective with larger numerical aperture is advantageous in differentiating the Bxa, Bxo, and optic normal figures.

The rotation angle may be used effectively to distinguish between different types of centered figures of the same crystal species observed with the same optical equipment. In Bxa, Bxo, and flash figures from differently oriented sections of the same crystal, the rotation angle is greatest in the Bxa, less in the Bxo, and least in the flash figure, as shown by Fig. 11-15. Very large or very small rotation angles are clearly

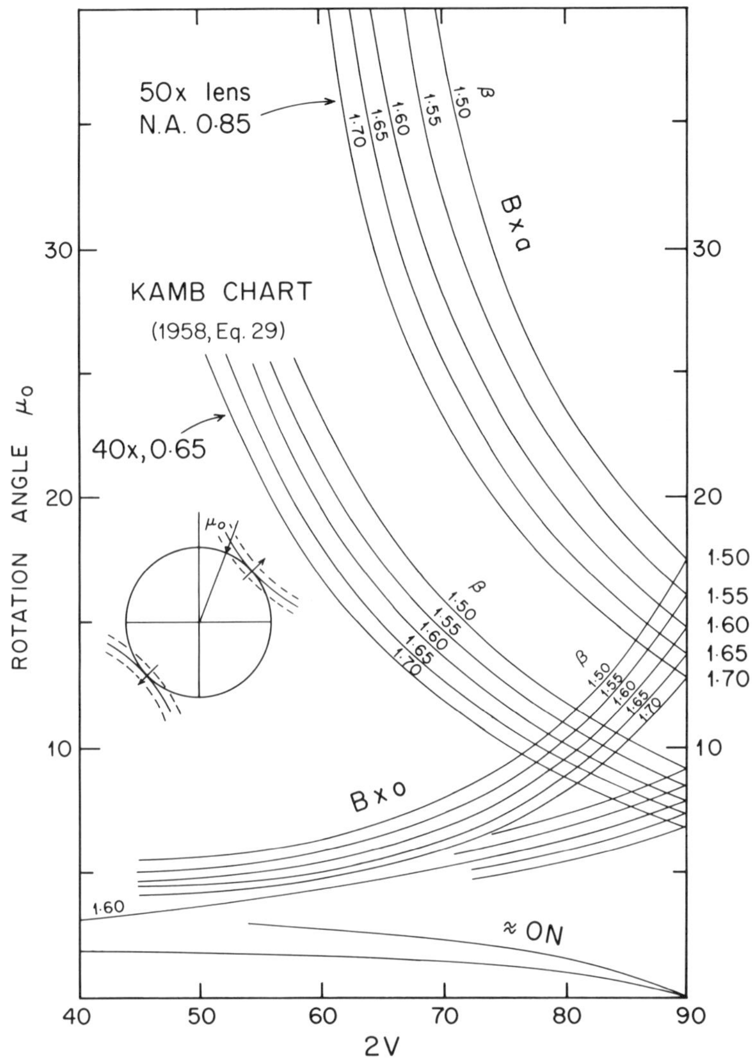

Fig. 11-15. Rotation angle as a function of 2V for the various 2-symmetry plane figures, after Kamb (1958). The curves illustrated are for a crystal of β = 1.5–1.7 and objectives of N.A. = 0.65 and 0.85.

indicative of Bxa or optic normal figures, respectively. Figure 11-15 shows that for N.A. = 0.85 there is very little difference in the rotation angle for Bxo and flash figures unless 2V is more than 50°. One may most easily distinguish the three types of figures when 2V is around 70°.

1-SYMMETRY PLANE FIGURES

General Description

Sections normal to a single symmetry plane of the indicatrix show 1-symmetry plane figures, of which there are two major categories, counter-rotating (CR) and same-rotating (SR) 1-isogyre figures. Two-isogyre 1-symmetry plane figures may belong to either category. All 1-symmetry plane figures have the property that at four positions of stage rotation a straight isogyre is coincident with a crosshair and simultaneously coincident with the trace of its symmetry plane. At other positions of rotation the isogyre is curved except in the special case of a centered optic axis figure from a crystal with 2V = 90°. The isogyre may leave the field. *One-symmetry plane figures are valuable because their crystals yield the principal value of refractive index of the vibration direction normal to the symmetry plane,* and the centered optic axis figure is especially valuable for determination of both optic sign and β.

CR and SR Figures

Single isogyres in figures from sections normal to the optic plane rotate counter to the direction of stage rotation (CR). Single isogyres in figures from sections normal to either of the other two symmetry planes rotate in the same direction as the stage is rotated (SR). The reasons for the difference in behavior may be deduced from the isogyre movement in the Bxa figure, as shown in Fig. 11-16. The figure contrasts the isogyre rotation in sections normal to the optic plane (CR) and to the Bxa-Y plane (SR). Attention is directed to one branch of the hyperbola as it moves outward from the crossed position. In Fig. 11-16, the small circles represent the microscope field showing a single-isogyre 1-symmetry plane figure as part of the parent Bxa figure. In Fig. 11-16(a)–(c), the microscope field is on the optic plane, marked by an arrowhead. As the stage is rotated clockwise (note movement of arrowhead) the isogyre sweeps counterclockwise. Note especially the motion of the north end of the isogyre near the arrowhead during the first few degrees of rotation, (a) to (b) in Fig. 11-16. A very slight amount of rotation from an extinction position (a) suffices to show the counterclockwise motion of the tip of the isogyre.

Figure 11-16(d)–(f) shows the very same rotation positions of the same Bxa figure, but with attention directed to the Bxa-Y symmetry plane (arrowhead). As the stage is rotated clockwise, the isogyre in the limited microscope field sweeps clockwise also, hence the name "same-rotating." Again, note especially the motion of the end of the

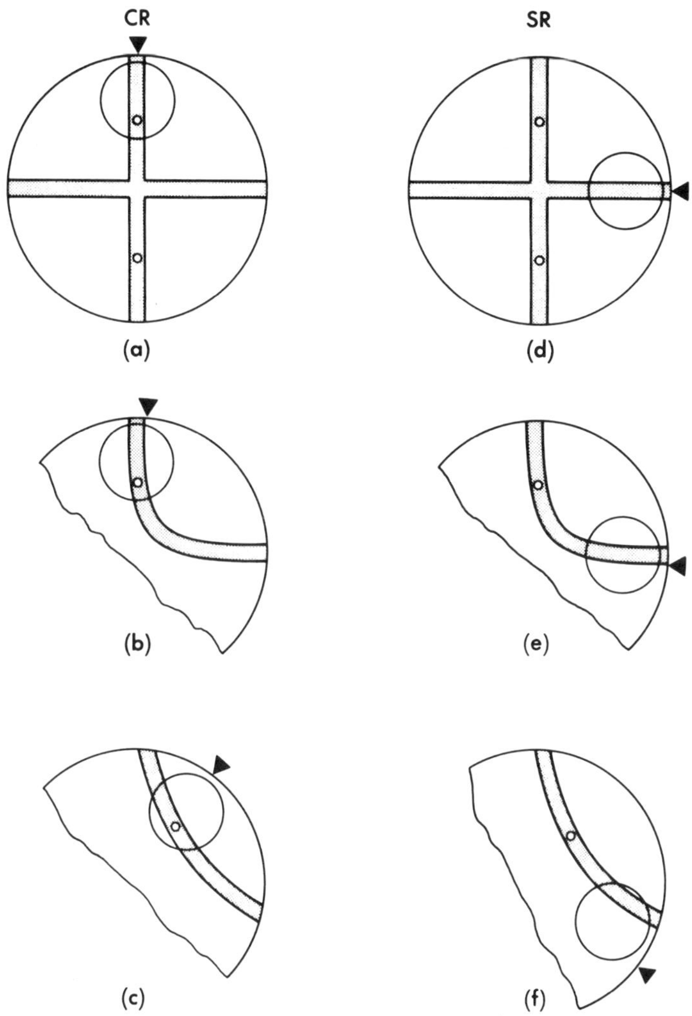

Fig. 11-16. Counter-rotating (a–c) and same-rotating (d–f) isogyres in different parts of the same Bxa figure. The small circle (the microscope field) remains on the trace of a symmetry plane: the optic plane in (a–c) and Bxa-Y plane in (d–f).

isogyre near the arrowhead during the first few degrees of rotation, (d) to (e). Compare especially (e) and (b) in this respect.

The relationship between the rotation sense of the 1-symmetry plane figures and the motion of the single isogyre common to both figures is summarized in Fig. 11-17. Two positions of the isogyre are shown superimposed, and small circles representing the microscope fields of CR and SR figures are included in the same diagram. Motion of the

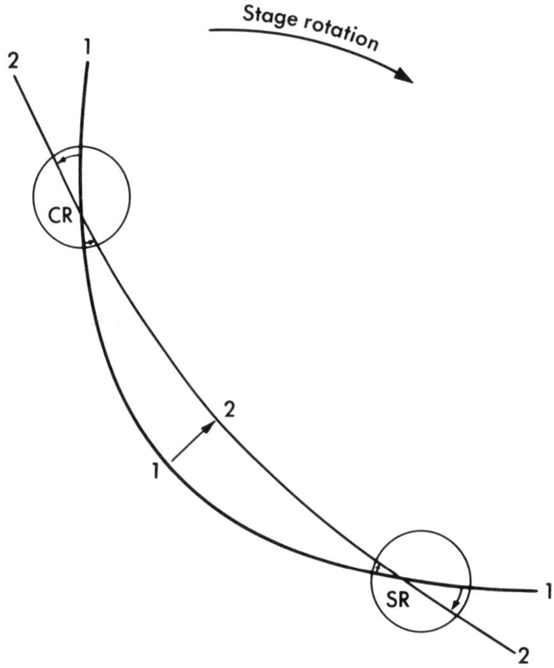

Fig. 11-17. Exaggerated straightening motion of a Bxa isogyre from position 1 to position 2, showing the rotation sense of CR and SR figures as the stage is rotated clockwise.

isogyre from position 1 (thick line) to position 2 (fine line) produces opposite rotation senses in the two 1-symmetry plane figures.

The behavior of CR and SR isogyres with stage rotation through a full quadrant is illustrated in Figs. 11-18 to 11-20. In these figures, the large circles show centered Bxa interference figures. The small circle, within the large one, is the conoscopic view that one sees when looking through the microscope. It shows the 1-symmetry plane figure, which has been scaled down to show its relationship to the parent Bxa figure. The small circles beneath the large ones are redrawn from those in the Bxa figure to show the position of the isogyre in 1-symmetry plane figures at different positions of stage rotation. In Figs. 11-18 through 11-20, note that in a Bxa figure the first motion of an optic plane isogyre as it leaves the crossed position is counterclockwise as the stage is rotated clockwise. The first motion of the isogyre not coincident with the optic plane is clockwise, as shown in Fig. 11-17.

Figure 11-18 shows the relationship of an SR figure to the parent Bxa; note that the original SR isogyre leaves the field and is replaced by a second one entering from the southwest.

Figure 11-19 shows a particular type of CR figure (between the optic axis and Bxa) in which the end of the isogyre away from the melatope whips very rapidly around with a small rotation of the stage.

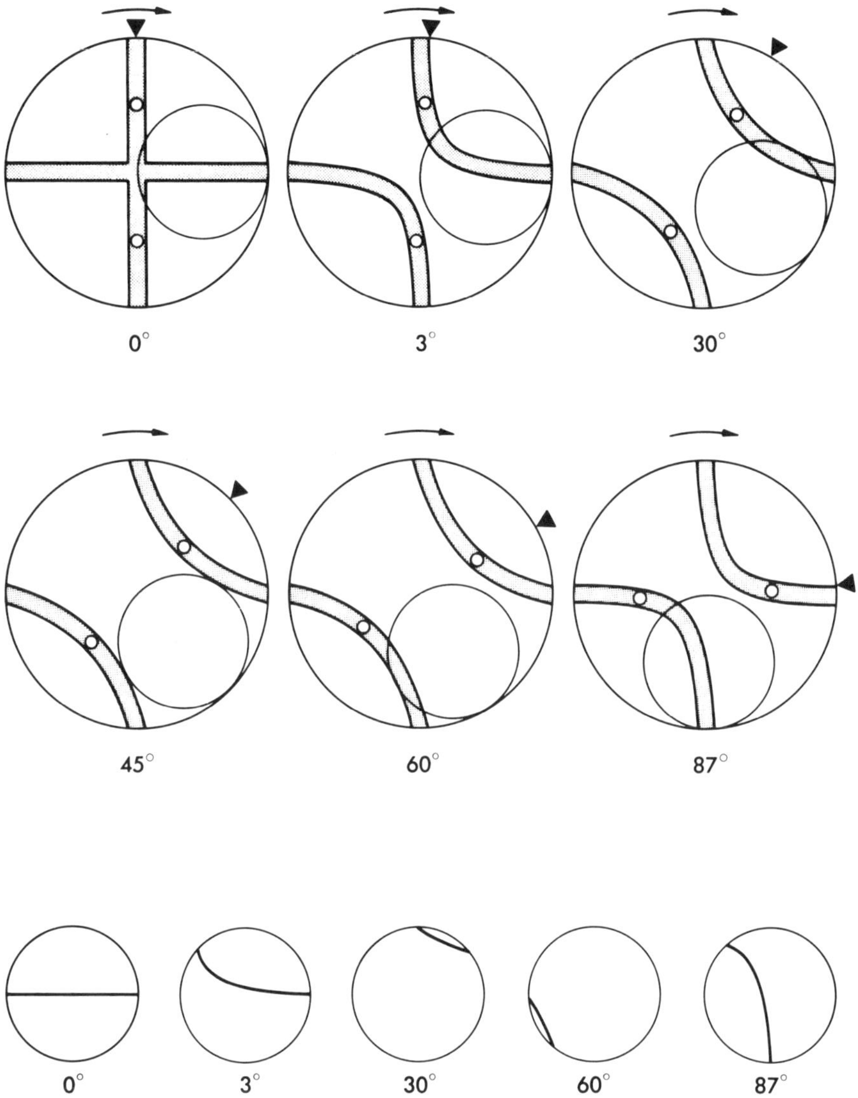

Fig. 11-18. Top: derivation of SR figure from Bxa figure, showing behavior of the SR figure (small circle) through a rotation of about 90°. Bottom: small circles repeated from top part represent interference figures as seen.

Figure 11-20 shows a CR figure (between the optic axis and the Bxo) with a less frantic motion.

Bxa figures having color curves that indicate the location of the melatope may be used to study the differences in rotation sense

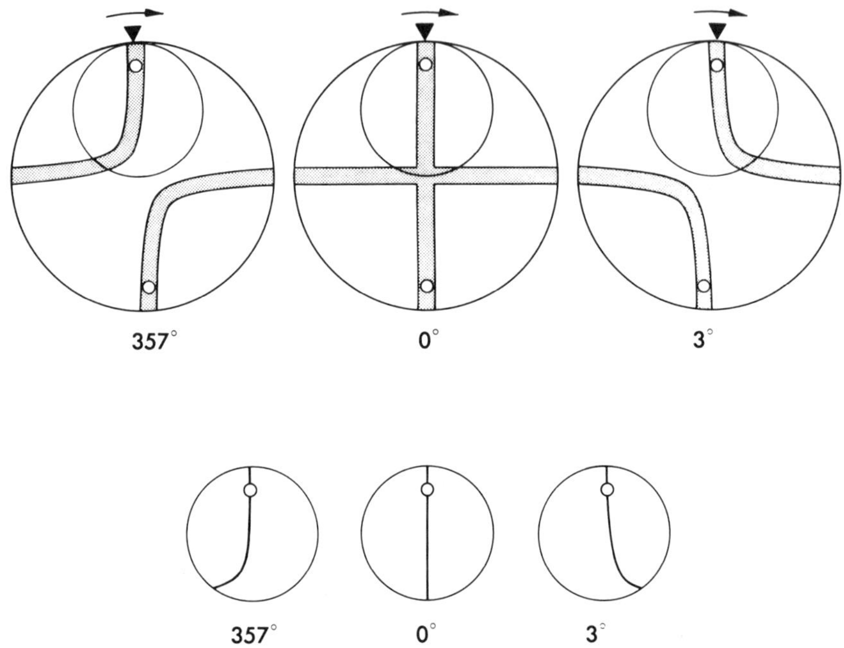

Fig. 11-19. Derivation of CR (counter-rotating) figure from the Bxa figure. The center of the CR figure lies between an optic axis and the Bxa. Arrows show direction of stage rotation.

of CR and SR figures. Such figures may be obtained from thick sheets of muscovite or a section of aragonite.

Optic Axis Figure

The optic axis figure is formed from a section normal to the optic axis. A single isogyre rotates about the melatope, which emerges in the center of the field as illustrated in Fig. 11-21. Because the plane normal to the section is the optic plane, one may predict that it will be a counter-rotating 1-symmetry plane figure. The isogyre is straight and bisects the field every 90° of rotation and has maximum curvature at the 45° positions. The smaller the curvature of the isogyre at the 45° position of rotation, the greater the optic angle of the crystal. If 2V is 90°, the isogyre is essentially straight at all positions of rotation. Vibration directions at the 45° position (Fig. 11-22) are the same as those in the vicinity of a melatope in a Bxa figure at 45°. The trace of the optic plane at the 45° position cuts straight across the curved isogyre at the melatope, and at the 0° or 90° position coincides with the isogyre (and with a crosshair) in the centered figure, as illustrated in Fig. 11-21. The

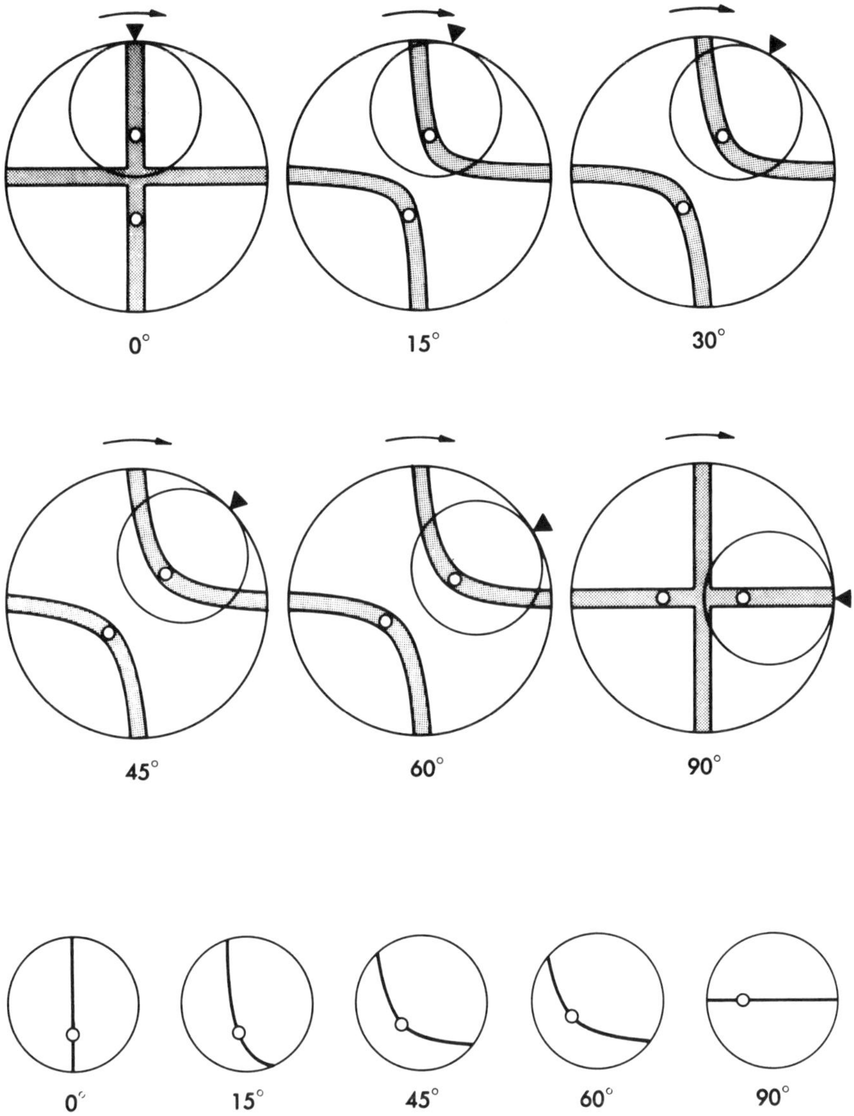

Fig. 11-20. Derivation of a second type of CR figure from the Bxa figure. The center of the CR figure lies beyond the optic axis from Bxa. Arrows show direction of stage rotation.

vibration vectors at the center of the figure and in the crystal viewed orthoscopically are Y in all directions.

If the optic angle is large, color curves are nearly circular, nearly concentric about the melatope, and analogous to the curves in the vicinity of the melatopes in a Bxa figure. Grains that will produce an

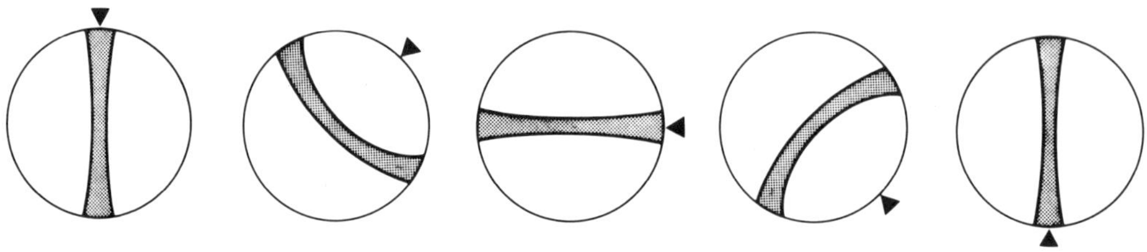

Fig. 11-21. Behavior of an optic axis figure with clockwise stage rotation. Arrowhead lies on trace of optic plane.

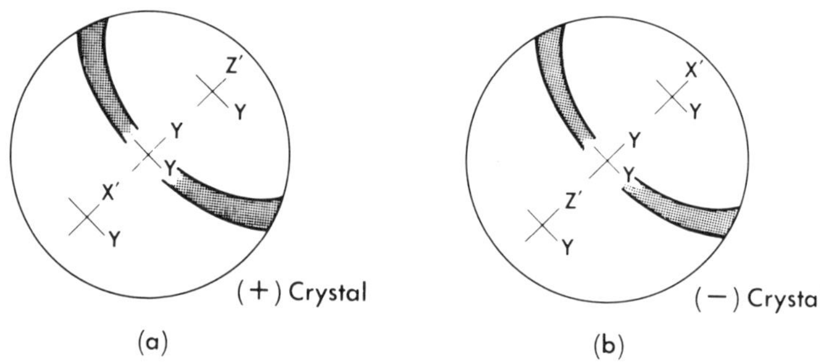

Fig. 11-22. Vibration directions in optic axis figures, positive (a) and negative (b) crystals.

optic axis figure can be provisionally identified when viewed orthoscopically with crossed polars, because they show complete extinction or remain gray or light gray or more rarely alternate from bluish to reddish at all positions of rotation.

> Centered or uncentered optic axis figures can be readily observed from cleavage flakes of babingtonite or bytownite.

2-Isogyre 1-Symmetry Plane Figures

Certain uncentered 2-isogyre figures are in fact 1-SP figures in which an isogyre coincides with a crosshair four times on stage rotation. They are almost as useful as 2-SP figures. They have an attribute that aids recognition: the isogyres of these 1-SP figures are still symmetrically disposed about either a NE–SW or NW–SE line at the 45° position of rotation. The trace of the symmetry plane in the indicatrix produces a line of symmetry in the interference figure. When an isogyre is rotated into coincidence with a crosshair, the crosshair marks the position of the symmetry plane and one of the vibration directions X, Y, or Z coincides with the other crosshair as with all 1-SP figures. Two such figures are illustrated in Figs. 11-23 and 11-24. Prismatic cleavage flakes of the mineral barite produce an interference figure like that of Fig. 11-24.

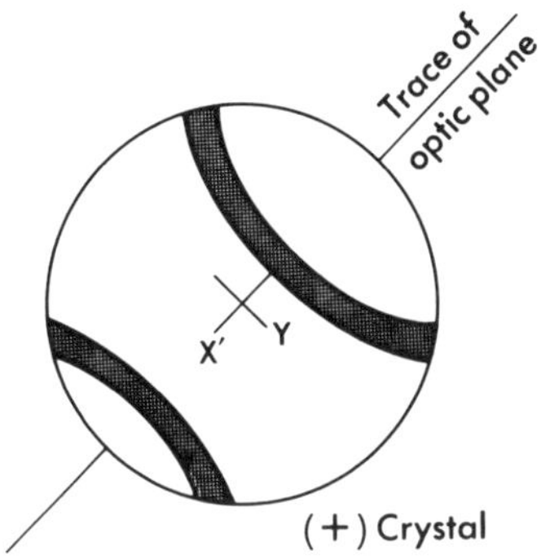

Fig. 11-23. Vibration directions in a 1-symmetry plane (uncentered Bxa) figure, positive crystal. The symmetry plane normal to the section is the optic plane.

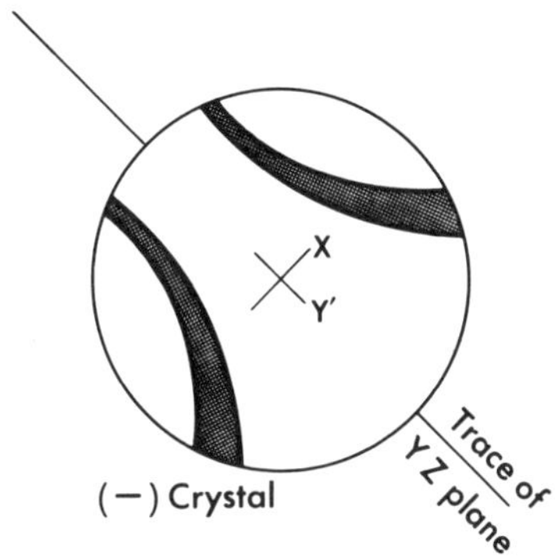

Fig. 11-24. Vibration directions in a 1-symmetry plane (uncentered Bxa) figure, negative crystal. The symmetry plane normal to the section is the YZ plane.

0-SYMMETRY PLANE FIGURES

Interference figures from sections normal to no symmetry plane of the indicatrix are of varying utility. Those which are only slightly uncentered versions of 2-SP or 1-SP figures are practically as good as the centered figures, the only restrictions being that refractive indices will depart slightly from their full value. On the other hand, 0-SP figures very far from any symmetry plane are useless. Clearly, the utility increases as the 0-SP figures become closer to 1-or 2-SP figures.

Many 0-SP figures can be used for optic sign determination; these are usually uncentered optic axis and Bxa figures. None is useful for precise values of refractive index. However, many are useful for setting limits on one or more refractive indices. One should make the best use of 0-SP figures, since they are encountered more frequently than any other figure. On the other hand, one should not be beguiled into spending more time than justified with a figure that is enigmatic and therefore useless.

EFFECT OF DISPERSION ON INTERFERENCE FIGURES

Dispersion of the vibration directions and refractive indices affects the isogyres and to a less noticeable extent the color curves in biaxial interference figures. The black isogyres are bordered by color fringes when vibration directions associated with different wavelengths (particularly at the extreme ends of the visible spectrum) are not coincident. These fringes are best seen in Bxa or optic axis figures at the 45° position. The fringes are wide in figures from crystals with strong dispersion. If dispersion is extreme, the isogyres are colored rather than black due to the lack of coincidence of isogyres for different wavelengths of the visible spectrum. Specific types of dispersion are discussed in a later chapter.

12 Identification of Biaxial Crystals

PROPERTIES USED IN IDENTIFICATION

Many properties may be used to identify biaxial crystals. The scheme used and the order in which the properties are determined depend to some extent on the nature of the tables to be used. With many of these, knowledge of the optic sign and the three refractive indices is of prime consideration, but while these properties are being measured, one often gains knowledge of other useful properties such as birefringence, pleochroism and absorption, dispersion, crystal habit, cleavage, twinning, and optic orientation. The last four properties are discussed in Chapter 13. Although this chapter deals largely with optical properties, it should be realized that in practice optical and crystallographic properties are so closely interrelated that they may best be determined simultaneously.

AXIAL CHARACTER

Biaxial crystals are those with two optic axes. They crystallize in either the orthorhombic, monoclinic, or triclinic system. Included also in the biaxial class are some crystals normally isotropic or uniaxial that have been subjected to external stress or have a composition that departs from normal to the extent that the structure is distorted. The mineral beryl, for example, is not infrequently biaxial if it contains a high percentage of rare earths. Calcite is often found to be biaxial.

About 46 percent of the entries in Winchell and Winchell's tables (1964) of inorganic substances are biaxial; Larsen and Berman (1934) found that about 62 percent of the entries in their tables of nonopaque minerals were biaxial, and about 64 percent of the entries in Fleischer et al. (1984) are biaxial. Of these, about 36 percent are biaxial negative and 28 percent positive. The major non-opaque rock-forming minerals are biaxial with a few exceptions; outstanding ones include quartz, garnets, rhombohedral carbonates, feldspathoids, and several accessory minerals such as apatite, zircon, and the spinels.

In several solid (crystalline) solution series of biaxial crystals the optic angle becomes zero for certain compositions, so that one or more members of the series are uniaxial at some wavelength of light. This situation is illustrated by the solid solution series lithiophilite-triphylite (orthorhombic) in Fig. 12-1.

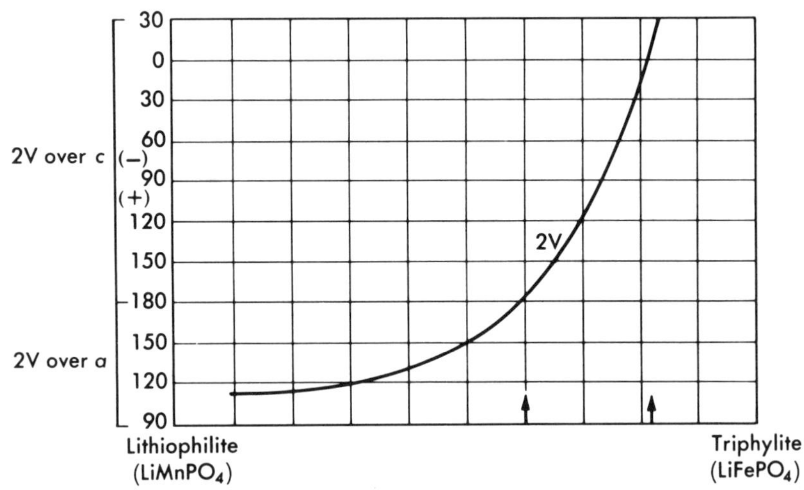

Fig. 12-1. Size of the optic angle, 2V, for the series lithiophilite-triphylite (Chapman, 1943).

Identification

A crystal may be identified as biaxial if the interference figure is unmistakably different from any uniaxial figure. This is true in several instances. If the isogyres form a cross at the extinction position and if on rotation of the stage the branches of the hyperbola separate but do not leave the field, the figure is biaxial (Bxa). If a figure shows only one isogyre that is never coincident with a crosshair or that rotates on a point in the field, the figure is biaxial (0-SP or optic axis). If the isogyres have borders of red or blue because of dispersion, the figure is surely biaxial. Other interference figures are unambiguously uniaxial; for example, if a cross remains with its center in the field during rotation the figure is uniaxial. Some interference figures are ambiguous. If both isogyres leave the field, the figure is a 2-SP figure, either uniaxial flash or biaxial Bxa, Bxo, or ON. The figures most easily confused are the very uncentered uniaxial optic axis figure and the very uncentered Bxa figure of small 2V. In both, if the optic axis or the melatopes remain outside the field, the isogyres remain almost straight and parallel to a crosshair at all positions of stage rotation. Such a figure appears to be uniaxial, although it may not be.

There are other features that distinguish uniaxial from biaxial crystals but all have exceptions or are difficult to apply. One of the most useful rules (to which there are exceptions) is that the crystal is biaxial if extinction measured against a cleavage direction in a cleavage fragment is inclined. Inclined extinction is very uncommon in cleavage fragments of uniaxial crystals and is common though not ubiquitous in cleavage fragments of biaxial crystals.

PLEOCHROISM AND ABSORPTION

There are a few differences between absorption and pleochroism in uniaxial and in biaxial crystals. In uniaxial crystals there are two, and in biaxial crystals three absorption axes at right angles to each other. In uniaxial crystals the absorption axes coincide with the principal vibration directions, whereas in biaxial crystals this is necessarily so only for those vibration directions that coincide with crystallographic directions. It follows that in orthorhombic crystals the absorption axes coincide with X, Y, and Z, the vibration directions. In monoclinic crystals the vibration direction that is parallel to the *b* crystallographic axis is coincident with an absorption axis, but the other vibration directions are not necessarily so. In triclinic crystals the principal vibration directions do not necessarily coincide with the absorption axes although their departure from coincidence is rarely noticeable. It is convenient to use a uniform frame of reference for the description of absorption and pleochroism, and these are characteristically described with reference to the X, Y, and Z vibration directions whether or not the absorption axes are coincident with the vibration directions. In some instances, though rarely for inorganic compounds, data may be available that indicate the angular relationship between the principal absorption axes and the vibration directions.

An absorption formula relating X, Y, and Z will express the relative absorption for white light in crystals in which pleochroic colors are not very different in wavelength:

Mineral	Pleochroism	Absorption
Chondrodite	X = yellow Y = pale yellow Z = colorless	X > Y > Z

Pleochroic colors and absorption may serve to point out peculiarities in chemical composition of members of a mineral group. In both the amphibole and pyroxene groups the members rich in Na have greater absorption than those without Na. The pleochroic colors of the Na-rich members are commonly blue or green. Violet pleochroic colors in pyroxene suggest a high titanium content. In some mineral groups, an increase in iron content is accompanied by increased intensity of pleochroic colors, usually yellow or brown for ferric iron and green for ferrous iron, as in common biotite.

Determination of the pleochroism and absorption formula of a biaxial crystal requires that the color and color intensity observed without crossed polars be correlated with vibration directions. This is done by identifying the vibration direction in grains that form appropriate interference figures. The vibration direction, the associated color of which is being sought, is oriented parallel to the vibration direction of the lower polar by turning to extinction; the associated color is then observed without crossed polars. The strategy followed in finding and identifying vibration directions is similar to that employed to determine indices of refraction. Pleochroism and absorption, once determined, may be useful in rapidly identifying vibration directions in grains in immersion mounts.

Quantitative measurements of pleochroism and absorption are seldom used in identification, but Mandarino (1959) shows them to be feasible and potentially useful.

BIREFRINGENCE

As in uniaxial crystals the birefringence of the *crystal* is the difference between the greatest and least indices of refraction:

$$B = \gamma - \alpha \tag{12-1}$$

The principal partial birefringences are $\gamma - \beta$ and $\beta - \alpha$. The birefringence of any *section* is the difference between the indices of refraction of the two components vibrating in the plane of the section. Birefringence of biaxial crystals is described as weak, moderate, strong, very strong, and extreme, as indicated in Table 9-1. Direct determination of the birefringence of a section has already been discussed in Chapter

6. The birefringence of a crystal, $\gamma - \alpha$, is that of an ON section, which may be recognized because it forms a centered flash figure. In immersion mounts, the birefringence can be calculated from the refractive indices determined as described below. Since errors in both γ and α may be additive, the calculated birefringence is usually less accurate than that measured directly.

Birefringence may be measured in thin sections by measuring retardation and thickness and applying equation 5-2. Thickness may be measured by measuring the retardation of a neighboring crystal of known birefringence and orientation. Further remarks on the measurement of retardation and birefringence are given in Chapter 6.

Estimation of Birefringence. A very rough approximation of the birefringence of an unknown crystal in an immersion mount or thin section is indicated by the number of color curves in an optic axis figure as compared with the number of color curves in the optic axis figure of another crystal of known birefringence. Both crystals must be of similar thickness and the same optical system must be employed in both instances. Comparison of the number of color curves in the optic axis figures from grains of approximately equal thickness (in thin section or immersion mount) of bytownite feldspar (B approximately 0.01), augite (B approximately 0.025), and common olivine (B approximately 0.038), will illustrate the effect of birefringence on the number of color curves.

A more accurate estimation of birefringence may be made from the retardation Δ at the edge of an optic axis figure. It is also necessary to know the thickness of the section t, the optic angle 2V, and the angular radius θ of the interference figure. Then from Winchell (1965):

$$B = \frac{\Delta}{t} \frac{\cos\theta}{[\sin\theta \ \sin(\theta + 2V)]} \qquad (12\text{-}2)$$

The value of θ in this case is obtained directly by setting $\theta = V$ in equation 11-1, hence $\sin\theta = \text{N.A.}/\beta$. The value of the second fraction in equation 12-2 is relatively insensitive to 2V in the range 2V = 40°–80°. For example, for $\beta = 1.7$ and N.A. = 0.85 ($\theta = 30°$), the value of this term lies between 1.843 at the end of the range of 2V, and 1.732 in the middle of the range, and the difference between these two values has a small effect on the estimated birefringence, as shown by the following illustration. Assuming a standard thin section thickness $t = 3 \times 10^4$ nm, the term Δ/t equals 0.018 for an interference color of first-order red at the margin of the field, and if 2V is anywhere in the range 40°–80°, B = 0.032±0.001. Common hornblende is a familiar example.

Equation 12-2 can also be used to control the thickness of a thin section, e.g., for a marble lacking quartz as a guide for the straw-yellow interference color. In such a case the desired interference color at the margin of a centered optic axis figure of calcite (2V = 0, B = 0.172) will be 2.75 orders (high third-order yellow) for N.A. = 0.85, or 1.5 orders (second-order yellow) for N.A. = 0.65.

The principal partial birefringences ($\gamma - \alpha$) and ($\beta - \alpha$) have some diagnostic utility, especially as a basis for calculating refractive indices, and 2V (equation 10-6). They may sometimes be used in the identification of minerals in thin sections where refractive indices cannot be determined. The partial birefringences may be determined from recognizable interference colors shown by sections that form centered Bxa or Bxo figures, if the thickness of the section is known or may be estimated. They may also be calculated from refractive indices when these can be determined.

OPTIC SIGN

Direct Determination

The optic sign is (+) if Bxa is Z and (−) if Bxa is X. Determination of sign depends on knowing the orientation of Bxa or Bxa′ in a grain or interference figure, and identifying this as X (or X′) or Z (or Z′) by means of an accessory plate. This may also be done indirectly by determining which vibration direction, X or Z, is the Bxo; the other vibration direction of the two, Z or X, must then be the Bxa.

There are several possibilities for locating Bxa or Bxa′ in interference figures.

1. The Bxa lies in the direction in which the isogyres of the *flash figure* or the *Bxo figure* have left the field at the 45° position of rotation (Fig. 12-2).
2. Bxa′ lies in an approximately radial[1] direction on the concave side of the curved isogyre of a *Bxa figure* or 1-SP figure at 45° from the extinction position (Fig. 12-3).
3. Bxa′ lies approximately radial everywhere in a *Bxa figure* at the crossed position.

There are some instances in which the Bxo direction is known. The important one is:

[1] Strictly radial along the trace of the optic plane; see Fig. 11-6.

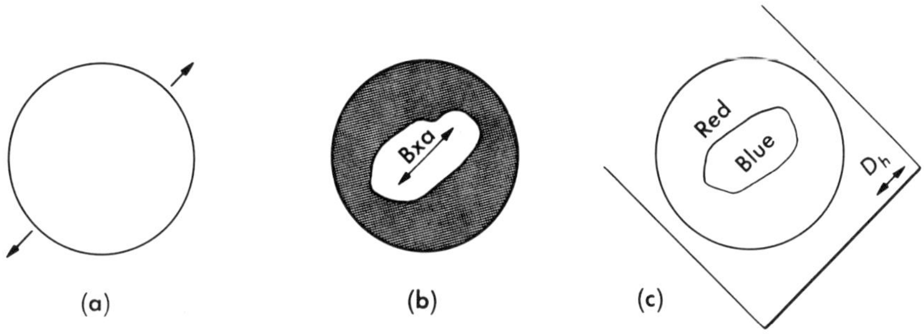

Fig. 12-2. Determination of optic sign from grain showing Bxo or flash figure. (a) Isogyres left NE–SW. (b) Orthoscopic view. (c) Gypsum plate added: colors rise, Bxa is D_h, crystal (+).

In the center of the *Bxa figure*, the Bxo direction is the one in which the isogyres have left or appear to be leaving the field at the 45° position. This is the direction pointing to a melatope on the convex side of an isogyre.

There need be no set system for finding the sign of the crystal. One may find it useful, however, to follow a few simple but exact rules as to what area to observe after an accessory plate is introduced. The rules are logically designed so that in the area to be observed the colors rise for a positive crystal and fall for a negative crystal when the accessory plate is inserted. The exact rules that must be followed so one can say "colors rise for a positive crystal" are stated here.

Rule 1: *For all figures but the Bxo and flash figures.* Observe the concave side of the isogyre at the 45° position, so rotated that the concave side is to the northeast. If there are two isogyres, either observe the concave side of one of these or observe the northeast quadrant of the cross. Insert the accessory plate. If the colors rise, the crystal is positive. This rule is valid only if the melatope remains in the field.

Rule 2: *For Bxo and flash figures.* Rotate the figure 45° from the crossed position so that during rotation the isogyres leave the field in the northeast and southwest quadrants. Then *observe the grain* orthoscopically. Insert the accessory plate. If the colors rise, the crystal is positive.

These rules are illustrated in Figs. 12-2 and 12-3. They work equally well for uniaxial and biaxial figures.

It is to be emphasized, however, that one may just as well disregard

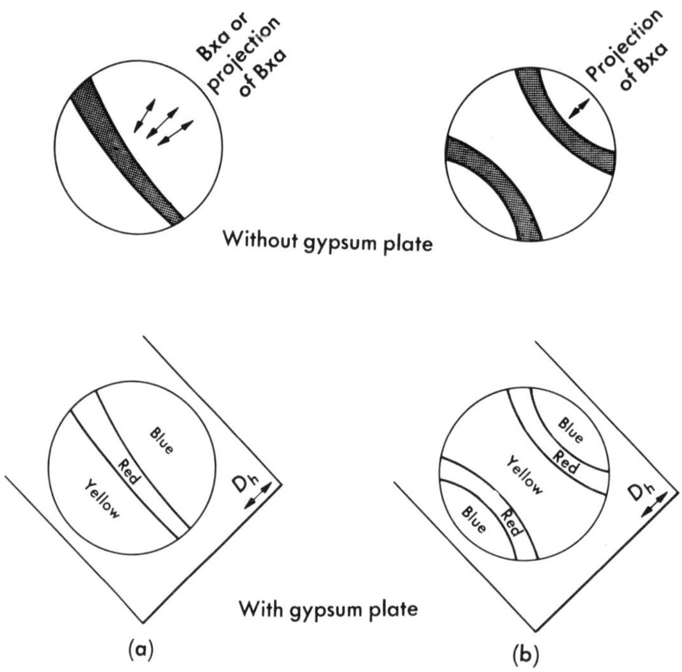

Fig. 12-3. Determination of optic sign at the 45° position from (a) uncentered optic axis figure or (b) Bxa figure. The crystal is positive.

these rules. For example, one may observe the *central* area of a Bxa figure after the isogyres have left the crossed position moving to the northeast and southwest. The direction in which they have moved is the Bxo direction. If the colors rise on insertion of a plate, this direction is the high index direction. This is the Bxo so the Bxo direction is Z. The crystal is negative. *Correct reasoning will lead to the right answer without the use of shortcut rules.*

Determination from Refractive Indices

The relationship between refractive indices and the optic angle has been presented earlier (page 166). Any of these equations may be solved for V by substitution of the values of the principal refractive indices. The sign of the crystal is therefore determined, (+) if 2V around γ is less than 90°, (−) if greater than 90°. The more general rules (see page 166) are:

When β is nearer α than γ, the crystal is (+).
When β is nearer γ than α, the crystal is (−).
When β is about as far from α as γ, these rules cannot be used.

OPTIC ANGLE

The size of the optic angle 2V is a function of the refractive indices α, β, and γ as expressed in equations 10-2 through 10-9. The optic angle may be calculated from the refractive indices using these equations, or may be determined graphically from nomograms of the equations (e.g., in Tröger, 1979, or Fleischer et al., 1984).

The optic angle may be measured directly with a universal stage or spindle stage by setting the optic plane vertical and rotating from one optic axis to the other. With the universal stage, a refraction correction is required, and even the corrected values are subject to uncertainties, as noted by Wyllie (1959), Munro (1963), and Kamb (1962); the latter paper furnishes theory and helpful advice on when and what corrections are to be used. With the spindle stage, the optic angle is routinely obtained by computation from measured extinction angles and partial refractive indices (Bloss, 1981).

Estimation

In practical crystal identification, 2V is very often estimated rather than measured. An accurate measurement is often not needed for identification, and relatively small variations in chemistry and structure within a single species may cause considerable variation of the optic angle. In short, the exact value of 2V is very often not diagnostic of a species, and sometimes not even diagnostic of composition within a solid solution series if structural modifications are unknown, as, for example, in the feldspars. On the other hand, accurate knowledge of 2V may, if the composition is known, furnish valuable structural information about such crystals as feldspars. The precision with which 2V is to be determined must therefore be tailored to the goals of the investigation.

A few additional facts are pertinent to the estimation of 2V. If two isogyres are present in a *centered optic axis* figure, 2E is less than 1/2 the angle subtended by the figure, corresponding to 2V of 30°–35°, depending upon the β refractive index (see equation 11-1). If only one isogyre is present in a centered optic axis figure, 2V therefore exceeds 30°–35°. One specific type of uncentered interference figure suggests that 2V is very small or zero, an isogyre that is straight and parallel to a crosshair at all positions of rotation. This figure has been noted earlier as being indistinguishable from an uncentered uniaxial optic axis figure whose cross center is outside the field. Biotite oriented in a thin section neither parallel nor normal to the cleavage forms such a figure.

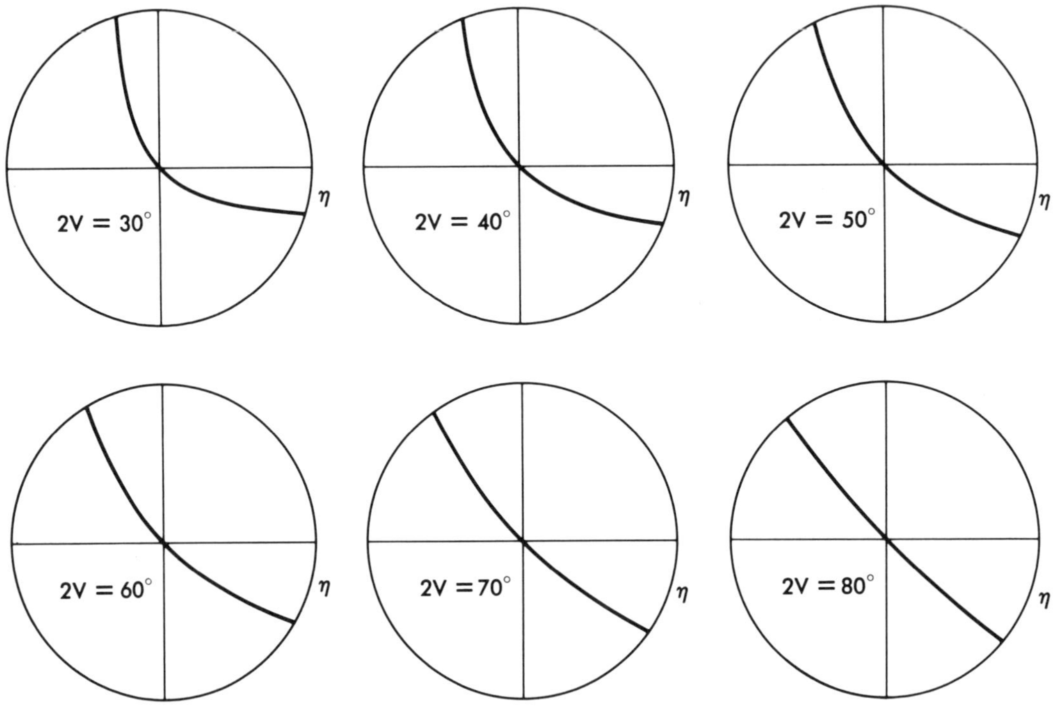

Fig. 12-4. Curvature of median line of isogyre of optic axis figure, as a function of 2V. The angle η is about equal to V.

Estimation from Isogyre Curvature. 2V may be estimated from the curvature of the single isogyre of a centered optic axis figure. The maximum curvature at the 45° position is a function of 2V, ranging from a right-angle bend at 2V = 0° to a straight line at 2V = 90°. The approximate curvature of the median line of the isogyre for selected values of 2V is shown in Fig. 12-4, after Wright (1905). Kamb (1958) has shown that Wright's equation is fortuitously correct enough for most practical purposes. Kamb also points out the useful fact that η, the angle between the tip of the isogyre and the crosshair, is about equal to V (see Fig. 12-4).

Centered optic axis figures from grains of the following minerals in thin section or immersion mounts illustrate the relation between isogyre curvature and 2V:

Common olivine: slightly less than 90°

Anorthite: about 77° (basal cleavage fragments are approximately normal to an optic axis)

Augite or diopside: about 60°

Hemimorphite, about 46°

When 2V is close to 90° it may still be possible to detect a slight curvature by inserting the ocular with a crosshair NW–SE along the isogyre, if the figure is well centered. The eye can detect a small deviation from parallelism of the straight crosshair and the isogyre. This test can frequently be used to determine that common olivine with 2V near 85° is negative.

Mallard's Method and Related Methods

Without a spindle or universal stage, 2V may be determined by measurement of the distance between the melatopes in a centered Bxa interference figure in which the isogyres do not leave the field. This is Mallard's (1882) method, which is of course limited to crystals with 2V less than about 65° if N.A. = 0.85 or 50° if N.A. = 0.65 (equation 11-1). The distance between the melatopes, 2D, in a Bxa figure at the 45° position is related to 2E by Mallard's formula:

$$D = K \sin E \tag{12-3}$$

K is *Mallard's constant,* which depends upon the optical equipment used; 2D is measured from a Bxa figure at the 45° position using a micrometer ocular. Mallard's constant is determined for a specific microscope, objective, and ocular combination from a Bxa section of a crystal of known 2V. A Bxa section of aragonite is frequently used to measure *K* because, for this mineral, the values of 2V = 18°, $\beta = 1.681$ are nearly constant from sample to sample. If E and β are known, 2V is calculated from Snell's law, which from equation 1-11 can be expressed as $\sin E = \beta \sin V$. A graph relating E, V, D, and *K* is furnished by Winchell (1946) and Tröger (1979).

A method that eliminates *K* by using the ratio $d = 2D/2R$, in which 2R is the diameter of the interference figure, is described by Tobi (1956), who furnishes a chart relating 2V, 2E, *d*, and β.

The error in determination of 2V by Mallard's method was examined by Fairbairn and Sheppard (1945). It is ordinarily small for accurately centered figures if 2V is between 20° and 60°, but increases rapidly below 2V = 20° as shown in Fig. 12-5.

> Cleavages of muscovite, heulandite, topaz, biotite, and prehnite may be used to illustrate the measurement of 2V by Mallard's method.

Other methods of 2V measurement using the movement of isogyres can be used. The rotation angle of Kamb (1958), discussed in Chapter 11 (Fig. 11-5), furnishes a basis of 2V measurement if the N.A. of the objective and the mean refractive index of the crystal are known (equation 11-2). The method can be used beyond the limits of the Mallard method, i.e., if the isogyres leave the field, and can if necessary be used with the Bxo figure, although with greatly diminished accuracy. Estimates of 2V and refractive index can sometimes be refined against one another to advantage.

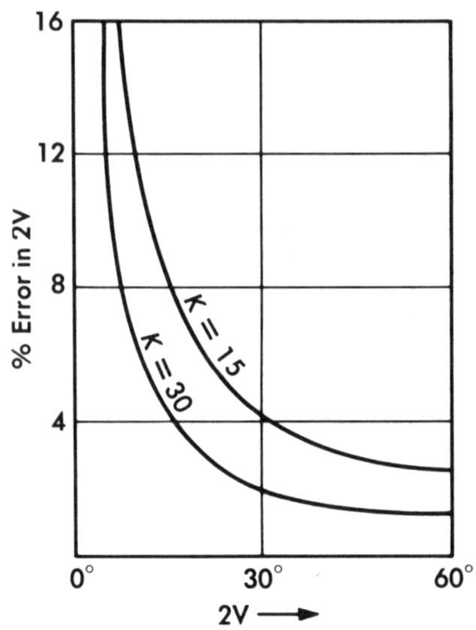

Fig. 12-5. Accuracy of the Mallard method of 2V determination. K is Mallard's constant. The curves refer to β known to ± 0.001 and 2D measured to within ½ scale division.

REFRACTIVE INDEX

Introduction

Refractive indices form the basis for most crystal identification schemes by optical means when material is available for immersion mounts. They are also used for composition determination within natural and synthetic solid solution series. In the course of determining refractive indices, one usually observes other useful properties of the crystal.

To measure a refractive index of a biaxial crystal by the immersion method, a grain must be found with the required vibration direction in the plane of the stage, and this vibration direction must be set parallel to the polarizer direction so that the refractive index that is determined will be a principal index. In thin section, refractive indices are estimated rather than measured, but the same principles apply.

Determination in Immersion Mounts

A vibration direction to be used for refractive index tests may be identified and oriented by means of interference figures alone. However, there is no point in orienting a vibration direction unless there is a good chance that the crystal is immersed in a liquid that matches the crystal index sought. Furthermore, the precise identity of the interference figure often need not be known; knowledge that the figure is a 2-SP or a specific kind of 1-SP usually suffices. For these reasons it is inefficient to concentrate upon a search for ideal figures rather than a search for higher or lower refractive indices. In this book, emphasis is placed upon random search methods for finding a match for α, β, and γ between crystal and liquid, and on using interference figures only to confirm the orientation of vibration directions.

Random Search Methods. As with uniaxial crystals, a random search is usually by far the most efficient means of chasing refractive indices through a series of immersion liquids. Biaxial crystals have three indices to be determined. It is efficient to find provisional limits for α and γ first, and then to seek a value for β. The reason for this strategy is that one of the mounts used for α and γ may by chance also provide an exact match for β. If an optic axis figure or a counter-rotating 1-SP figure is found, then the value of β is present in the grain and can be tested for a match with the oil. If no such figure is easily found, but the value of 2V can be estimated from another figure, the expected value of β can be calculated from the limiting α and γ (equation 10-7 or 17-1), and then the critical interference figure can be sought in the most likely matching oil, as discussed below.

For γ, mounts are prepared with liquids of higher index. These are searched under oblique illumination to see if any grain in any orientation still has a higher index than the liquid (shadow near or yellow color fringe). If so, a new mount is prepared immediately with a higher liquid, the old mount being put aside for reference. The random search is continued until a mount is found that shows no crystals higher than the liquid in any orientation. This mount should then contain one or more grains that match the liquid (orange-red color fringe) at some rotation position. Until this time there has been no need to use crossed polars or to observe an interference figure. Now the polars are crossed to see if the matching crystal matches at an extinction position. If so, an interference figure is taken to see if it is a 2-SP (centered) figure or an SR 1-SP figure. If it is either of these, γ has been found, as can be confirmed by analysis of the figure, discussed below.

The α refractive index is found by an analogous procedure, searching for grains lower than the index of the liquid until none appears in the mount, at which point one or more grains should match the liquid at an extinction position and should show a 2-SP or SR 1-SP figure.

In practice, one usually overshoots α and γ in the random search, since it is more efficient to approach alternately from high and low values than to approach from one direction alone. When testing for limiting refractive indices, the search for grains with index beyond the oil can be hastened by removing the lower polar, as with uniaxial ϵ.

Pronounced cleavage may prevent or inhibit grains from lying with the required vibration direction parallel to the stage, in which case one may at least place an upper limit on the value of α and a lower limit on the value of γ. The effects of pronounced cleavage are discussed more fully below.

Confirmation of Gamma and Alpha. If either γ or α matches the liquid in some grain, the grain will show either a flash figure, one of the other two 2-SP figures, or an SR 1-SP figure, as listed in Table 12-1. When the random search method has been used, one need not know the exact identity of the figure, but only its symmetry. In any of the 2-SP figures, the isogyres leave in the direction of α or γ.

Flash figures furnish both α and γ. The other 2-SP figures furnish α or γ when the optic plane is set parallel to the polarizer, usually N–S. To set the optic plane in this position, observe the line connecting the melatopes or the direction in which the isogyres have left at the 45° position, then set that line or direction N–S by a 45° rotation to extinction. Bxa figures furnish α if the crystal is (+), γ if the crystal is (−). Bxo figures furnish γ if the crystal is (+), α if the crystal is (−).

All SR 1-SP figures furnish either α or γ, and one non-principal index. Either X or Z lies N–S when the isogyre coincides with the E–W crosshair, since the isogyre marks the trace of a symmetry plane which is not the optic plane. Which of the two vibration directions lies normal to the symmetry plane is easily determined: set the isogyre E–W, rotate the stage 45° to the right, thus setting the symmetry plane NW, view the grain orthoscopically, and insert an accessory plate. If the colors rise, Z is now NE; if the colors fall, X is now NE. The identity of the vibration direction is usually known beforehand if the random search method has been used.

Refinement and Confirmation of Beta. If optic axis sections are abundant or at least not prohibited by cleavage, β is easily found by using different liquids until grains cut or broken approximately normal to an optic axis match the liquid. To confirm that the match for β is valid, the interference figure from the matching grain should be checked to

Table 12-1
Interference Figures whose Grains Furnish Exact Values of α, β, γ.

Refractive Index	Interference Figures
α	Flash figure (ON)
	Either Bxa if crystal (+) or Bxo if crystal (−)[a]
	Some SR 1-SP figures (SP = YZ plane)[b]
β	Optic axis figure
	Any other CR 1-SP figure
	Bxa
	Bxo
γ	Flash figure (ON)
	Either Bxa if crystal (−) or Bxo if crystal (+)[a]
	Some SR 1-SP figures (SP = XY plane)[b]

[a]Any 2-SP figure furnishes either α or γ.
[b]Any SR 1-SP figure furnishes either α or γ.

see that the isogyre rotates on a point at the center of the crosshairs. If the melatope describes a circle around the center of the crosshairs, either a new grain should be found or some allowance made for uncertainty in the recorded value of β.

If optic axis sections are lacking, other strategies may be used (Table 12-1). Aside from the optic axis figure, three other types of figure show that the optic plane is vertical, and therefore indicate grains that furnish β. Any CR 1-SP figure places Y north–south when the isogyre (hence the trace of the optic plane) is east–west. In both the Bxa and Bxo figures, Y lies 90° from the trace of the optic plane, which is marked at the 45° rotation position by a line connecting the melatopes, or by the direction in which the isogyres have left the field. The figures whose grains furnish β are listed in Table 12-1.

Once α and γ are known, β may be sought more efficiently by knowing the optic sign and approximate 2V. If 2V is near 90°, β is about halfway between α and γ. If the sign is (−), β is nearer γ than α, and if (+), β is nearer α than γ. How near may be estimated qualitatively from 2V, or quantitatively from 2V and equation 10-8 (page 166). With such knowledge, a form of random search can again be used in the vicinity of β, with the aid of interference figures.

Effects of Pronounced Cleavage. Cleavage may prevent one or more of the vibration directions from lying parallel to the stage, in which case one or more refractive indices may not be measurable by routine methods. For purposes of identification, the problem may not be severe if interference figures are not badly uncentered. In such cases, a useful limiting value of the index in question can often be stated, or an estimate made of its probable value by consideration of how far a figure is uncentered and of the birefringence. If the birefringence is weak, the effect of approximate orientation is less drastic than if the birefringence is strong.

Cleavage fragments of orthorhombic or monoclinic crystals are usually parallel to one or two principal vibration directions. If one or more of the principal vibration directions is not parallel to a cleavage plane, it may be very difficult to find the appropriately oriented fragment for study. This difficulty is encountered particularly in platy and rod-like cleavage fragments such as those of mica, chlorite, amphibole, and pyroxene.

Gelatin mounts, mentioned on page 45, provide the simplest means of overcoming preferred orientation due to cleavage. Use of the spindle stage (Bloss, 1981; Fleischer et al., 1984) overcomes orientation problems completely.

It has been suggested that fragments of glass be introduced in the immersion mounts to prop up the platy fragments or that sections of soft crystals be cut perpendicular to the offending cleavage with a sharp blade, placing the resulting slice in the immersion mount in the position desired. Slight pressure on one edge of the cover-glass commonly suffices to orient all but platy grains. An excess of liquid will make the material more mobile.

The vibration directions of cleavage fragments of many triclinic minerals are not those associated with the principal refractive indices. The partial refractive index on the cleavage may be effectively determined in the few instances in which these values have been recorded. Partial indices are particularly useful in determining the composition of members of the plagioclase series for which the values of α' and γ' on the two cleavages have been plotted, as shown by Tsuboi (1923, 1934) and Morse (1968); see Tröger (1979).

Mixed Species in a Sample

In the discussion above it has been tacitly assumed that only a single unknown crystal species is present in the sample. If this is not the case, confusion may easily result. Often the characteristic type of interference figure, or birefringence, or difference in relief, or general appearance can be used to distinguish among species in a mount. Disagreement between values of refractive index from grain to grain, when good confirming interference figures are present, is usually diagnostic of mixed species. More explicit hints are given in Chapter 17, where identification of crushed whole rocks in immersion is discussed.

Estimation in Thin Section

As with uniaxial crystals, estimation of refractive index in thin section is restricted to comparison of the unknown with neighboring crystals of known identity or with the mounting medium. Principal refractive indices can be compared only when the section is cut by chance normal to one or more symmetry planes. One usually compares partial indices. The purpose of estimation is usually to eliminate one or more suspected candidates, and a comparison of one or more partial indices may suffice very well for this.

Some colorless minerals easily confused in thin section may be distinguished on the basis of refractive index, usually by observation of the Becke line against quartz, plagioclase, or the mounting medium, presumed near 1.54. For example, untwinned cordierite, which is easily mistaken for quartz, may be distinguished by its distinctly higher mean refractive index. K feldspar shows low relief and a faint pinkish hue against albite and quartz when a section is scanned with the illumination stopped strongly down.

DISPERSION

Introduction

The relationship between certain types of dispersion and crystal symmetry is elegant, and useful when it can be observed. In practice, one recognizes chiefly rhombic dispersion, which produces weakly colored fringes along isogyres near the melatopes at the 45° position. One also notes whether the fringe nearer the bisectrix is red or blue, and then compares the result with information in the literature on the suspected crystals. Dispersion of this or a similar type is present in most biaxial crystals, but is easily noticeable in relatively few and strikingly noticeable in still fewer. As a result, strikingly noticeable dispersion is usually a strong direct clue to the identity of the crystal.

Dispersion is seen more easily in thick grains than in thin, hence is ordinarily more obvious in immersion than in thin section. It is also much more easily seen in light from the sky, which has a stronger blue component than filtered tungsten light.

Dispersion causes incomplete or colored extinction in some crystals, because red and blue light either do not travel the same paths, or do not vibrate in the same plane.

Dispersion of the Refractive Indices. Each of the refractive indices of biaxial crystals decreases as the wavelength of light increases. Determination of this dispersion is made only when variable monochromatic light is used.

The magnitude of refractive index dispersion in biaxial crystals is not unlike that of uniaxial crystals. Extreme dispersion is illustrated by the mineral sphene (titanite), weak dispersion by the mineral natrolite:

Dispersion expressed as $n_{\mathrm{Tl}} - n_{\mathrm{Li}}$:

	γ	β	α
Sphene (extreme)	0.025	0.023	0.022
Natrolite (weak)	0.007	0.006	0.005

The table illustrates that the dispersion is of similar magnitude for each refractive index of the crystal, but tends to diminish with refractive index. The average dispersion of minerals rises with refractive index, as given by equation 16-12; this relation is useful in the application of the dispersion method to the rapid determination of refractive index.

Dispersion of the Optic Angle (Rhombic Dispersion)

This is also known as *dispersion of the optic axes* or *axial angle dispersion*. It is the only type of common dispersion found in orthorhombic crystals, but it occurs also in monoclinic and triclinic crystals. It is not always possible to distinguish rhombic dispersion from the more complex dispersion of lower symmetry discussed below, and for this reason the observed effect is often tabulated as though it were rhombic dispersion without regard to detailed identification.

Rhombic dispersion results from dispersion of the refractive indices. It can be seen from equation 10-4 (page 166) that the optic angle for one wavelength will not equal that for another unless by chance all indices vary with wavelength in such a way that β maintains the same relation to α and γ. The crystal has an indicatrix of different size and shape for any different wavelength of light. Blue and red light are chosen as extremes for illustrative purposes, since these are the regions of the spectrum most easily seen by the eye.

Minerals are likely to show strong rhombic dispersion when they contain as essential constituents U, Sb, P, V, Fe, Mn, or Ti.

Rhombic dispersion produces color fringes along the isogyres near the melatopes in a Bxa figure. The effect is symmetrical about a line normal to the trace of the optic plane, as shown in Fig. 12-6. The optic angle for blue light may be either greater or less than the optic angle for red light (Fig. 12-7).

The relative position of the blue and red fringes is just opposite to

Fig. 12-6. Rhombic dispersion: color fringes in Bxa figures at the 45° position. R and B stand for red and blue fringes, respectively.

the position of the melatopes for the respective colors in the crystal, because a blue fringe marks the extinction position for red light, and vice versa. The angle in the crystal is symbolized as follows: the angle for red is *r* (from Greek *rho*, ρ); the angle for blue is *v* (from Greek *upsilon*, υ). Thus, if blue fringes appear outside red fringes, the optic angle for red is greater than that for blue ($r > v$), and vice versa. This

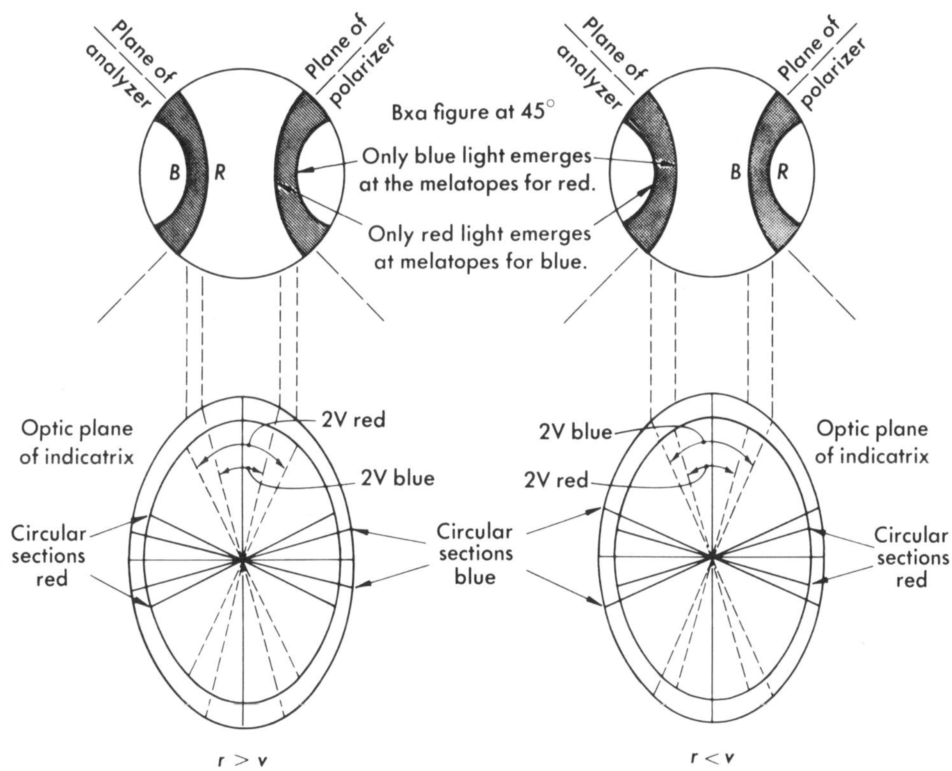

Fig. 12-7. Explanation of rhombic dispersion: optic planes of two indicatrices for red and blue light, with interference figures above.

leads to the following rules with respect to the concave side of the isogyre:

If the red fringes are on the concave side, $r < v$.

If the blue fringes are on the concave side, $r > v$.

No fringes are visible when the interference figure is in the crossed position unless dispersion of the vibration directions is present.

In describing the difference in size of 2V for red and blue, the adjectives weak, distinct, strong, very strong, and extreme are commonly employed. These adjectives refer to the ease with which the difference in angle can be observed in the Bxa interference figure using white light. Numerical values of the difference in r and v equivalent to these adjectival designations are not standardized.

The width and clarity of the fringes in the Bxa interference figure at the 45° position are measures of the strength of the dispersion as described in Table 12-2. Other adjectives used to describe strength of dispersion such as "marked" and "perceptible" are self-explanatory.[2]

Table 12-2
Strength of Dispersion

Dispersion	Description of Fringes	Shown by Cleavages of:
Weak	Fringes barely perceptible	Epsomite, $r > v$
Distinct	Fringes easily seen	Anhydrite, $r < v$
Strong	Fringes obvious, wide	Niter, $r < v$
Very strong	No black isogyre at melatope	Glauberite, $r > v$
Extreme	Melatopes for different wavelengths so very different that no overlap occurs and there is no black isogyre.	

In orthorhombic crystals, in which symmetry prohibits dispersion of the vibration directions, fringes are caused only by dispersion of the optic angle. Because fringes are not modified by dispersion of vibration directions, dispersion of the optic axes can be identified in crystals known to be orthorhombic from an optic axis figure, as in Fig. 12-8. If the crystal is not known to be orthorhombic, interpretation of such a fringe on a single isogyre is risky, since another type of dispersion may be involved. If two isogyres are present in a Bxa figure,

[2] The fringes become clearer in interference figures from thicker grains. This complicates quantitative appraisal. Sheets of muscovite ($r > v$) of several thicknesses will illustrate the dependence upon thickness.

Fig. 12-8. Rhombic dispersion in a centered optic axis figure at the 45° position. R = red fringe, B = blue fringe, so $r < v$.

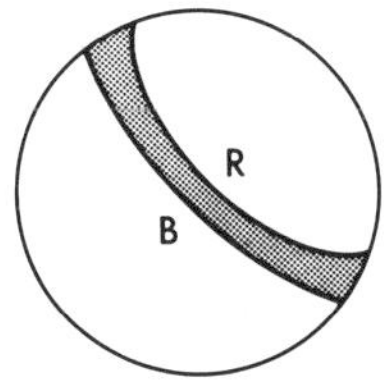

rhombic dispersion may be recognized in some monoclinic and triclinic crystals.

Dispersion of the Vibration Directions

In monoclinic and triclinic crystals the position of the vibration directions relative to the crystallographic axes may be different for different wavelengths. This difference is dispersion of the vibration directions. A principal vibration direction that is coincident with a crystallographic axis is coincident for all wavelengths, and hence has no dispersion. For this reason there is no dispersion of the vibration directions in orthorhombic crystals (if crossed axial plane dispersion[3] is excepted). Dispersion of the vibration directions not coincident with the unique *b* axis can occur in crystals of the monoclinic system, and there may be dispersion of all three vibration directions in triclinic crystals.

Monoclinic Crystals. Dispersion of the vibration directions in a monoclinic crystal may be one of three kinds, as illustrated in Fig. 12-9. The dispersion is always compatible with $2/m$ symmetry: the dispersed vibration directions always lie in the (010) plane of the crystal. Any of the three symmetry planes of the indicatrix may lie in the (010) plane. The type of dispersion is named from the position of the optic axes for different colored light, which is dictated by which symmetry plane of the indicatrix lies in (010). The names are:

Inclined if optic plane = (010), Y = *b*

Horizontal if Bxa—Y plane = (010), Bxo = *b*

Crossed if Bxo—Y plane = (010), Bxa = *b*

[3] Crossed axial plane dispersion is due to the optic plane being coincident with one crystallographic plane for some wavelengths and coincident with another plane at right angles for other wavelengths. Bxa figures are then quite abnormal when viewed in white light. This type of dispersion is not common, but is famous in the mineral brookite, TiO_2.

When dispersion is inclined, there is a single optic plane for all colors of light; vibration directions and optic axes are dispersed within the optic plane [Fig. 12-9(a)].

In horizontal and crossed dispersion (dispersion of the optic plane), the position of the entire optic plane varies with the color of light. All the possible optic planes intersect along the *b* axis of the crystal. The inclination of the optic plane varies relative to *c* with color. If Bxa lies in the (010) plane and Bxo is *b* (horizontal dispersion), the optic axes for red and blue light are symmetrically displaced across the trace of the *b* axis. If Bxo lies in the (010) plane (crossed dispersion) and Bxa is b, the pairs of optic axes for red and blue light are related to each other by a rotation about *b*, but not by a symmetry plane.

The relation between horizontal and crossed dispersion can be appreciated by considering the optic planes shown in Fig. 12-9(b), and then imagining that the optic angle increases till the *b* axis becomes the acute bisectrix. This is simply an exchange of Bxo and Bxa, and it produces the crossed dispersion shown in Fig. 12-9(c).

The three types of monoclinic dispersion of the vibration directions are recognized in Bxa figures by the position and symmetry of pairs of fringes, as shown in Fig. 12-10. In the interference figures, the color symmetry line is the trace of the optic plane for inclined dispersion, and the normal to the trace of the optic plane for horizontal dispersion. Figures showing crossed dispersion have no symmetry line, only a center of symmetry. In horizontal and crossed dispersion, fringes can sometimes be seen at the crossed position as well as at the 45° position. The color fringes at any rotation position result from the difference in position of the melatopes and isogyres for red and blue light. Unless dispersion is extreme the isogyres for all colors coincide and are black except near the melatopes where the colored borders appear.

Dispersion of the vibration directions is also indicated by incomplete extinction, different for red and blue light. In a section parallel to (010), and to a lesser extent in other sections except those parallel to *b*, the traces of the vibration directions in the plane of the section do not coincide. This causes a reddish or brownish color at the position of extinction for blue wavelengths and a bluish color at the position of extinction for red (Fig. 12-11, ferroaugite from Renfrew, Ontario; data from Wülfing, 1895). This is most noticeable if the crystal is colorless and if the interference colors are first-order gray or white and therefore do not obscure the effect.

The strength of dispersion of the vibration directions is described qualitatively from observations of the width of the fringes, or the difference in the angle between the trace of a cleavage and the

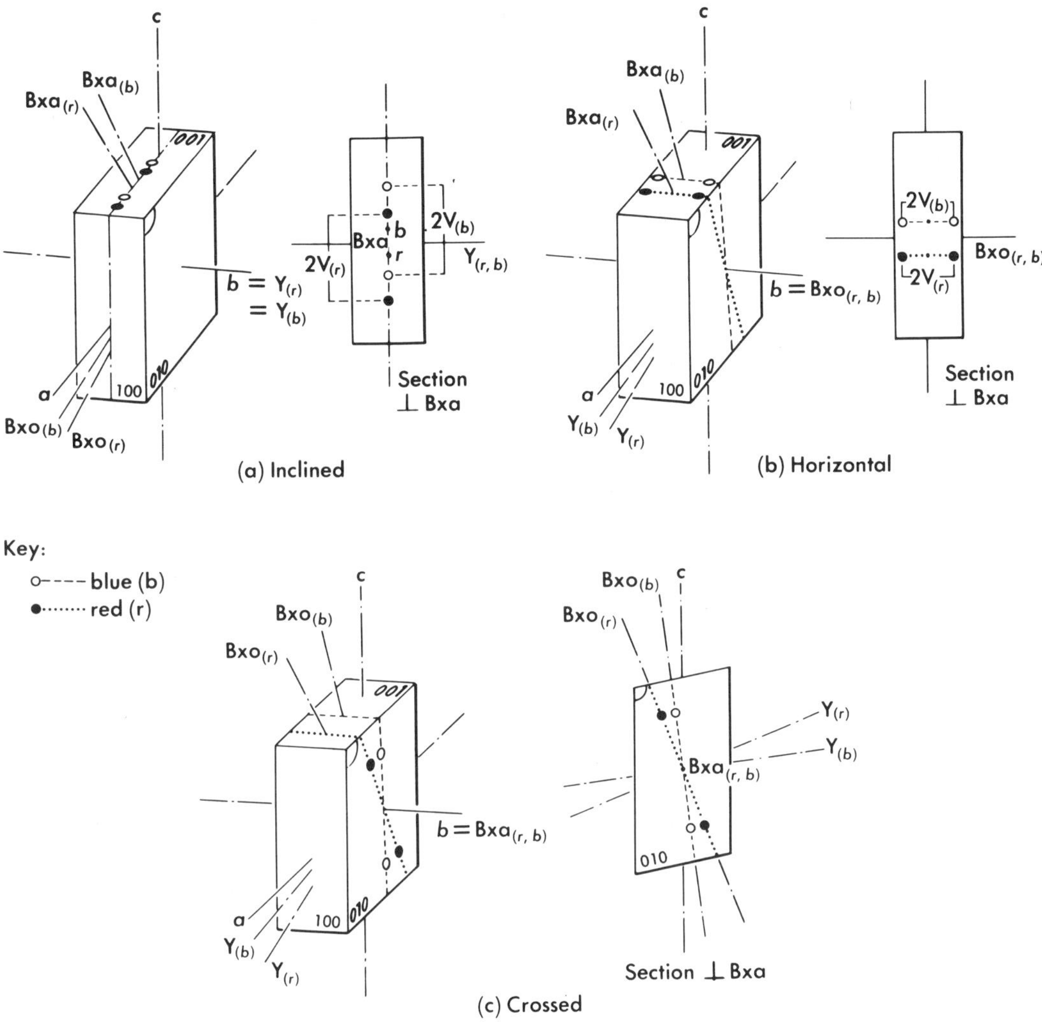

Fig. 12-9. Dispersion of the vibration directions in monoclinic crystals. (a) Inclined dispersion; (b) Horizontal dispersion; (c) Crossed dispersion.

extinction position for red and blue in the (010) plane. The extinction angle for red may be greater or less than for blue. Cleavage fragments of heulandite that show white or gray interference colors exhibit different positions of extinction for red and blue. Since dispersion is commonly small and extinction angles are commonly measured only to the nearest degree, and since the blue and red may be obscured by the interference color, quantitative measurement of the extinction angles for red and blue on (010) are impractical in routine examination.

Determination of which extinction angle is greater can sometimes

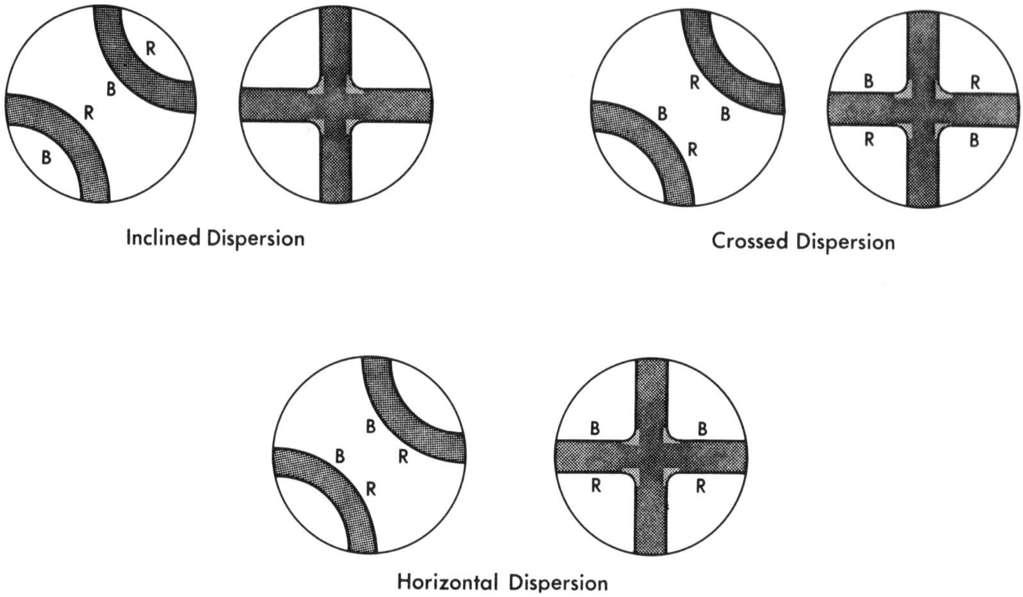

Fig. 12-10. Monoclinic dispersion fringes in Bxa figures. R, red; B, blue.

be useful confirmatory evidence in crystal identification. Borax crystals exhibit crossed dispersion. Inclined dispersion is evident in most monoclinic pyroxenes. The difference in extinction for r and v is evident in sections parallel to (010) in which the interference colors are white or gray. When rhombic dispersion is added to the dispersion of the vibration directions, the effect may be to make the observed fringes near one optic axis very much more dispersed than near the other optic axis, as distinctively seen in titanaugite (Tröger, 1979).

Triclinic Crystals. Dispersion of the vibration directions in triclinic crystals is more complicated, for there may be dispersion of all three vibration directions. To the extent that dispersion is very much stronger with respect to two of the three vibration directions, the nomenclature applicable to monoclinic dispersion is employed. If the dispersion affects all three axes, it is known as *combined dispersion*. The fringes in a Bxa figure of a triclinic crystal showing combined dispersion are not symmetrical with respect to the center of the figure or the trace of the optic plane or the plane normal thereto. Extinction angles that differ for red and blue are also indications of dispersion in triclinic as in monoclinic crystals.

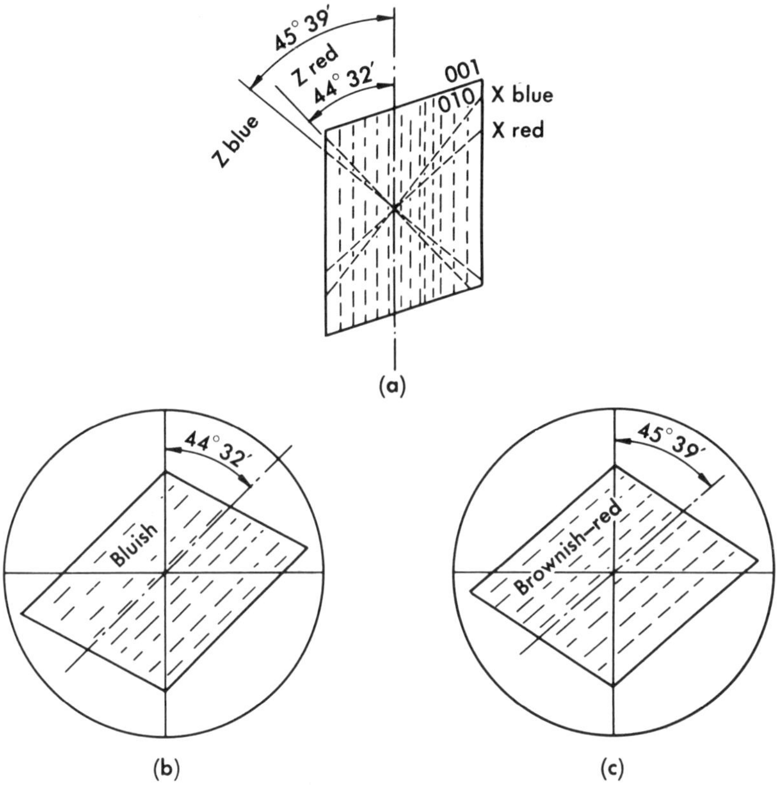

Fig. 12-11. Extinction of white light in a crystal of ferroaugite lying on (010) showing dispersion of the vibration directions. (a) Orientation diagram for red and blue light; (b) Extinct for red; (c) Extinct for blue.

General Remarks

Difficulties may arise in determining dispersion. Inclined dispersion cannot be distinguished in the Bxa figure if 2E is so large that the isogyres leave the field, for only at the 45° position of rotation can the fringes be seen clearly. Abnormal extinction or fringes adjacent to the black isogyre frequently suggest dispersion of the vibration directions, but do not permit identification as crossed, horizontal, or inclined.

WORKED EXAMPLE

A worked example of the determination of a biaxial crystal is offered at this point because it is mainly concerned with optical properties rather than the optic-crystallographic properties (optic orientation) treated in some detail in the next two chapters. As before, we present the observations in white light first.

White Light

The crystal was found to show color fringes in a 1.616 oil. The following observations were made:

1.616 oil:
  Crystal clear, colorless
  Flat tops common: one good cleavage?
  Many Bxa figures: (+) 2V 62° (N.A. 0.85 lens, eq. 11-1, p. 185)
  Brownish red color fringe—α low by about .002, from Fig. 3-15.
  Orange red color fringe on same grain with Bxa figure—β on
  Some orange to bright yellow color fringes—γ as much as 0.006 high, from Fig. 3-15.

The interpretation requires further comment. The Bxa figure gave a positive optic sign because colors rose on the concave side of the isogyre in the NE quadrant when a gypsum plate was inserted. This meant that Z, the acute bisectrix and high index direction, was normal to the grain, and X and Y were therefore parallel to the stage. The two refractive indices giving color fringes in this grain were therefore full values of α and β. The data so far suggest $\alpha = 1.614$, $\beta = 1.616$. The bright yellow color fringes suggest, with the help of Fig. 3-15, that the highest index is near $1.616 + .006 = 1.622$.

1.622 oil:
  Lots of brown color fringes and shadow opposite—these are old low refractive indices near 1.616
  Some orange-red color fringes—got SR1SP figure with Z normal to isogyre so γ on

The γ index is confirmed to be 1.622. The birefringence B is $1.622 - 1.614 = +0.008$. By reference to Tröger's endpapers, the nearest candidates are topaz at B = 0.010, n = 1.61, and Mg-Fe chlorite at B = 0.008, n = 1.613. The mineral is colorless, so it can't be chlorite. It is therefore probably topaz. A perfect basal cleavage with Z = c is reported, and is consistent with the observation of many Bxa figures on flat topped grains. Further investigation shows that the 2V is just right for topaz with a bit of OH, as the table gives 66° for the pure fluorine end member and 44° for the OH end member. Reference to USGS 1627 shows a whole flock of biaxial positive minerals with β near 1.616 ± 0.002. However, the only colorless one with $B < 0.02$ is

REFRACTIVE INDEX WORKSHEET

SAMPLE NO. TP-1

	ALPHA α (NLOW)	BETA β (OMEGA)	GAMMA γ (NHI)
Oil:	1.616	1.616	1.622
Shuttle reading	128 mm	112 mm	115.5 mm
Lambda match = λ_0	626 nm	586 nm	595 nm
k_D, from table	−0.1697	0.0168	−0.0284
t match	23.8 °C	23.8 °C	22.6 °C
dn/dt (bottle)	− 4.1 E-4	− 4.1 E-4	− 4.1 E-4
$n_D^L = n_{25} + [(t-25)(dn/dt)]$	1.61649	1.61649	1.62298
Δ^L(F-C) (bottle or graph)	0.0238	0.0238	0.02439

(Bxa, same grain — alpha and beta)

Crystal:

Calculate Δ^S(F-C) = 10↑(5.9 log n - 3.2) or = exp (5.9 ln n -7.368):

Provisional Dispersion, Δ^S(F-C)	0.011	0.011	0.011
$n_D^S = n_D^L + k_D(\Delta^L - \Delta^S)$	1.6143	1.6167	1.6226

(If warranted, recycle with refined values of the crystal dispersion)

(Book says 0.008 but 8 points give 0.011 ± 0.001, so stet.)

$\gamma - \beta$ = 5.9E-3 B = 8.3E-3 $Vx \approx \arcsin \sqrt{(\gamma - \beta)/B}$ = 57.5

$2V_{calc}$* 65 $2V_{obs}$ 62 from figure Centered Bxa

Sign + Cleavages 001 prob. Color, etc. colorless

Optic orientation Bxa = c ? Crystal system Orh?

Distinguishing features low B, many Bxa figs., H = 8!

* wt % F = 19.8 from γ

* Ribbe + Rosenberg, Am. Min. 56, 1816. (First 2 eqn's of Table 2 are scrambled.)

Mineral Name Topaz Composition X_F = 0.96 : $Al_2SiO_4(F_{1.9}OH_{.1})_2$

Observer Morse Date 20 Jan '88

Fig. 12-12. Worked example of biaxial crystal identification using the worksheet shown in Fig. 16-4.

topaz, with an entry at 1.616, 1.618, 1.625. These indices are consistently higher than our observations. Tröger says that the β index may range all the way from 1.609 for the pure fluorine end member to 1.637 for varieties rich in OH, so individual entries could be all over the map. We now refine the data with the dispersion method and refine our interpretation of the composition.

Dispersion Method

The example is summarized in Fig. 12-12, based on our worksheet for crystal identification, Fig. 16-4. Two readings (for α and β) are available from the same Bxa grain in the 1.616 oil. A separate reading is made for γ in the 1.622 oil. The calculations are done just as before for the isotropic and uniaxial examples at the ends of Chapters 4 and 9. The figures for the oil are the same in the first two columns except for the lines related to wavelength.

The provisional dispersion of 0.011 is obtained from the formula discussed in Chapter 16 and shown on the worksheet. (The book gives 0.008, but it so happens that eight independent data points for this particular sample, TP-1, give a value of 0.011±0.001 for the dispersion, so we retain the provisional value.) The refined refractive indices for the crystal are 1.6143, 1.6167, and 1.6226, each a bit higher than the estimates in white light, but not different by more than 0.0007.

The partial birefringence $\gamma - \beta$ is 0.0059 and the total birefringence is $B = 0.0083$, hence the calculated V_X is 57.5°. If we double this and take it away from 180 we have $+2V = 65°$. This compares very favorably with the estimate of 62° from the centered Bxa figure. The refractive indices are therefore consistent with the observed 2V. As a capstone to our identification, we put a few grains between two dry glass slides and rub them together. They scratch the glass easily, so the hardness of 8 is very credible. No other nearby mineral in the tables has a hardness greater than 6, so the test is definitive.

A determinative diagram for (F-OH) topaz was published by Ribbe and Rosenberg (1971). Our value for γ corresponds to 19.8 weight percent F, using the equation furnished for γ by these authors (note that α and β appear to be interchanged in their other two determinative equations). The mole fraction of fluorine is given by $X_F \approx .04854$ wt% F − 0.00485, or, for 19.8 percent F, 0.96. Of the two fluorohydroxyl positions in this sample, 1.9 are filled with fluorine. We have therefore not only identified the mineral, but also found its composition in terms of (OH, F) without laborious chemical analysis.

13 Crystallographic Relations of Biaxial Crystals

SIZE, SHAPE, AND ELONGATION

The size, shape, and elongation of biaxial crystals are determined essentially the same way as for uniaxial crystals. The sign of elongation is defined somewhat more completely. Crystals may be elongated parallel to either X, Y, or Z, or to none of these. Crystals elongated parallel to Y are considered to have an indeterminate (±) sign of elongation. The rules for the sign of elongation are therefore:

Euhedral crystals with length parallel, or nearly parallel to the vibration direction:

Z, have + elongation ("length slow" of some authors)
Y, have ± elongation
X, have − longation ("length fast" of some authors)

Interference figures are observed to determine the identity of the vibration directions in elongate grains. Crystals of orthopyroxene or amphibole provide examples of + elongation.

Some biaxial crystals are flattened in one plane (elongate in two directions). The *sign of flattening* is defined as follows:

Crystals in which the vibration direction perpendicular, or nearly perpendicular to the plane of flattening is:

X, exhibit + flattening (e.g., micas)
Y, exhibit ± flattening (e.g., stilbite)
Z, exhibit − flattening

In order to identify the vibration direction normal to the section, interference figures from flattened plates are observed.

> Elongation and flattening properly refer to the forms of crystals and not cleavage fragments, because cleavages do not always reflect the growth habit of the crystal. Nevertheless, the terms can usefully be applied to cleavage fragments if reference to cleavage is clearly stated. The main point of caution is not to confuse cleavage and crystal habit.

OPTIC ORIENTATION

Optic orientation is the relationship between the principal optical directions X, Y, and Z and the crystallographic axes *a*, *b*, and *c*. It may be determined completely for the crystal or for one or more selected planes. The optic orientation of a biaxial crystal is not a fixed relationship as in uniaxial crystals, for which c = E in all cases. To the extent that crystallographic directions in biaxial grains are unknown, as is commonly the case, the optic orientation is determined provisionally.

A statement of the relationship between a vibration direction and a crystal axis partially describes the optic orientation. It may also be necessary to identify the crystallographic plane containing the vibration direction and the crystal axis. A complete statement requires that the relationship of X, Y, and Z to *a*, *b*, and *c* be indicated. The complete orientation is usually derived by the synthesis of the partial determinations in several crystallographic planes. The procedure varies with the crystal system. The name of the crystallographic directions in the plane of a section may be indicated by a recognizable euhedral crystal face, cleavage, or composition plane of a twin. The directions of vibration in the plane of the section are the extinction positions. The vibration directions in and normal to the section are identified by the interference figure formed; they may be principal (X, Y, Z) or partial (X′, Y′, Z′) vibration directions.

Extinction Angle Measurement

Measurement of the extinction angle gives quantitative expression to the relationship in the selected plane between a vibration direction and a crystallographic direction. The plane selected is usually a cleavage, or a plane normal to one or two cleavages. When a section lying parallel to the microscope stage is normal to a cleavage, the cleavage trace is a fine line that does not change position upon change of focus. The

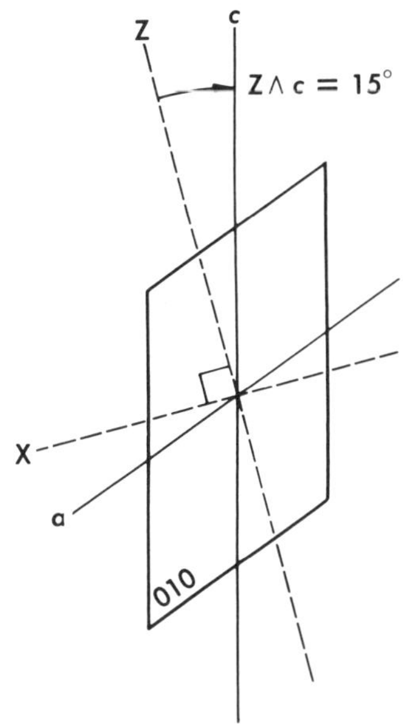

Fig. 13-1. Measurement of extinction angle in (010) section of a monoclinic crystal.

extinction angle, which is the angle between a vibration direction and a crystallographic direction, is expressed in a formula as, for example, Z∧c = 5°, in which the symbol ∧ is read "makes an angle with". The same information is sometimes written γ:c = 5°, or Z:c = 5°. In every section two vibration directions lie in the plane of the section, whereas frequently only one crystallographic direction can be recognized, and this may or may not be a crystallographic axis. One vibration direction lies nearer to the crystallographic direction than the other. It is most usual to record the angle between the crystallographic direction and the vibration direction which is less than 45°. If *c* and some other crystallographic direction can both be recognized in the plane of the section the extinction angle is commonly measured from the *c* direction (Fig. 13-1).

The vibration directions in cleavage sections may be partial rather than principal vibration directions. If so they are designated as Z′, the high index vibration direction, and X′, the low index direction. The two are distinguished by means of an accessory plate.

Extinction is usually measured in sections that include one or more of the crystallographic directions *a, b,* or *c* in the plane of the section. If, however, the crystallographic direction is the projection of a crystallographic direction in the plane such as the trace of cleavage, the Miller index (see Chapter 14) of the intersecting plane is cited in the extinction formula and reference is made to the plane in which the extinction is measured. In some instances the intersecting plane is cited in the extinction formula even if the plane of the section is parallel to a crystallographic axis, so as to indicate which line of reference was used in measuring the extinction angle. The basal cleavage of the plagioclase feldspar, albite, provides an example. In this mineral, in the (001) section, the vibration direction X related to the lower index α' makes an angle of 4° with the trace of the (010) cleavage, which contains the *a* axis. This may be expressed:

$$X' \wedge (010) = 4^\circ \text{ on } (001), \text{ or } X' \wedge a = 4^\circ \text{ on } (001).$$

Using ordinary equipment, the extinction angle is usually measured only to the nearest degree. The color of the isotropic field and the grain at extinction may be matched more closely if both are red than if both are black. Measurement with gypsum plate inserted will bring this about.

Sign of Extinction. In monoclinic crystals the extinction position from which the extinction angle is measured may lie in either the acute or obtuse angle between the crystallographic axes *a* and *c*. An example is shown in Fig. 13-1. Some authors have called the sign of the extinction angle positive when measured from a vibration direction in the obtuse angle, whereas others have called it negative. The confusion in nomenclature is best avoided by indicating whether the extinction angle in the (010) plane of a monoclinic crystal is or is not in the obtuse crystallographic angle beta. The positions of *a* and *b* in an (010) section, and hence the angle beta, may be determined from the euhedral crystal outline or from (001) and (100) cleavages, as indicated in Fig. 13-2.

A rule that many authors support determines whether in the triclinic feldspars the extinction angles on the (001) and (010) cleavages will be designated as plus or minus. The rule, known as *Schuster's convention* (1880), is that an extinction angle measured in the *obtuse* angle between the two crystallographic axes that lie in the plane of the section is *positive*, and in the *acute* one is *negative*. The sign of the extinction angle can rarely be determined because the position of more than one crystallographic axis is usually not known unless the grain is euhedral. In the (010) section of a triclinic feldspar the trace of the crystallographic axis *a* is marked by the trace of the (001) cleavage and the *c* axis is parallel to the face making an angle of approximately 114° with *a*. In the (001) section the composition plane of albite twinning marks the position of the *a* axis and *b* is parallel to the trace of an external face making an angle of approximately 94° with *a*.

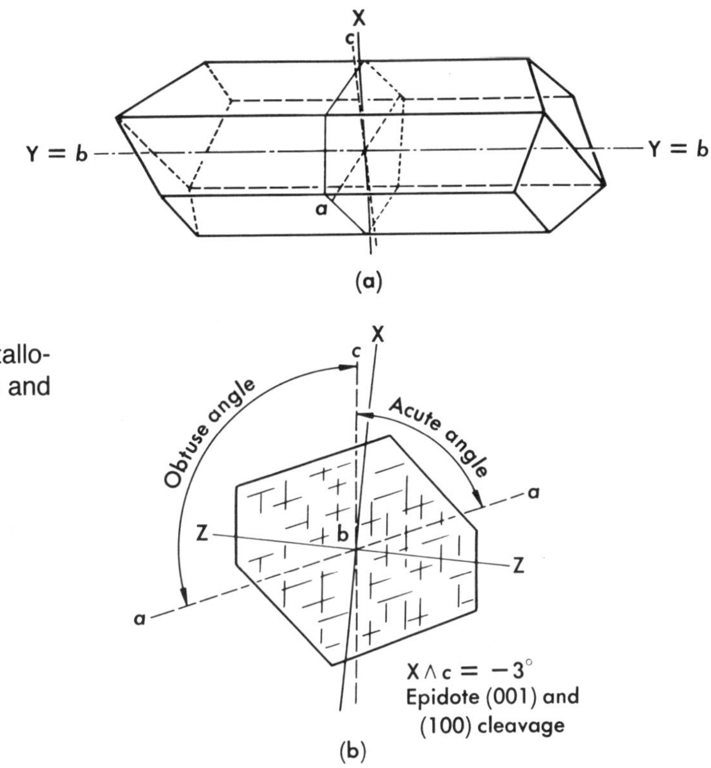

Fig. 13-2. Location of crystallographic axes from cleavage and external form.

Types of Extinction in Biaxial Crystals

The type of extinction shown by a biaxial crystal depends to some degree upon the crystal system, and hence in favorable cases may be used to identify the crystal system. The peculiarities of each crystal system are discussed here and summarized in Table 13-1.

Orthorhombic System. In orthorhombic crystals, because X, Y, and Z are parallel to *a*, *b*, and *c*, it follows that extinction in any pinacoidal or prismatic section is either symmetrical or parallel. Extinction is symmetrical in basal pinacoidal sections of crystals with prismatic cleavage. It is parallel in side pinacoidal and prismatic sections of crystals with pinacoidal or prismatic cleavage.

In all sections not parallel to a crystallographic axis, angles measured from pinacoidal cleavage traces may be small (if *c* is the Bxa) or moderate (if *a* or *b* is the Bxa). The magnitude of the maximum orthorhombic extinction angle for various angles of inclination of the section from the vertical zone is graphed in Fig. 13-3. Note that the extinction angle also varies with 2V. Such extinction angles would be encountered if a

Table 13-1
Extinction in Sections of Biaxial Crystals[a]

Crystal System	Sections Perpendicular to X, Y, Z	Sections Perpendicular to Cleavage	Cleavage Fragments	In a Field of Diversely Oriented Grains
Orthorhombic	Parallel or symmetrical	Parallel or symmetrical	Parallel or symmetrical	Sections parallel to *a*, *b*, or *c*, parallel or symmetrical[b]
Monoclinic	Parallel, symmetrical, or inclined	Parallel or symmetrical in some, inclined in some	Inclined on cleavage not parallel to *b*[c]	Parallel or symmetrical in sections parallel to *b*; all others inclined
Triclinic	Inclined	Inclined	Inclined	Inclined

[a]It is assumed that crystallographic reference directions such as cleavage traces are present in each section observed.
[b]In all sections nearly parallel to *c*, the extinction angle is small if the Bxa is *c*.
[c]Parallel or symmetrical on cleavage parallel to *b*.

crystal fragment lay on a pyramidal (*hkl*) cleavage (for example) in immersion. Because pinacoidal and prismatic cleavages are far more common than pyramidal cleavage, orthorhombic extinction angles are rarely encountered in immersions, but of course are quite generally encountered in thin section. One must not make the common mistake of thinking that an obvious extinction angle like 5° precludes orthorhombic symmetry, for it does not, as Fig. 13-3 indicates. Orthorhombic symmetry is, however, highly unlikely if all crystals of the same species in a thin section show extinction angles.

Monoclinic System. Extinction is inclined in most sections. It is symmetrical or parallel (depending on the orientation of the cleavage) only in sections parallel to the *b* axis. This follows from the fact that in all monoclinic crystals one of the principal vibration directions coincides with the *b* axis: It is possible, though not usual, for either *a* or *c*, but not both, to coincide by chance with a vibration direction. This is nearly the case in orthoclase. In this mineral the extinction angle on (010) measured against basal cleavage is only 5°. The generalities regarding extinction are summarized in Table 13-1. A study of extinction in sections perpendicular or parallel to cleavage ordinarily helps in distinguishing the crystal system. There are limitations to such methods, however, as well illustrated by monoclinic crystals. If no cleavage is

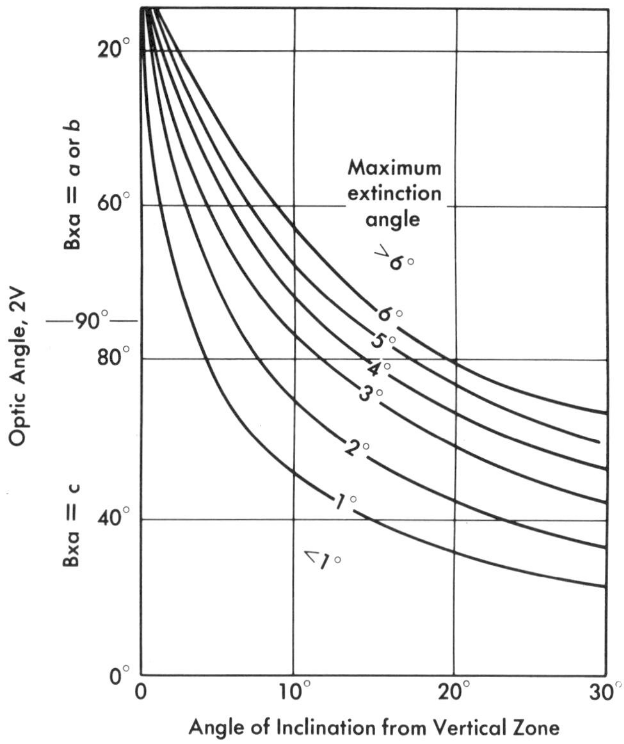

Fig. 13-3. Maximum extinction angle in oblique sections of orthorhombic crystals, after Harker (1901).

present parallel to *b*, no cleavage fragment will show parallel extinction and the crystal will be assumed to be triclinic. Similarly, if the only cleavages are parallel to the *b* axis, the extinction on the cleavage will be parallel and the crystal will appear to be orthorhombic. Such problems can usually be resolved with the aid of interference figures. If there is more than one cleavage, one of which is parallel to *b* and another not parallel to *b*, there will be no mistaking the monoclinic character of the crystal.

Triclinic System. Extinction in all sections of triclinic crystals is inclined except in the few sections in which there is chance coincidence of a vibration and crystallographic direction. The limitations of course prevail: there must be cleavage cracks or other visible crystallographic directions from which extinction angles may be measured. Furthermore, the type of extinction in grains of diverse orientation is not effective in determining the crystal system if the angles between the *a*, *b*, and *c* axes depart very little from those in a crystal of higher symmetry. When this is so the extinction angles are very small so that inclined extinction may appear to be parallel and lead to an erroneous judgment as to the crystal system. The general rules also fail if, in a plane of chance orientation, the extinction angle is equal to the angle between the traces of two cleavages. In a section nearly parallel to (001) of the triclinic mineral rhodonite, for example, the extinction appears to be symmetrical and the crystal might be suspected to be monoclinic. Extinction in grains of diverse orientation may help suggest the crystal system if these limitations are appreciated.

Special Uses of Extinction Angles. Extinction angles in specific planes are sometimes useful in distinguishing members of an isomorphous series. The plane chosen should be one in which the extinction angle varies largely with changes in composition and only a small amount as a result of slight departure of the orientation of the section from the desired plane. Extinction angles in several planes in the triclinic plagioclase feldspar series will determine the composition of the plagioclase in terms of the albite and anorthite end members. It has been noted by Goransen (1926) that even a considerable error in the orientation of a plagioclase section normal to both (001) and (010) cleavages changes the extinction angle very little and that the extinction angle in this section varies rapidly with composition of the sodic plagioclases. Hence this plane, the so-called "*a*-normal" plane, is useful in determining plagioclases. This is in contrast to the extinction angle on the (001) cleavage which, for the 0–50 percent An range of plagioclase composition, varies little with An content and is affected relatively strongly by errors of orientation. Extinction angle curves appear in standard reference works such as those of Tröger (1979) and Jones and Bloss (1980).

Expression of the Optic Orientation of a Crystal

Because of the different relationships between optical and crystallographic directions in each of the three biaxial systems, the crystal system may be tentatively identified, as discussed later in this chapter. When this is done one may formulate one or more possible expressions for the optic orientation that do not violate the limitations imposed by the crystal system. The procedure for determining the complete optic orientation differs for each system.

Orthorhombic System. The crystal axes are mutually perpendicular, and coincide with the optical directions X, Y, and Z in one of six possible combinations. The optic orientation for an orthorhombic crystal is completely specified by a set of equalities indicating parallelism of vibration and crystallographic directions, for example, X = *a*, Y =

b, Z = *c*. The six relationships form the set of all possible ways of ordering *a*, *b*, and *c*, as follows:[1]

X	Y	Z
a	b	c
a	c	b
b	a	c
b	c	a
c	a	b
c	b	a

In order to arrive at the orientation formula of an orthorhombic crystal, vibration directions are identified from centered Bxa, Bxo, and flash figures and related to crystallographic directions identified in the grains observed orthoscopically. Cleavages are parallel to or symmetrical with respect to the vibration directions in sections giving centered figures. An example of the relationship between an interference figure, morphology, and cleavage is shown in Fig. 11-2.

The chief problem is that of identifying crystallographic directions. There are several features illustrated in Fig. 13-4 and enumerated below that may be useful in provisional identification. To each there are frequent exceptions.

If crystals are acicular in habit [Fig. 13-4(a)], *c* may provisionally be assumed to be the direction of elongation; if platy [Fig. 13-4(b)], to be the direction normal to the plate.

If twinning of the aragonite type [Fig. 13-4(c); see below] is present, the direction bisecting the angle between two (110) composition planes in a basal section is the *b* axis. If the composition planes are sharp, the section is normal to them and hence normal to *c*.

If cleavage can be provisionally identified, assumptions can be made regarding the crystallographic directions in certain instances. If cleavage is prismatic and not at right angles, *b* can be assigned to the acute angle [Fig. 13-4(d) left]. If nearly at right angles, no assumption is possible [(d) right].

If cleavage is prismatic, and in thin section a grain shows only one cleavage trace (lies in the prism zone), the cleavage trace is the direction of the *c* axis [Fig. 13-4(e)].

[1] The occasional instance in which X, Y, and Z do not coincide with the same crystallographic directions for all wavelengths introduces additional relationships. One of the six combinations may apply for one part of the spectrum and another for the remaining wavelengths. This is crossed axial plane dispersion, described in a footnote on p. 219.

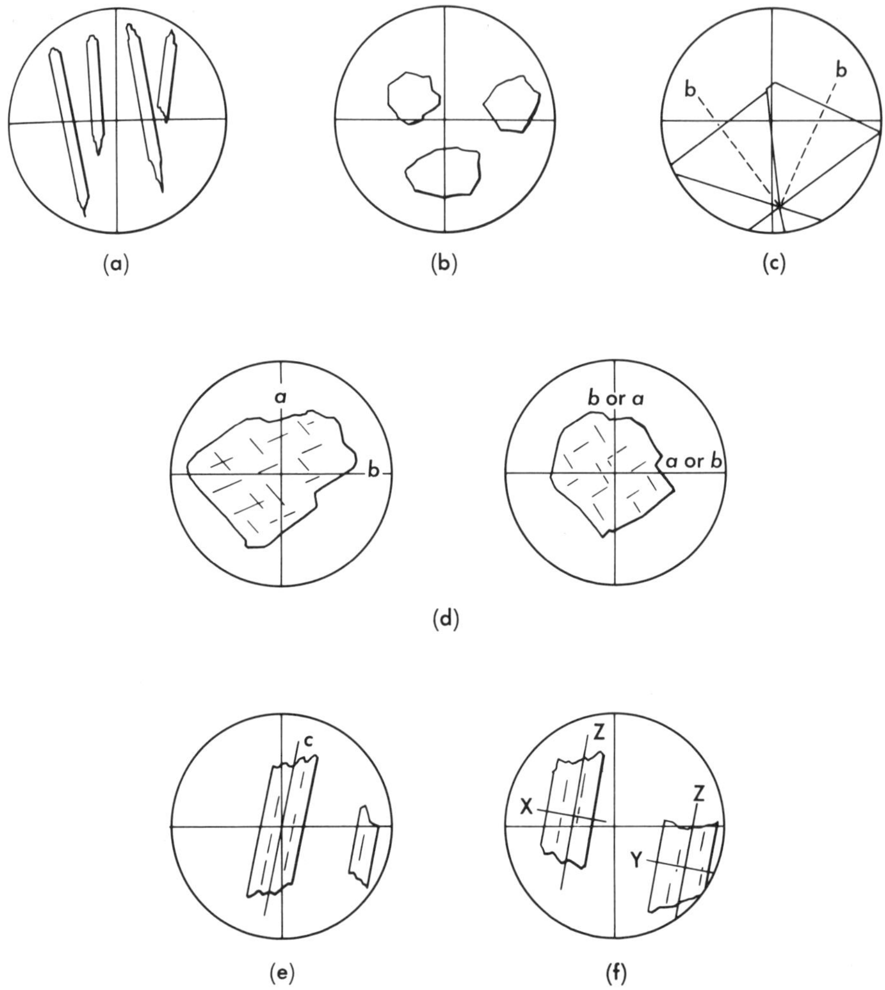

Fig. 13-4. Assumptions regarding crystallographic directions in orthorhombic crystals.

If two pinacoidal cleavages are present [Fig. 13-4(f)], they may be assumed provisionally to be (001) and (010) and the *a* axis will be common to both cleavage planes (e.g., Z = *a* in the figure).

Although identification of the crystallographic directions in orthorhombic crystals is frequently uncertain, one or more of the sets of possible orthorhombic orientation formulas can usually be provisionally assigned. Alternate formulas are considered to the extent required by the assumptions regarding the identity of crystallographic directions. Other optical properties will suggest the identity of the crystal and incorrect alternatives may be discarded.

Monoclinic System. In crystals of the monoclinic system the crystallographic axes *a* and *c* are perpendicular to *b* but not to each other. The optical directions X, Y, and Z are mutually perpendicular. The crystallographic axis *b* always coincides with one of the vibration directions X, Y, or Z and the other two vibration directions lie in the plane of *a* and *c*. The optic orientation of a monoclinic crystal is described by indicating which vibration direction coincides with *b*, and the angular relationship between *c* and one of the two vibration directions in the *ac* plane, the (010) plane. The optic orientation is recorded in two expressions: X, Y, or Z = *b* and X, Y, or $Z \wedge c = n°$. It is clear that much depends on the features in the (010) plane.

The (010) plane is one of the three planes (but generally the only cleavage plane) from which a centered figure is obtained. It is *the one* of the three that shows inclined extinction. Having identified a (010) section, one then names its interference figure according to which indicatrix direction coincides with *b*. This direction will be the Bxa, Bxo, or optic normal. The first part of the monoclinic orientation expression may now be written, perhaps Y = *b*. The two vibration directions in the plane of the section can now be identified with an accessory plate. It remains only to relate one of these principal vibration directions to a crystallographic direction by measuring the appropriate extinction angle. When this is done, the orientation expression is completed by an appropriate statement, such as $Z \wedge c = 15°$.

One cannot always follow a straight line procedure and identify the (010) section, the extinction angle in the section, and the vibration direction normal to the section. There may be no crystallographic direction in the (010) plane to use in measuring the extinction angle. Cleavage other than (010) may make it difficult to locate a (010) plane in an immersion mount. It may be impossible to identify the figure formed from a (010) section with certainty.

In attempting to avoid these difficulties with respect to the (010) section one may make use of monoclinic dispersion if present to indicate which vibration direction is coincident with the *b* crystallographic axis. As explained earlier (Fig. 12-9) if the Bxa figure shows:

Crossed dispersion, Bxa = *b*
Horizontal dispersion, Bxo = *b*
Inclined dispersion, Y = *b*

(The optic sign of the crystal identifies the Bxa and Bxo as X and Z or as Z and X, respectively.) These relationships are most useful when a Bxa figure can be obtained more easily than a figure from a (010) section.

If the extinction angle can be determined on a prismatic cleavage rather than the desired (010) face, one may calculate the extinction angle on (010) if 2V and also the angle between (010) and the cleavage are known. In a few instances the value of the extinction angle on the cleavage is recorded with other data. For the amphiboles, for example, all of which have approximately the same cleavage angle, and for which the extinction angle is a characteristic property, data have been compiled by Tertsch (1938) relating 2V and extinction on (110) and (010). These are conveniently summarized in Tröger (1979).

If the (010) section with centered figure is not found in a thin section one may still estimate the extinction angle on (010) by determining the maximum extinction angle shown by sections that lie in the vertical zone. Grains of crystals in which all cleavage lies in the vertical zone are identified in thin section by the lack of cleavage traces in more than one direction. The maximum extinction angle in the vertical zone is either equal to or larger than the extinction angle on (010) depending upon whether $2V_{ac}$ is around *c* or not. It has been shown by Daly (1899) that, for crystals in which $2V_{ac}$ is around *c*, the maximum extinction angle in the vertical zone is the extinction angle on (010) and, if not around *c*, the maximum angle occurs in a prismatic section between (010) and (100). If $2V_{ac}$ is not around *c* the maximum angle is useful only to tell one that the extinction angle on (010) is smaller because the maximum angle in the vertical zone is not recorded with other optical data. If $2V_{ac}$ is around *c* the maximum angle found will likely be very near the true one. The angle does not vary greatly from the maximum in sections not quite parallel to (010). Whether $2V_{ac}$ is around *c* is determined by noting the direction of the Bxa in an interference figure obtained from a section in the vertical zone. This is compared to the trace of the *c* axis marked by the cleavage.

Triclinic System. In crystals of the triclinic system none of the crystallographic axes is perpendicular to any other. The optical directions X, Y, and Z do not coincide with crystallographic directions except by chance. The optical orientation is described by citing the positions at which crystallographic directions and optical directions intersect a sphere with center coincident with the center of symmetry of the crystal. The crystallographic directions so located are usually the normals to prominent pinacoidal faces; the optical directions are usually X, Y, and Z and the two optic axis directions A and B.

Two angles serve to describe each position. The angles are usually designated as φ, the angle in a clockwise direction from the zero line of longitude on the equatorial meridian, and ρ, the angle of inclination to the vertical axis. It is to be noted that not all authors employ these designations for the angles or these conventions regarding the sense of the angular measurements. These data are easily plotted on a stereographic projection (see Chapter 14) and are particularly useful if a spindle or universal stage is employed for making the appropriate measurements.

TWINNING

Provisional identification of a twin law and determination of the relationship between vibration directions and the composition planes of twins often provides useful evidence in biaxial crystal identification. Some twin laws are characteristic of particular mineral groups and in some crystals the prevalence of twinning is suggestive evidence of identity. Twinning may also aid in identifying the crystal system in some instances.

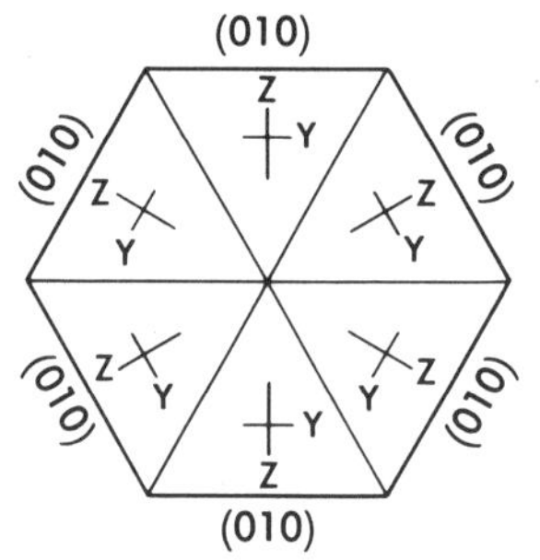

Fig. 13-5. Section of twinned cordierite (orthorhombic) cut parallel to (001). The composition planes are (110), and the optic plane is radial in each segment.

Orthorhombic System

Twinning on the prism face, which has a prism angle near 60°, is a common form of twinning in the orthorhombic system. It is particularly apparent between crossed polars in sections nearly parallel to (001). It is characteristic of cordierite (Fig. 13-5) and aids in identifying this mineral. It also occurs in the mineral aragonite, and is sometimes called twinning of the aragonite type.

Monoclinic System

Twinning according to the Carlsbad law, in which the composition plane is (010), is common in orthoclase and also in the triclinic plagioclases. Polysynthetic twinning parallel to the base is not uncommon in pyroxenes and in the monoclinic members of the humite group, chondrodite and clinohumite. Stellate twins of the type found in cordierite are also common in the monoclinic humites. The optical properties of the humites and olivine are similar, but olivine is not twinned, so that twinning is an aid in identification. Twinning with composition plane (100) is found in some monoclinic pyroxenes, for example diopside, and in titanite (sphene), which also exhibits lamellar polysynthetic twinning on the pyramid (221).

Triclinic System

The most important triclinic minerals are the plagioclase feldspars. Twinning and extinction angles referred to the twin planes are espe-

Fig. 13-6. Polysynthetic twinning in plagioclase [albite law, composition plane (010)]. Thin section of slowly cooled rock, Kiglapait intrusion, Labrador; crossed polars. Width of field about 3 mm.

cially useful in determining the identity of the members of this group. There are three common twin laws, albite, Carlsbad, and pericline, and many less common laws. Certain identification of the twin law requires use of the universal stage. Examples of plagioclase twinning are shown in Figs. 13-6 and 13-7.

Albite twinning is the most common of the polysynthetic types. The twin lamellae are continuous. They are parallel to, and therefore not visible in, (010) cleavages, but they do appear in (001) sections. This

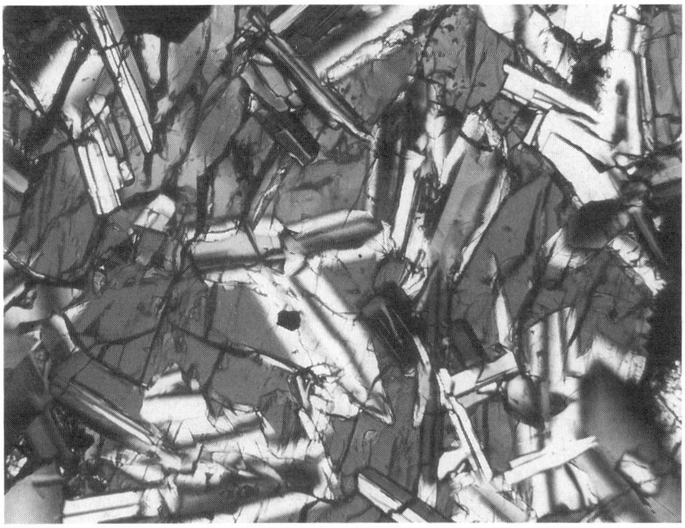

Fig. 13-7. Plagioclase twinning in ophitic lunar basalt: laths of plagioclase enclosed by augite (uniform gray) and ilmenite (black). Width of field about 1 mm; X polars. Sample from Mare Tranquilitatis, Apollo 11 lunar landing site; photo courtesy of John S. Dickey, Jr.

difference distinguishes (001) from (010) cleavage fragments. Pericline twins, also polysynthetic, are common. The lamellae are less continuous than albite twin lamellae and are often tapering. The composition plane of the pericline twin (the rhombic section) is nearly parallel to the basal pinacoid in plagioclases of intermediate composition. In microcline the composition plane of pericline twinning is almost perpendicular to (001) and parallel to *b*. The lamellae of these pericline twins are perpendicular to those of the albite twins in (001) sections and the traces are mutually perpendicular in most sections in the zone parallel to *b*. Both types of twin are present in microcline, creating a distinctive plaid appearance under crossed polars, as shown in Fig. 13-8.

CLEAVAGE AND CRYSTAL SYSTEM

It is often possible to identify the crystal system of a biaxial crystal by making use of the relationship between interference figures and cleavages. The scheme presented in this book makes use of the number of symmetry planes of the indicatrix normal to a cleavage. Cleavage

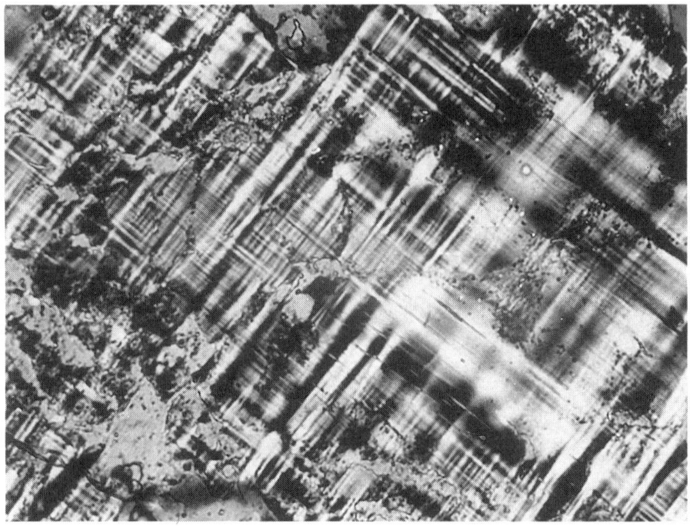

Fig. 13-8. Twinning in microcline, crossed polars. The intersection of albite and pericline twins creates the plaid appearance. Width of field about 1.5 mm.

and crystal system are identified simultaneously through the same series of observations. Unique solutions may not be achieved, particularly if an unusual type of cleavage is encountered, but the scheme yields useful results with many crystals.

> Cleavage is recognized by different means in thin section and immersion. In thin section, one recognizes cleavage, when it is approximately normal to the section, by the presence of parallel cracks (see Fig. 11-2). One cannot ordinarily recognize a cleavage parallel to the section. The scheme described here applies most effectively to immersion mounts. In immersion mounts, flat fragments (recognized by their uniform interference color) are assumed to be cleavage fragments, especially if many flat fragments show the same interference figure.

Types of Cleavage

One may assume that crystals belonging to the orthorhombic, monoclinic, and triclinic systems will have cleavage parallel to (001), (010), (100), or (*hk0*). Other cleavages are so much less common that they are neglected. One may further assume that among pinacoidal cleavages (001) is the most probable, (010) the next most probable, and (100) the least probable. Although these assumptions may lead to trouble, they are valid in most cases and hence valuable as working hypotheses.

In rare cases there may be morphological or other evidence suggesting the identity of a pinacoidal cleavage.

Prismatic cleavages are distinguished from pinacoidal cleavages by their extinction characteristics or their relationship to symmetry planes of the indicatrix. Because the prismatic cleavage form may produce intersecting cleavage traces in sections near the basal plane, it is important to avoid confusion between two intersecting pinacoidal cleavages and the two cleavage directions of the single prismatic cleavage form. All grains lying on a prismatic cleavage, whether *hk0* or $h\bar{k}0$, produce the same interference figure, which is not true of grains lying on two different pinacoidal cleavages.

Number of Cleavage Forms

The number of different cleavage forms present is of first importance. All cleavages of the same form will show the same interference figure. Straight edges on cleavage fragments are also significant. The forms (001), (010), and (100) in all three systems and (*hk0*) in the triclinic system have only two faces in the form, so they do not give rise to intersecting cleavage planes. However, (*hk0*) in both orthorhombic and monoclinic systems has four faces, so that cleavages of this form intersect each other. Hence, if only one cleavage is present, unless it is (*hk0*) of the orthorhombic or monoclinic system, flat-topped fragments will show no marginal straight edges. Conversely, straight margins on flat-topped cleavage fragments indicate the presence of more than one cleavage, unless the cleavage is (*hk0*) of the orthorhombic or monoclinic system.

Identification of Cleavage and Crystal System

Knowing the number of cleavages represented, one then turns to the problem of identifying each, and at the same time making a tentative identification of the crystal system. Each cleavage can be identified by its type of interference figure and by whether the extinction is parallel, symmetrical, or inclined. Some examples will help to illustrate the procedure.

> *Example 1*. Fragments have straight edges, hence either two cleavage forms or one prismatic cleavage form. None shows symmetrical extinction, hence the combination of basal and prismatic cleavage is not present. Fragments show different interference figures, hence there must be two pinacoidal cleavages or one pinacoidal cleavage parallel to *c* and a prismatic

cleavage. Interference figures are 0-SP figures, hence no cleavage is normal to a symmetry plane. The system is therefore triclinic, and the cleavages are tentatively (001) and (010).

Note that both cleavages would show inclined extinction unless by chance a vibration direction lay so close to a crystallographic direction that extinction appeared parallel. In most cases the inclined extinction would quickly suggest the triclinic system.

Example 2. Fragments have straight edges, hence either two cleavage forms or one prismatic cleavage form. Some fragments have symmetrical extinction, hence the combination of basal and prismatic cleavage is indicated. Some fragments (the ones with symmetrical extinction) show 2-SP figures, and others, with parallel extinction, show 1-SP figures. The highest biaxial symmetry is therefore indicated: the system is orthorhombic, and the cleavages are (001) and (*hk0*).

Note that the extinctions in this example would also be permissive of monoclinic symmetry, in which case the number of symmetry planes intersected by the two cleavages would be 1 and 0 rather than 2 and 1.

The possible combinations of cleavage and interference figure for crystals with one or two cleavage forms are given in Table 13-2. Crystals with three cleavages may be studied successfully by extrapolation from the table. The examples given above may be found in the first and second columns of the table, respectively. Any of the entries in the table can be derived by arguments similar to those used in the examples; the table is merely a convenience to simplify the deductive process, and to show in juxtaposition the likely alternatives when there is some doubt about any of the input data.

Doubt about the input data may arise from several causes. If appropriate straight edges are not present on the cleavage fragment, it is of course impossible to make use of the extinction criteria. This will be the case when only one pinacoidal cleavage form is present, as in column C of the table, or when a second cleavage is very poor. Some cleavages have similar properties, as, for example, the three pinacoidal cleavages of orthorhombic crystals. In this instance one makes use of the working hypothesis that (001) is the most likely and (100) the least likely. The same working hypothesis may be used with the two monoclinic pinacoidal cleavages that show 1-SP figures.

In distinguishing cleavages, one should when possible make use

Table 13-2
Cleavage and Crystal System, Determined from types of Interference Figures in Cleavage Fragments

Type(s) of Interference Figures from Cleavage Flakes	Cleavage Fragments (A)	(B)	(C)	(D)
0-SP	Triclinic pinacoidal		Triclinic pinacoidal	Monoclinic (hk0)
1-SP	Monoclinic (001) and (100)		Monoclinic (001) or (100)	Orthorhombic (hk0)
2-SP only	Orthorhombic any two of (001), (010), (100)		Orthorhombic either (001), (010), or (100) Monoclinic (010)	
1-SP and 0-SP	Monoclinic (hk0) and (100)	Monoclinic (001) and (hk0)		
2-SP and 1-SP	Orthorhombic[a] (hk0) and either (010) or (100) Monoclinic[b] (010) and either (001) or (100)	Orthorhombic (001) and (hk0)		
2-SP and 0-SP	Monoclinic (010) and (hk0)			

A: Two cleavage forms; neither shows symmetrical extinction, hence not both basal and prismatic.
B: Two cleavage forms; one shows symmetrical extinction, hence one basal, one prismatic.
C: One cleavage form without straight edges, hence not prismatic.
D: One cleavage form with straight edges, hence prismatic.
[a]If both show parallel extinction.
[b]If one shows inclined extinction.

of crystallographic directions indicated by the composition planes of twins whose identity is known, or by euhedral crystal faces.

Thin Section Procedures for Identification of Crystal System

In thin section the recognition of grains with cleavage lying parallel to the stage is not possible. The failure to observe cleavage cracks in minerals with one cleavage is not sufficient reason to believe cleavage is parallel to the stage. As a result one cannot use the generalizations of Table 13-2, which apply to cleavage fragments in determining the crystal system of a mineral in thin section. One can, however, recognize grains in thin section in which cleavage is *perpendicular* to the stage because cleavage cracks in such cases do not appear to move sidewise on raising or lowering the focus of the microscope. To make the test, oblique illumination cannot be used, so care should be taken that the field is evenly illuminated. In sections having cleavage perpendicular to the stage it is significant for the identification of the crystal system to determine whether a symmetry plane of the indicatrix is also perpendicular, as indicated by a 1- or 2-symmetry plane interference figure. If the trace of the cleavage in the grain and an isogyre in the figure are coincident with the same crosshair, a symmetry plane is parallel to the cleavage. Fig. 11-2 illustrates such a relationship. The limitations imposed upon the orientation of the biaxial indicatrix in the three crystal systems are such that some cleavages are parallel to symmetry planes of the indicatrix and others are not. A symmetry plane of the indicatrix is parallel to (001), (010), and (100) cleavage in the orthorhombic system and only to (010) in the monoclinic system. The other common cleavages in the three crystal systems (*hk0*) are not parallel to a symmetry plane of the indicatrix.

Information as to the cleavage and crystal system may also be provided by a section perpendicular to two cleavages and showing symmetrical extinction. If such a section shows a 2-symmetry plane figure, the two cleavages are prismatic and the crystal system orthorhombic. If such a section shows a 1-symmetry plane figure, the cleavages are prismatic but the crystal system is monoclinic.

14 Optic Orientation in Stereo

INTRODUCTION AND REVIEW

The visualization and representation of optic orientation is the nitty-gritty of practical optical mineralogy. A complete description is concisely made with the use of a stereographic projection. Perspective drawings can also give complete information and are often easy to understand. Drawings of principal sections, particularly the pinacoids, are also helpful and can be easy to construct. The identification of pinacoidal and other sections requires observation of the interference figure. We will illustrate several examples of optic orientation completely, with stereo nets, perspective drawings, sections, and interference figures. We begin with a brief review of morphological crystallography in stereograms of the holosymmetric classes of the five crystal systems that are optically anisotropic. The optical symmetry of a crystal must always be consistent with the holosymmetric class of the crystal system to which it belongs.

Crystallographic Review

The *operations* of crystallography include those of the *first sort,* such as translations, and such as rotations about diad, triad, tetrad and hexad axes; and those of the *second sort,* such as reflection across a mirror and inversion. The first stereogram of Fig. 14-1 shows the operation of inversion, symbolized ($\bar{1}$), in the holosymmetric class of the triclinic system. The operation shows that when the crystal is rotated 180° and flipped, the pole to a face that was on top is now on the bottom as though the face were enantiomorphously transferred through the inversion center. The *c* axis emerges at the center of the stereogram, here as in the general case. The trace of the *a* axis is indicated by a dashed north–south line, and the emergence of the negative end of the axis in the upper hemisphere is indicated. Similarly, the trace of the *b* axis is shown at an oblique angle to *a*, and the upper-hemisphere emergence of *b* is shown. No rotation axes or mirrors occur. The neighboring diagram shows the triclinic crystal axes and the names of the interaxial angles.

The second stereogram of Fig. 14-1 shows the holosymmetric class of the monoclinic system, $2/m$, whose class symbol proclaims a diad axis normal to a mirror. A face inserted on the top right hemisphere will immediately be repeated by the diad on the bottom right, and by

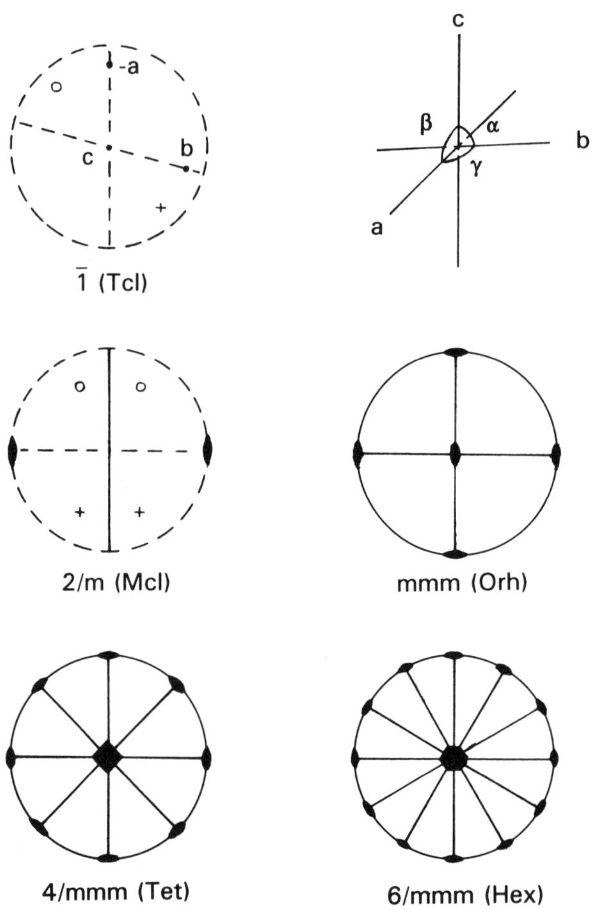

Fig. 14-1. Stereograms of the holosymmetric classes of crystals. Solid lines are mirrors. Poles to faces are indicated by + for the upper hemisphere and o for the lower hemisphere.

the mirror on the top left, so it will also appear on the bottom left, by the diad operation. Or one can follow the face back to the origin by the operations: rotation, reflection, rotation, reflection. The third stereogram shows the orthorhombic holosymmetry *mmm*, which is equivalent to 222. All three great circles are mirrors, normal to each of which is a diad axis. The last two diagrams show the holosymmetries of the tetragonal and hexagonal systems, both of which are optically uniaxial with E = c = 4- or 6-fold rotation axes.

A cleavage can be represented in stereo as a plane denoted by a great circle, or as a pole. Both are shown for the three pinacoidal cleavages in Fig. 14-2. The pole of (001), for example, is plotted 90° from the great circle containing the axes a and b. Since c is located at the center of the net and the pole of the cleavage lies a distance x away from c, the great circle lies a distance x away from the equator. The stereogram serves as a useful reminder that the pole of the (001) cleavage does not coincide with c in the triclinic system. Similarly, the pole of (010) is normal to the ac plane but not coincident with b, and p_{100} lies also on the equator, not at a.

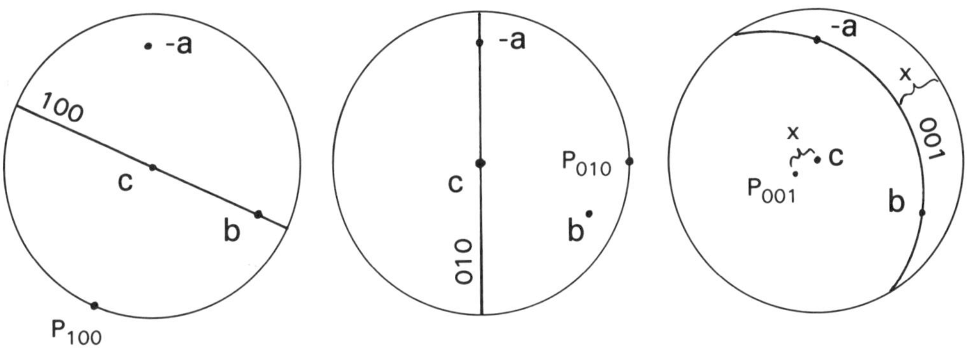

Fig. 14-2. Representation of a plane such as cleavage in a stereogram.

A kind of crystallographic axis that includes the crystal axes as well as others is the *zone axis,* e.g., [001], which is the axis of common intersection of all faces in the vertical zone. A *zone* exists for any set of faces whose mutual intersections are all parallel, and the zone axis is the line of intersection. The *zone index* [*uvw*] can be found by cross-multiplying the *Miller indices* (*hkl*) of any two forms that intersect. In this operation, the Miller indices are each written down twice, the first and last numbers are thrown away, and the remaining three pairs are cross-multiplied, observing the convention that down and to the right is positive, while down and to the left is negative. For example, the prism face (110) and the side pinacoid (010) intersect to form the vertical zone axis [001]:

```
1 ¦ 1   0   1   1 ¦ 0
  ¦   \/  \/  \/  ¦
  ¦   /\  /\  /\  ¦
0 ¦ 1   0   0   1 ¦ 0
_____________________
       [0   0   1]
```
(14-1)

For practice, one may verify that the zone axis formed by the intersection of (230) and (131) is [323].

Optical Conventions

Optical directions are usually plotted on the upper hemisphere of the stereo projection. The conventional symbols are shown for a fictional

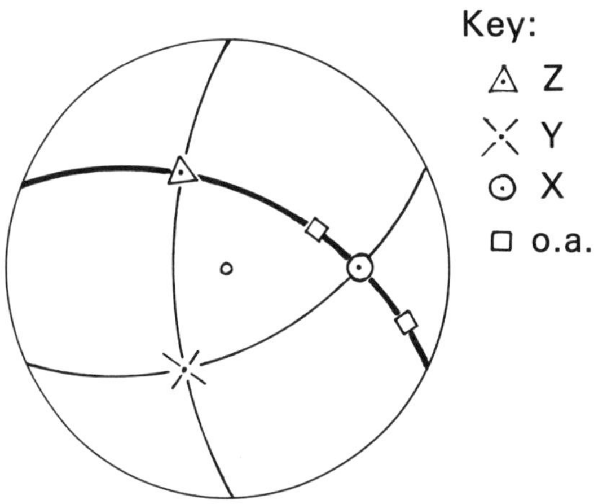

Fig. 14-3. Symbols for optical directions in a stereogram. The optic plane is emphasized.

crystal in Fig. 14-3. The *c* axis is at the center of the net. It is clear that the crystal is triclinic because none of the indicatrix axes plots on a crystal axis on the equator. The optic angle is acute about X, so evidently the crystal is negative, with a 2V of about 50°. The first indicatrix axis plotted is the pole to a great circle containing the other two axes, since all the indicatrix axes are orthogonal. The three great circles are the symmetry planes of the indicatrix.

TRICLINIC CRYSTALS

When the stereo nets of crystallographic and optical directions are combined, the result is a complete stereo depiction of the optic orienta-

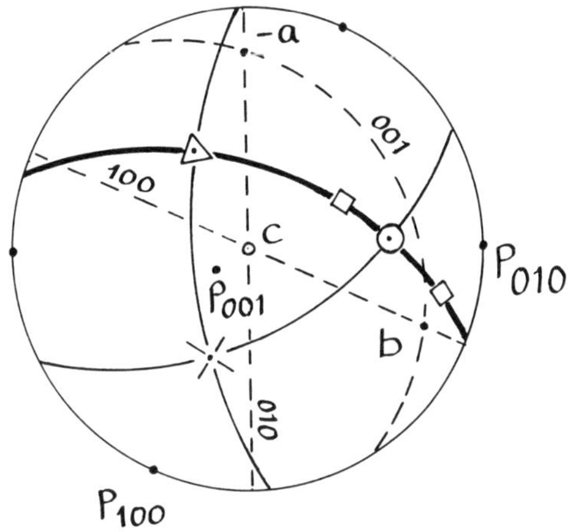

Fig. 14-4. Symbols for both optical and crystallographic directions in a stereogram.

tion. Fig. 14-4 is a combination of Figs. 14-2 and 14-3. Cleavage traces are shown as dashed great circles whose poles are indicated as heavy dots. Indicatrix axes and symmetry planes are now verified to have no special relation to the crystal axes and planes. The optic orientation could be described as "Y $\wedge$ *c* = Z $\wedge$ *c* = 50°, X $\wedge$ *b* ~ 30°, YZ plane $\wedge$ (010) ~ 28°." An idealized perspective view of these relationships is shown in Fig. 14-5. Although an optic axis lies close to *b*, it will not fall near the center of an (010) section because the pole to (010) lies rather far from the *b* axis (see Fig. 14-4).

The pinacoidal sections of this same fictional crystal are shown in the middle row of Fig. 14-5, and the interference figures are stylized in the bottom row. Several features of the pinacoidal sections are of note. The (100) section shows nearly parallel extinction relative to traces of (010), but this is fortuitous, because the Y indicatrix axis falls almost exactly between the pole to (100) and the *c* axis (Fig. 14-4). The parallelism of the extinction, by itself, might lead one to suspect a higher symmetry than triclinic, but this idea is not supported by the 0-SP figure seen on (100). The X′ direction makes an angle with the trace of *b*. The (010) section shows a small inclined extinction angle, which could be described as X $\wedge$ *c* small, provided that traces of (100) cleavage or faces could be seen to indicate the orientation of *c*. The YZ plane makes a moderate angle with the *a* axis. The interference figure from this section is also a 0-SP figure, but an off-centered optic axis figure in which a melatope is present. The (001) section shows a moderate extinction angle Z′ $\wedge$ *a*, and the interference figure, if recognizable at all, would be a 0-SP figure having some of the behavior of a Bxo or ON figure. In the pinacoidal sections, fine lines indicate cleavage traces,

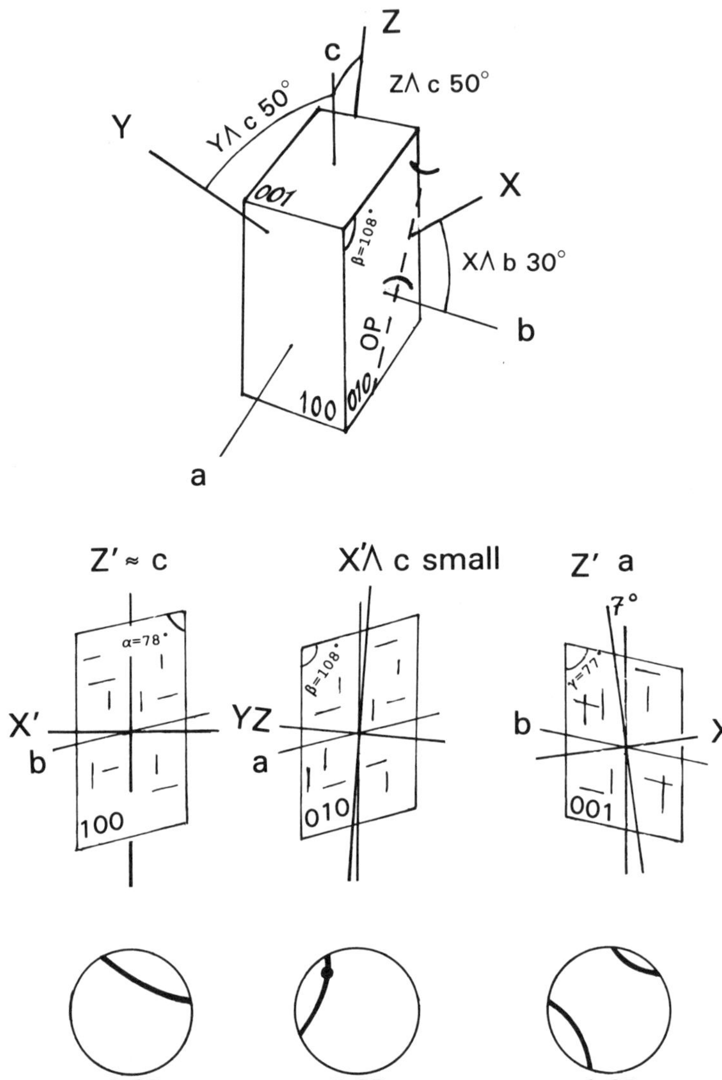

Fig. 14-5. Perspective (top) and pinacoidal (middle row) views, and interference figures (bottom row), of the crystal in Fig. 14-4.

as might be revealed by cracks in thin section or straight edges in grain mounts. The arrangement of the traces suggests the presence of (010), (001), and (100) cleavages.

In summary, triclinic crystals can be recognized by inclined extinction or accidental parallel extinction, combined with 0-SP interference figures. Often in crystal identification, these observations will be used to confirm that a triclinic candidate is consistent with the evidence. More rarely, if enough observations are made and the cleavages are good, one can infer the presence of triclinic symmetry and make a provisional statement of the optic orientation. A decisive statement of optic orientation would require comparison of optical directions either to crystal morphology or to X-ray diffraction studies that define the locations and identities of the crystal axes. Such comparisons can be made if the crystal is mounted on a spindle that can be interchanged between a spindle stage on the microscope and a goniometer head on the X-ray diffractometer.

Sections and Symmetry Planes

Sections of crystals may have rational indices, as for cleavages, or irrational ones, as for thin sections that randomly slice through a rock. When the Miller indices of a section or cleavage are known from morphology or structure, they usually become the focus of attention. However, in grains having poor cleavage or lying in thin section, the orientation of the crystal on the microscope stage is more directly referable to the symmetry planes of the indicatrix. If a crystal is cut or lies normal to no plane of symmetry, its *pole* lies within no plane of symmetry and the section is a *0-SP section.* It will show a 0-SP interference figure. By the same token, a *1-SP section* has its pole in a symmetry plane, and a *2-SP section* has its pole in the intersection of two symmetry planes.

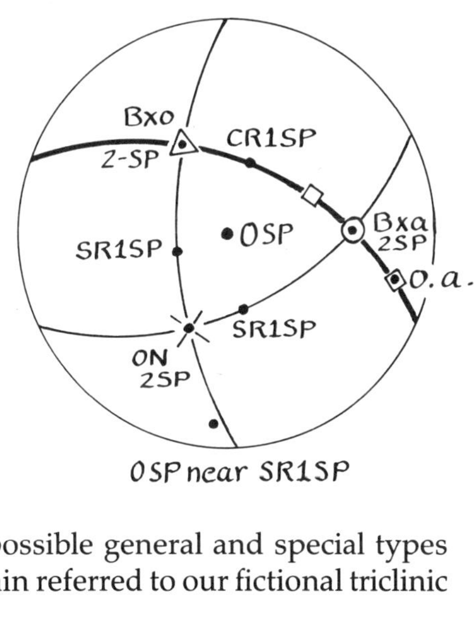

Fig. 14-6. Poles of symmetry-plane sections in the same triclinic crystal previously shown.

The locations of poles for all the possible general and special types of sections are shown in Fig. 14-6, again referred to our fictional triclinic crystal with the undistinguished optic orientation. There are an infinite number of 0-SP sections, but some are near one or more symmetry planes and can be described as off-centered versions of the higher-symmetry section. Of the three symmetry planes, two yield same-rotating figures and one yields counter-rotating figures; of the latter, a special case is the optic axis figure. There are three 2-SP sections, each with a pole at an indicatrix axis. Fig. 14-6 reminds us that whenever a stereo plot is available for the optical properties of a crystal, we can predict the general character (and sometimes the exact behavior) of an interference figure from the location of its pole relative to a symmetry plane or two. This information is usefully assimilated in terms of the most general case, that of triclinic crystals, which show nothing but 0-SP figures except by accident. In crystals of higher symmetry, figures of higher symmetry appear by constraint, not by accident.

TRICLINIC EXAMPLES

Aenigmatite is a soda-iron-titanium chain silicate found in some alkalic igneous rocks. It is triclinic prismatic, with good prismatic (110) cleavage. The indicatrix axis Y is near *b* and the optic plane near (010), as shown in Fig. 14-7. The optic angle is small and positive. The *c* axis bisects the angle between the indicatrix axes X and Z, although the stereogram and perspective drawing do not quite agree on the matter. Sections and interference figures are shown in the bottom part of Fig. 14-7. Judging from the stereogram, a section parallel to (100) would accidentally show a counter-rotating 1-symmetry plane (CR1SP) figure, with a small extinction angle between Z′ and the trace of *c* defined by intersections of the prismatic cleavage. The prismatic cleavage itself shows a 0-SP figure, possibly showing a melatope near the edge of the field. A section parallel to (001) shows a 0-SP figure that is very close to a CR1SP (optic plane) figure. The near-coincidence of the optic plane with *c* makes this triclinic crystal look monoclinic unless very careful measurements of extinction angles are made.

A more flagrantly triclinic mineral is the Ca, Fe hydroxysilicate skarn mineral **babingtonite**, shown in Fig. 14-8. The observed cleavages are shown in the upper right next to the perspective drawing. The nearest indicatrix axis to a crystal axis is X, about 40° from *c*. Large extinction angles will be seen on all cleavages except (110), for which the foreshortened Z axis is nearly in line with *c*. The better cleavage ($1\bar{1}0$) shows a 0-SP figure and an extinction angle X′:*c* = 37°. The perfect basal cleavage

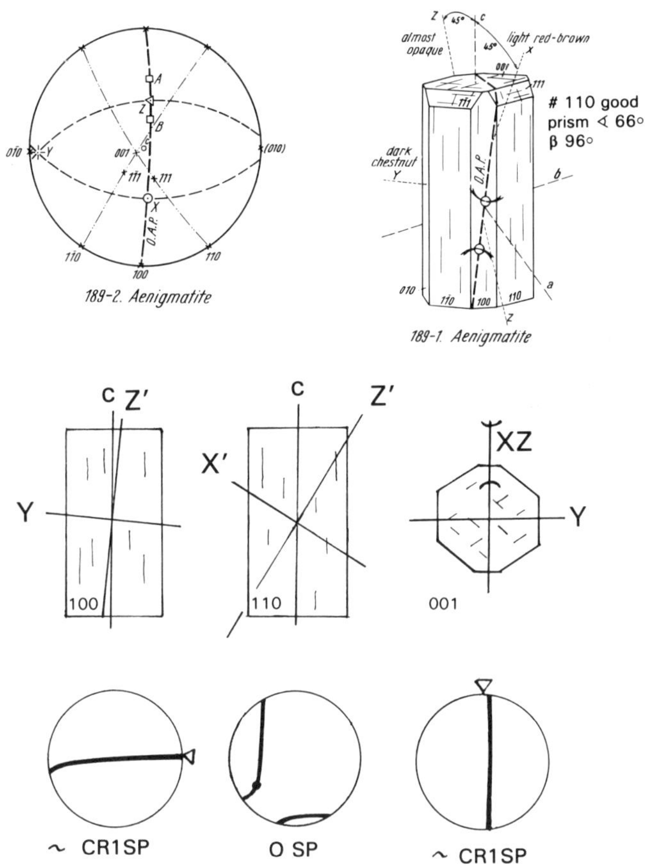

Fig. 14-7. Optic orientation of aenigmatite.

shows a 0-SP figure, but optic axis A is very near the center of the field, so the figure is an off center optic axis figure.

Rotating the Stereogram. The interference figures to be seen on selected sections or cleavages are not always obvious from an inspection of the stereogram. Babingtonite, where the indicatrix axes are far from the crystal axes, is a case in point. In this case, it was necessary to rotate the stereogram to the view seen in the section, and then to apply the planar approximation of the Biot-Fresnel rule to find the isogyres from the optic axes. The procedure of rotating the stereogram is shown in Fig. 14-9. The pole to $(1\bar{1}0)$ was first turned to the south and then rotated 90° to the center of the figure. All other elements were then carried 90° north along their respective small circles, as shown by the arrows in the figure.

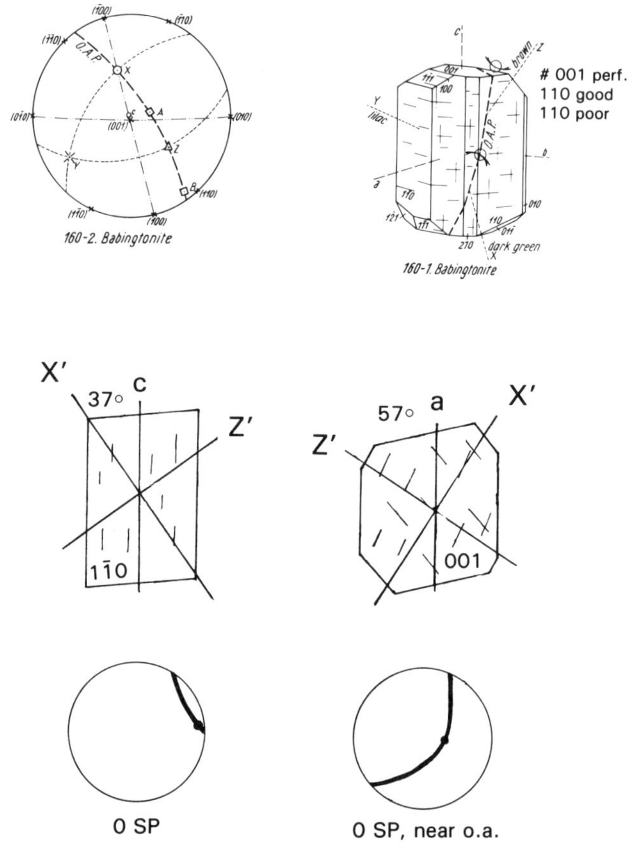

Fig. 14-8. Optic orientation of babingtonite.

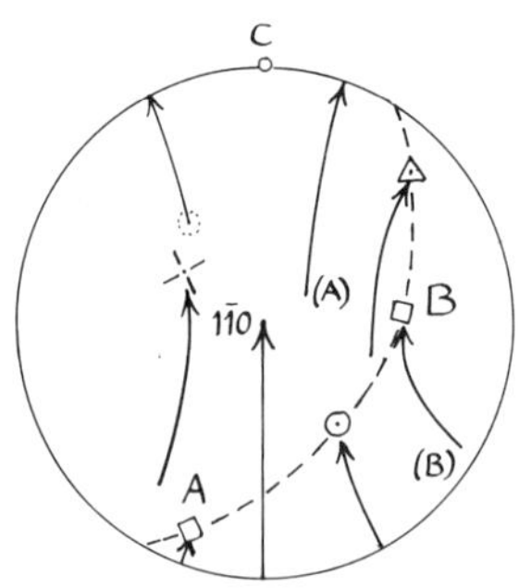

Fig. 14-9. Rotation of a stereogram to the plane of the section assumed to lie on the stage.

Optic axis B has moved farther into the stereogram, and close enough to the pole of the section that it may appear in the field if the objective lens has a large numerical aperture. Both the X indicatrix axis and optic axis A crossed the equator of the net and reappeared 180° away. A Biot–Fresnel construction from the optic axes located the isogyre seen in the lower left of Fig. 14-8.

A last triclinic example is furnished by the common metamorphic aluminum silicate **kyanite**, Fig. 14-10. The common cleavages and their angles to one another are shown in the upper right. The X indicatrix axis nearly coincides with the pole to (100), but is some distance from the *a* axis, shown in the stereogram. Y and Z lie sensibly in the (100) plane. The mineral is negative with a large 2V. The interference figure seen on the (100) cleavage is practically a centered Bxa figure. That seen on (010) is close to the YZ symmetry plane and shows a same-rotating isogyre. The pole to the basal cleavage nearly lies in the YZ plane, and the figure shows another off-centered, same-rotating isogyre.

MONOCLINIC EXAMPLES

The sheet silicates in general are typified by **talc**, shown in Fig. 14-11. The monoclinic symmetry constrains one indicatrix axis, here Z, to coincide with *b*. The crystallographic angle beta is 100°. By coincidence, the angle between X and *c* is about 10°, with the result that Y $\approx$ *a*. This is why sheet silicates show nearly or quite parallel extinction when viewed parallel to the perfect basal cleavage. Talc is negative (see section on the structural basis for the optic sign, Chapter 10, p. 168) with a small 2V. The basal cleavage shows a perfect Bxa figure, by accident of the X:*c* angle being equal to beta-90°. A prismatic section is nearly normal to the YZ symmetry plane of the indicatrix, and so it shows a same-rotating isogyre, nearly a SR1SP figure. A section parallel to (010) would show a Bxo figure (see stereo net).

The rare high-pressure, metamorphic, aluminous mineral **sapphirine** is illustrated in Fig. 14-12. The name originates from the deep blue to green to yellow pleochroic colors, which change to blue-green and yellowish brown with increase of the ferric/ferrous ratio. One distinct and two poor cleavages occur. The optic orientation is Y = *b*, Z:*c* 6–8°, as shown in the stereogram. The X indicatrix axis appears 6–8° from the equator in the north part of the net. The beta angle is 110°, so the ($-a$) axis would plot 20° inside the equator, and some 12–

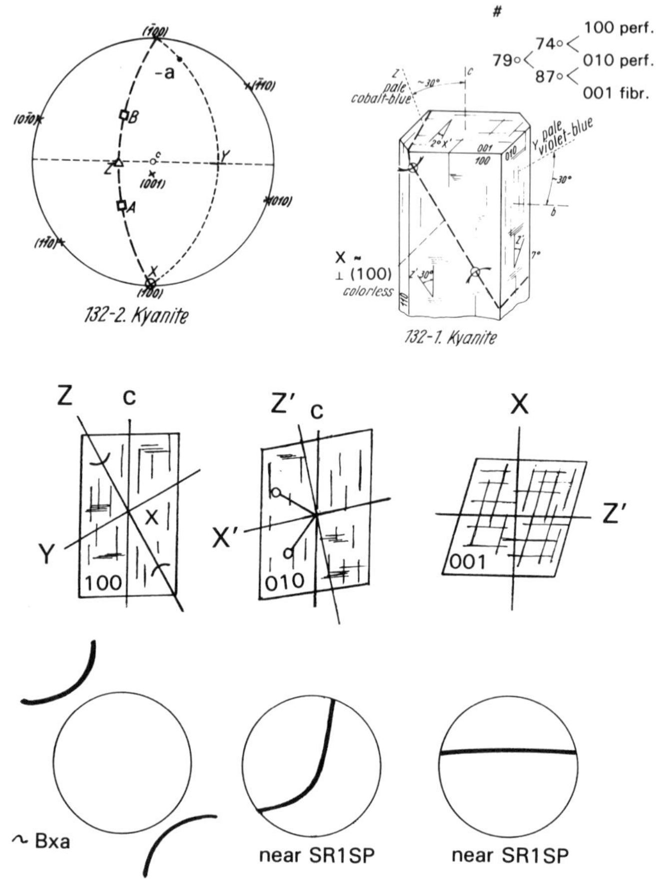

Fig. 14-10. Optic orientation of kyanite.

14° inside the X axis. The optic angle is large, negative. The cleavage sections and interference figures to be seen on them are shown in the lower part of the diagram. The figure on (100) is a CR1SP figure from the optic plane. The figure on (010) is an optic normal figure, and the (001) section shows another CR1SP figure, very close to a centered Bxo figure. All three refractive indices are accessible on cleavage flakes.

The calcium-titanium silicate **sphene (titanite)** is a common monoclinic mineral with distinctive wedge-shaped cross-sections and a very high refractive index. Its optic orientation is shown in Fig. 14-13. Like very many other monoclinic minerals, it has Y = *b*, but in this case Z:*c* is very large, 51°. Prismatic and pyramidal cleavages occur, but they are not well developed. The optic angle is small and positive, and the dispersion is strong, $r >> v$. The (110) cleavage shows a 0-SP figure

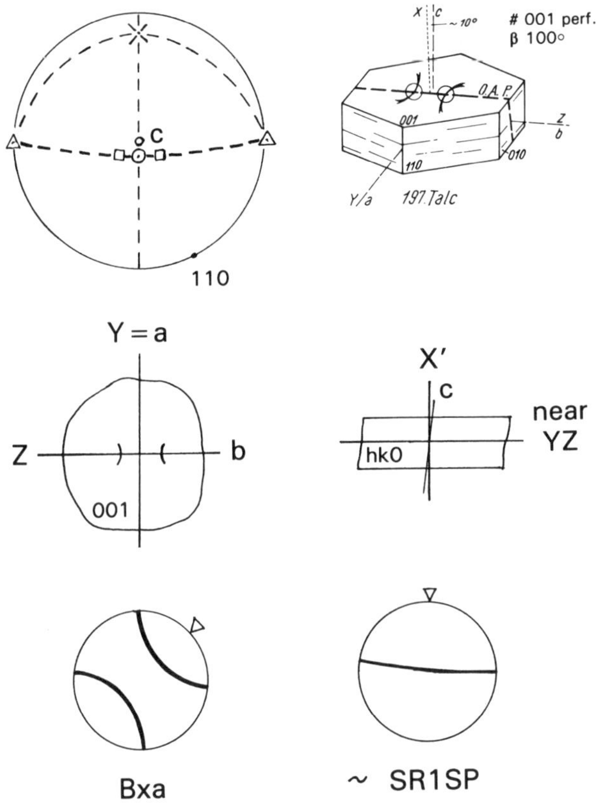

Fig. 14-11. Optic orientation of talc.

and a large extinction angle. An (010) section, if it can be obtained, shows a flash figure and the maximum extinction angle of 51°. The (111) cleavage nearly shows a same-rotating 1-symmetry plane figure that will confirm a close approximation to the alpha refractive index, and that shows the small positive 2V in an off center Bxa figure.

Pyroxenes are among the major rock-forming minerals and their orientations and intergrowths are so fundamental in the study of rocks that most practitioners learn them by heart. Most of the monoclinic pyroxenes are varieties of augite, of which the Mg end member diopside is the type example (Fig. 14-14). Good prismatic cleavages occur, as well as important (010) and (100) cleavages and a (100) parting in the vertical zone. The stereo net summarizes the orientation, with Y = *b* and Z:*c* large and variable. The optic sign is positive and the optic angle moderate. A (100) cleavage or parting shows parallel extinction and a CR1SP interference figure, yielding the beta refractive index. An

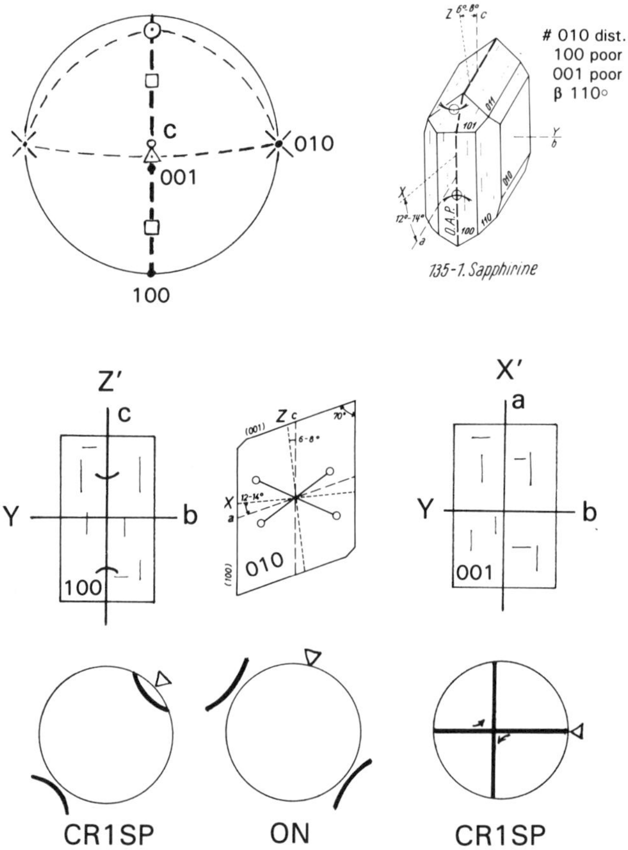

Fig. 14-12. Optic orientation of sapphirine.

(010) section shows an optic normal figure and all the critical angular relationships. The prismatic cleavage shows only 0-SP figures and a moderate extinction angle.

ORTHORHOMBIC EXAMPLES

The orthorhombic pyroxenes are referred to the same internal structure as the monoclinic ones. As in the example of **hypersthene** (Fig. 14-15), the optic plane is (100) and Z = *c*. The prismatic cleavage is denoted (210) because the orthorhombic symmetry occurs by an approximate doubling of the *a* cell edge with respect to diopside [more precisely, *a* (OPX) = 2*a* sin β (CPX)]. This relationship means that the prismatic cleavage, which has the same orientation as in diopside, cuts the OPX

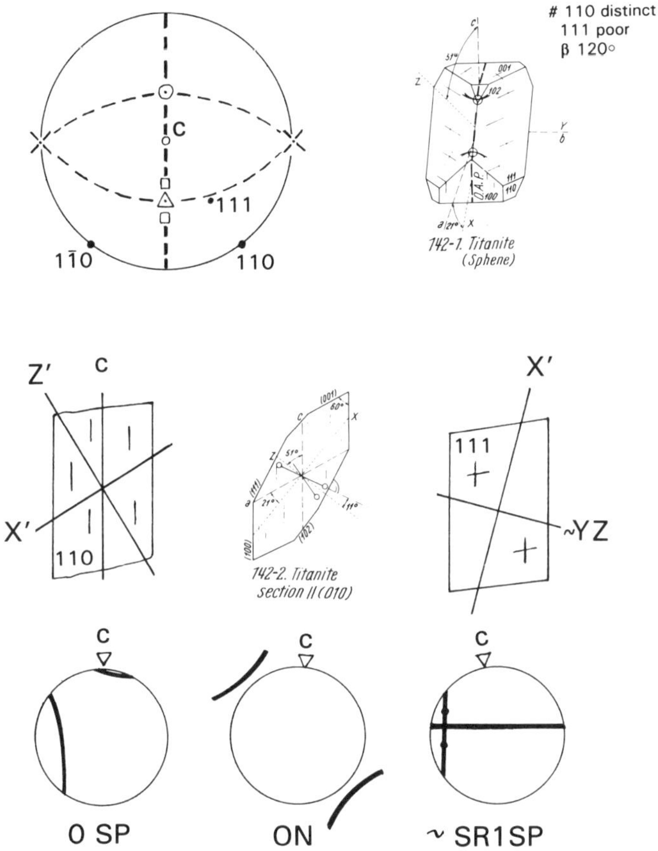

Fig. 14-13. Optic orientation of sphene (titanite).

a axis at 1/2 its length, yielding the Miller index 2. The (100) cleavage yields a flash figure and therefore both alpha and gamma. The prismatic cleavage yields a SR1SP figure and gamma. Both have parallel extinction, unlike the case of augite. An (001) section shows symmetrical extinction and the characteristic pyroxene cleavage near 92° and 88°. It shows a Bxo figure. The combination of parallel extinction and interference figures with two symmetry planes helps to establish the orthorhombic nature of this mineral in cleavage mounts.

Orthorhombic-Monoclinic Intergrowths. Intergrowths of both **pyroxenes** and of **amphiboles** occur along lattice planes that are closely matched in terms of their atom positions. The interfaces, called optimal phase boundaries, are often close to or equal to (100) and (001). The resulting intergrowths are chiefly exsolution lamellae, ranging from

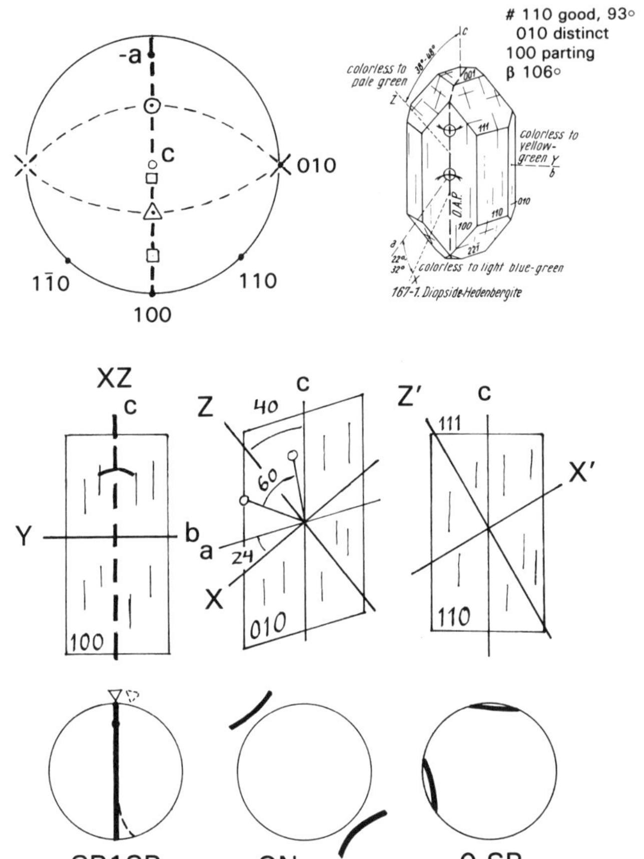

Fig. 14-14. Optic orientation of diopside.

submicroscopic or just barely detectable from their Becke lines at 1–2 microns (micrometers) wide, to thick plates many tens of μm thick. Some examples of pyroxene intergrowths are shown in Fig. 14-16. All have augite hosts. In part A of the figure, the lamella is hypersthene, and the (100) planes of both host and guest are rigorously parallel. In parts B and C of the figure, the lamellae are pigeonite, and the exsolution interfaces are best-fit planes near, but not equal to, (100) and (001). By reference to the previous drawings for augite and hypersthene, and from the data given in Fig. 14-16, one can draw for oneself the stereograms illustrating the optical and physical orientations of hosts and lamellae.

Counterparts to the intergrowths just discussed occur with orthorhombic hosts and monoclinic lamellae. The same rules of fit apply,

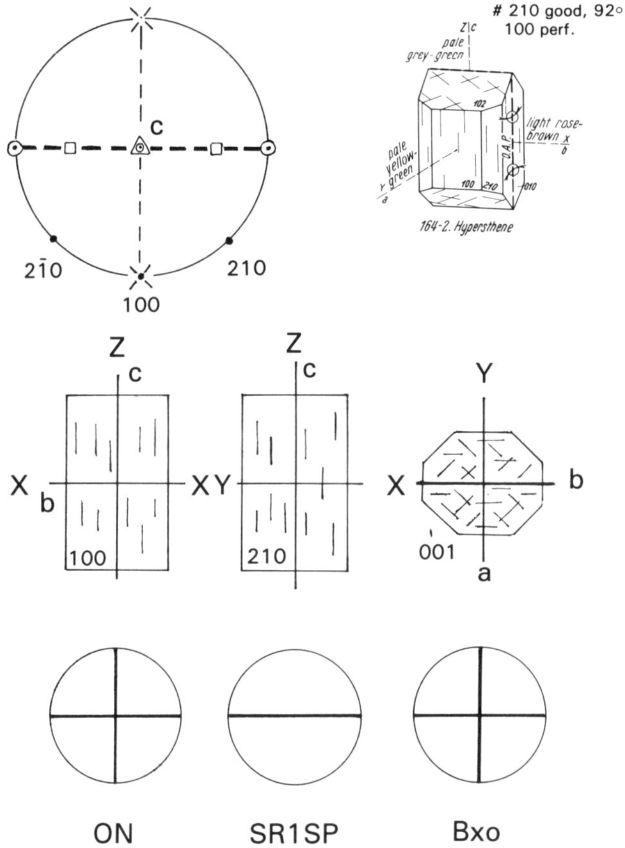

Fig. 14-15. Optic orientation of hypersthene.

so the orientations are similar. A most interesting type of intergrowth occurs when augite exsolves near (001) of the Ca-poor monoclinic pyroxene **pigeonite**, and the host then inverts to orthopyroxene, in some cases forming a matrix connecting many previous grains of pigeonite in a single OPX crystal (see frontispiece and cover). Important clues to the recognition of exsolution features are the extinction angles (about 45° for CPX, and parallel for OPX) seen on the common (010) sections, and the interference figures. The type and orientation of pyroxene intergrowth can furnish valuable information about the temperature and pressure of formation, as discussed by Robinson et al. (1971). Amphiboles have similar intergrowth relationships.

Among other examples of orthorhombic minerals may be mentioned the aluminum fluoro-hydroxy silicate **topaz** (Fig. 14-17) and the iron aluminum silicate **staurolite** (Fig. 14-18). The optic orientations are

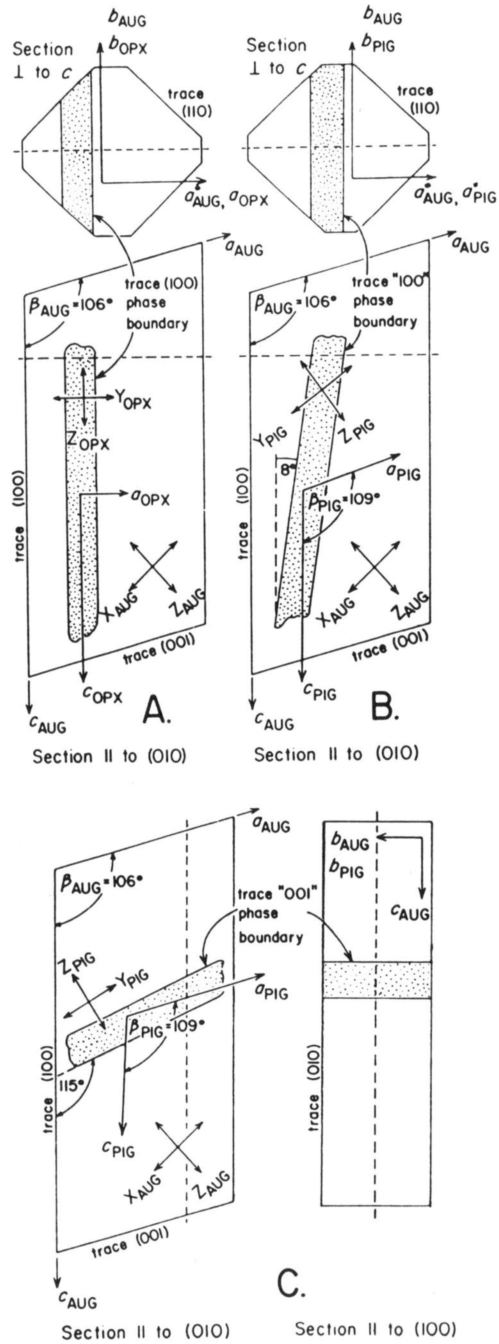

Fig. 14-16. Examples of pyroxene intergrowths, from Jaffe et al. (1975). (A) orthopyroxene in augite, (B) pigeonite "100" in augite, (C) pigeonite "001" in augite.

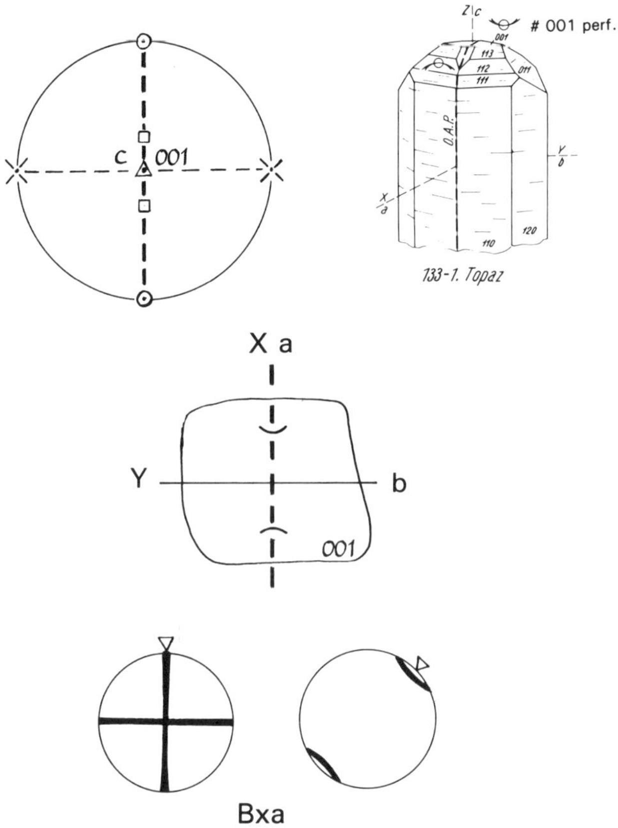

Fig. 14-17. Optic orientation of topaz.

straightforward, and the orthorhombic symmetry is evident from the parallel extinction in cleavage flakes and the frequent observation of interference figures with high symmetry, either one or two symmetry planes, but never zero.

FELDSPARS

Introduction

The rock-forming feldspars are the most abundant minerals in the earth's crust and they display some of the most elegant relationships among optical and crystallographic properties to be found in minerals. By simple optical measurements, one can determine both the composition and the

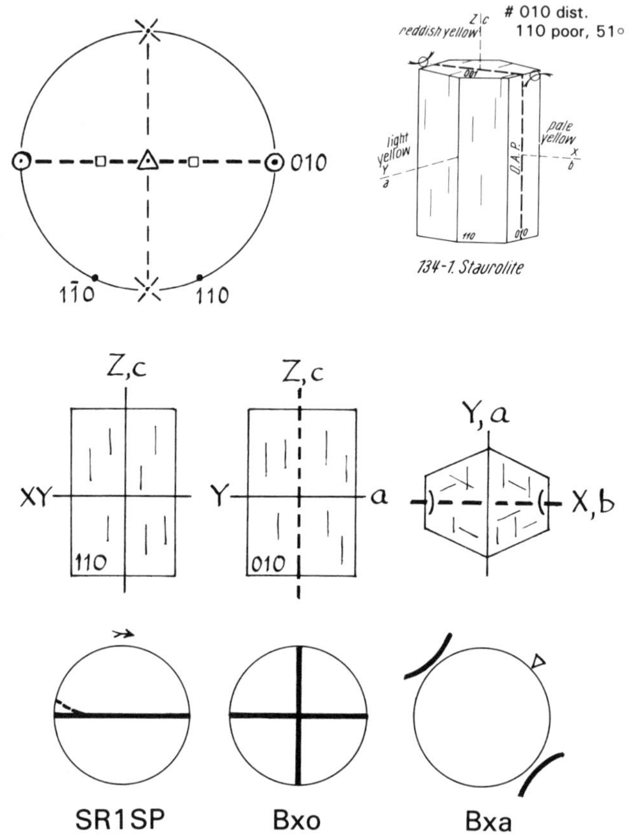

Fig. 14-18. Optic orientation of staurolite.

thermal history of many common feldspars. The major end members are orthoclase, Or, $KAlSi_3O_8$; albite, Ab, $NaAlSi_3O_8$; and anorthite, An, $CaAl_2Si_2O_8$. The substitutions along the two complete solid solution series are NaK_{-1} for the alkali feldspars and $CaAlNa_{-1}Si_{-1}$ for the plagioclase series; the solid solution between An and Or is very limited. It is useful to focus at first on the alkali feldspars, starting with orthoclase.

Orthoclase and Sanidine

The crystal optics of orthoclase are illustrated in Fig. 14-19. The mineral is monoclinic, with Z = *b*. Cleavages are listed beside the perspective drawing in the upper right. In the middle row, three sections are shown, and below are shown the corresponding interference figures. On (100) is seen a same-rotating 1-SP figure from the XY plane. The

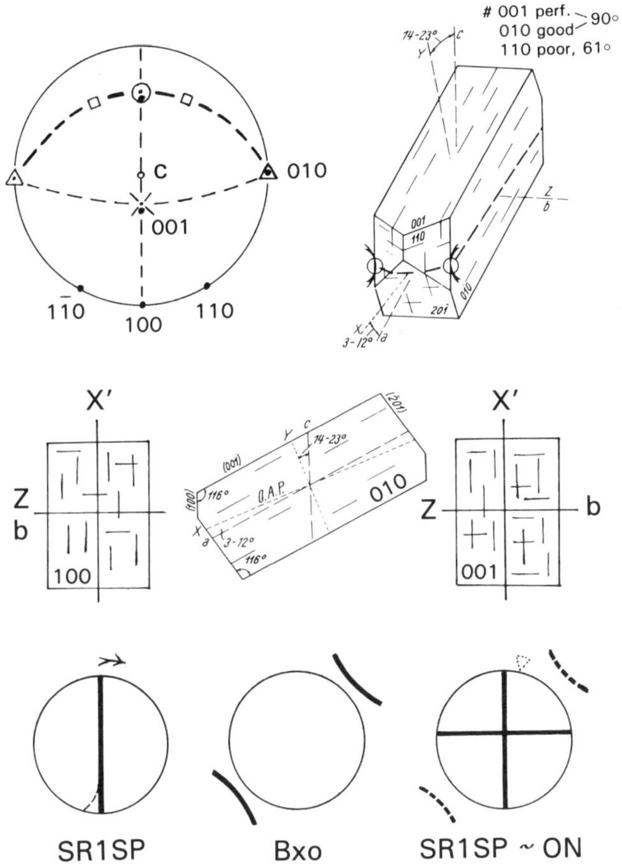

Fig. 14-19. Crystal optics of orthoclase.

(010) section shows a Bxo figure, and the (001) section shows a SR1SP figure that is very close to an optic normal figure. Flash figures from the perfect (001) cleavage should be abundant in grain mounts.

The contrasting crystal optics of **high sanidine** having the same composition as orthoclase are shown in Fig. 14-20. Here the optic plane is (010), perpendicular to where it was in orthoclase. The kinds of interference figures seen on the cleavages are the same as in orthoclase except that the identities of the symmetry planes are all switched to varieties of counter-rotating 1-SP figures, and an optic normal figure instead of a Bxo on (010). How does this switch of optic orientation occur, and what is the meaning of the mineral name "high sanidine"? These questions are answered in the next two sections; high sanidine is *disordered* sanidine.

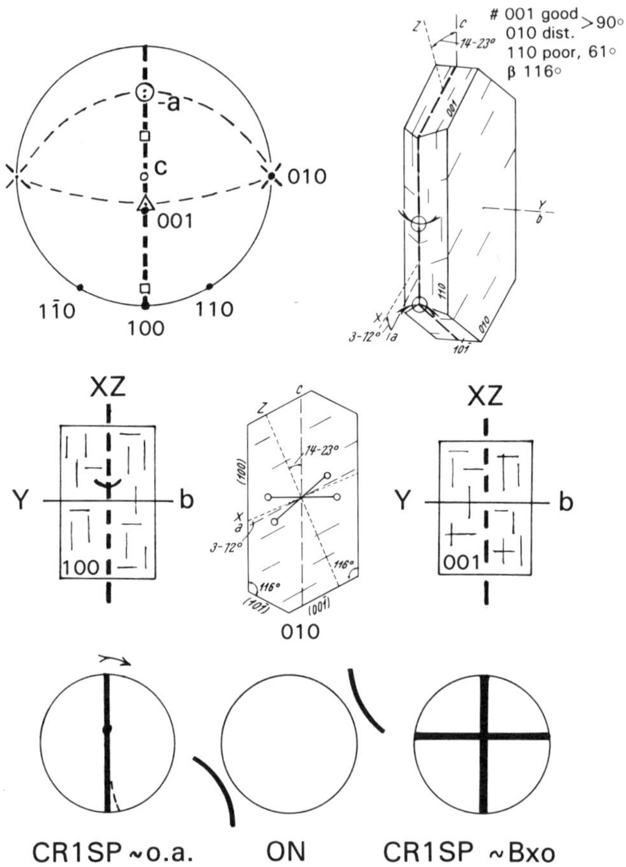

Fig. 14-20. Crystal optics of high sanidine.

Feldspar Ordering

The structure of the feldspar framework is illustrated in Fig. 14-21 in projections along the reciprocal lattice direction c^*.

The reciprocal lattice is constructed from the direct crystal lattice by a consideration of the planes of atom positions passing through the origin of the cartesian coordinate system of axes x, y, z or a, b, c. Each set of lattice planes is characterized by a unique orientation (hkl) and interplanar spacing (d). To each set of lattice planes erect a perpendicular. Give this perpendicular a vector length $1/d$. Place a point at the end of this vector. The orientation and interplanar spacing of this lattice plane are now both uniquely identified by this point. The collection of all such points constitutes the reciprocal lattice. It has lattice axes a^*, b^* and c^*, and corresponding axial lengths. The view along c^* of feldspars happens to be the felicitous one for discussing order-disorder.

The framework consists of TO_4 tetrahedra (in which T stands for either Si or Al), all linked at each corner to other tetrahedra. Among the tetrahedra are octahedra that contain the large cations K, Na, and Ca. In Fig. 14-21, the tetrahedral sites are emphasized by labels. There are four kinds of these sites, called T_1(m) and T_1(o), and T_2(m) and T_2(o). The four sites near the middle of the unit cell in the upper figure of Fig. 14-21 suffice to tell the story of all possible sites. One of these is black and the rest are open. The black site represents an Al atom and the rest are Si positions; the formula is evidently based on an $[AlSi_3O_8]$ group. There is no mirror plane parallel to the *a* crystallographic axis, because the Al sits only in the one site. The symmetry is triclinic, lacking the mirror required for monoclinic crystals. This figure depicts a completely *ordered* structure in which Al and Si occur only on specific sites. If we now heat the crystal up so that all the atom positions get bigger because of thermal motion, the crystal can no longer distinguish between the Al and Si atoms, which are similar in size. The atoms may occasionally break their local bonds and rearrange themselves by random diffusion. If the temperature is high enough, the Al atom can be found with equal probability in each of the four T sites. Since there are four sites and only one Al atom, the probability of finding the Al at any site at any moment is simply 1/4. This structure is a random or *disordered* structure.

The state of order between the two limiting extremes is easily described by the probability of finding the Al atom in any of four positions, or by the sum of Al atoms in any specific subgroup of tetrahedra. For the ordered structure, the probability of finding Al is zero on three sites and 1.0 on the fourth. For the T_1 sites alone, the sum of the atoms in T_1 is $1 + 0 = 1.0$. For the disordered structure, the sum of Al atoms in T_1 is $0.25 + 0.25 = 0.5$. A function "Sum T_1" (shorthand for the sum of Al atoms on T_1 sites) therefore runs from values of 1.0 for perfect order to 0.5 for perfect disorder. There are many alternative ways of describing the state of order with a numerical parameter, some redundant and others having special utility for special purposes. All of them are based on counting the number of Al atoms among various sites. In general, disorder signifies high temperature. The end members of ordering in K feldspar are high sanidine (in lavas) and **maximum microcline** (in metamorphic rocks). The "maximum" in this name refers to maximum triclinicity, which occurs at maximum order.

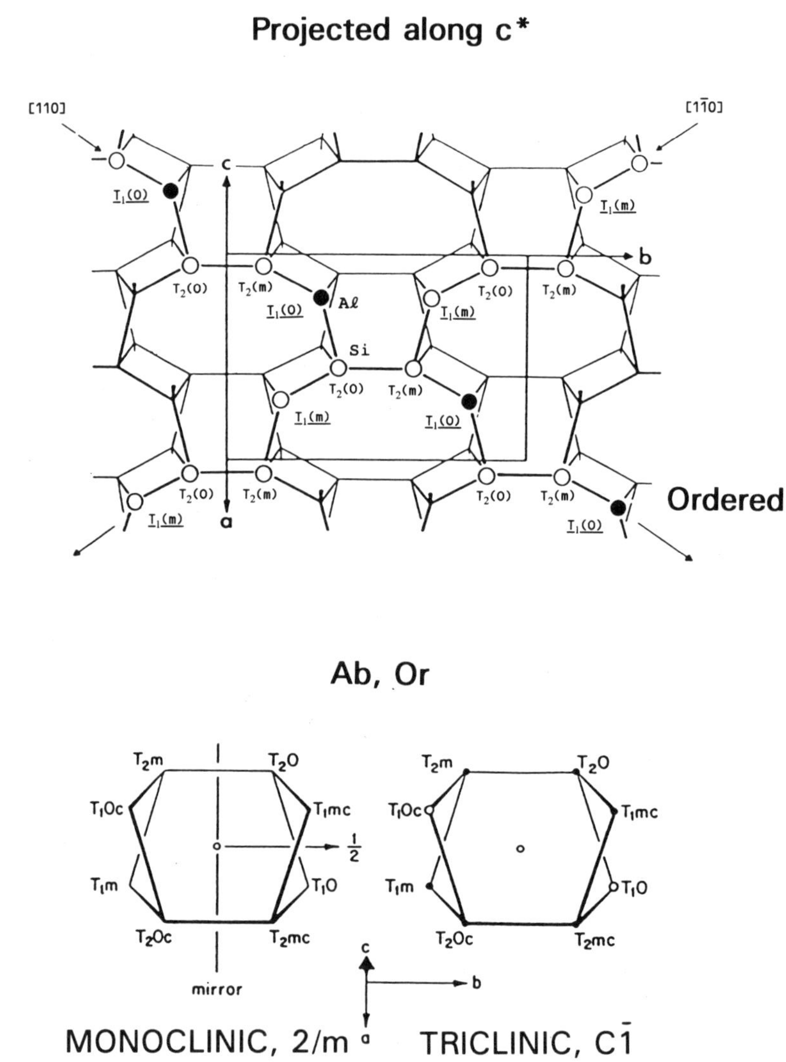

Fig. 14-21. Structure of the feldspar framework projected along c^* (from Ribbe, 1983a).

Effect of Ordering on Optics

The average position of the Al atom in the K feldspar structure has an effect on the local cell dimensions that, in turn, has an inverse effect on the refractive index. As a result, the measurement of certain optical properties can bring information about the state of ordering. The optic angle, 2V, is particularly sensitive to the state of order.

A determinative diagram for ordering in K feldspar is shown in Fig 14-22. The abscissa is an order parameter like that described above. High sanidine is on the left, and maximum microcline on the right. (The scale at the top represents the number of Al atoms encountered in a traverse along the *b* or *c* axes.) In the top part of the diagram is plotted the optic angle, which goes through zero as the optic plane switches from (010) to an orientation approximately normal to (010). Details of this switch are to be discussed, but first it makes sense to look into the origin of the switch and the change of optic angle through zero.

Refractive indices and cell edges are plotted in the bottom part of Fig. 14-22. As ordering occurs, the cell edge *a* changes very little, whereas *b* and *c* change a great deal, and in opposite senses. The cell edge *b* contracts, while *c* expands. This behavior occurs in direct response to the average location of Al, which swells the crystal structure locally. The refractive indexes also vary linearly and in opposite senses, if they are labelled according to the nearest crystal axis (Bloss, 1985), rather than slavishly according to their relative magnitudes. The refractive index associated with the vibration direction nearest the *c* axis decreases with order, opposite to the cell edge, which increases. This behavior is an inevitable consequence of the inverse relationship between refractive index and density (e.g., the Gladstone-Dale relation), but now we see that the relationship holds along specific directions within a crystal, not merely for the bulk crystal. Compare also the antithetical relationships between n_b and *b* and between n_a and *a*. Note also that as n_b and n_c cross, they change identity in terms of β and γ.

The optic plane switch. The optic orientations of both high and low sanidine are shown on a stereogram in Fig. 14-23. The conventional symbols for Y and Z are missing, and instead there are dots labelled n_b and n_c. The X indicatrix axis is shown in the ordinary way as a circle: it plots close to $-a$ (see Figs. 14-19 and 14-20). The different positions of the optic plane are shown for high and low sanidine. Along the "low" optic plane are optic axis positions labelled 1, 2, and 3. These show that the optic angle collapses to zero along the "low" optic plane,

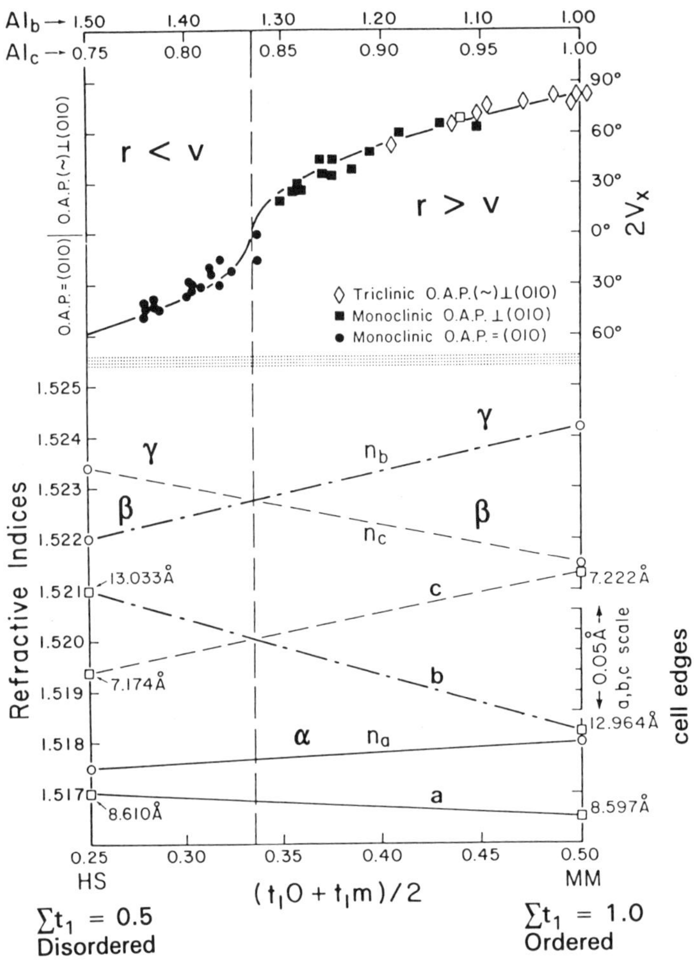

Fig. 14-22. Determination of ordering in K feldspar (after Su et al., 1984).

as the D-line value of the refractive index near *c*, n_c, overtakes the value of n_b. New positions of the optic axes 4 and 5 are shown along the "high" optic plane, as n_c rises away from n_b. The sequence 1–5 is the sequence from right to left in Fig. 14-22.

In the middle part of Fig. 14-23 are shown three sections normal to the X indicatrix axis for low, neutral, and high sanidine. These sections show how the isogyres in the interference figure approach each other as the optic angle decreases, and then depart from each other along the new optic plane as the optic angle increases. The sections show the equivalent of stages 1, 3, and 5 in the stereogram above. All the relations illustrated above are taken for a single wavelength of light, the D line.

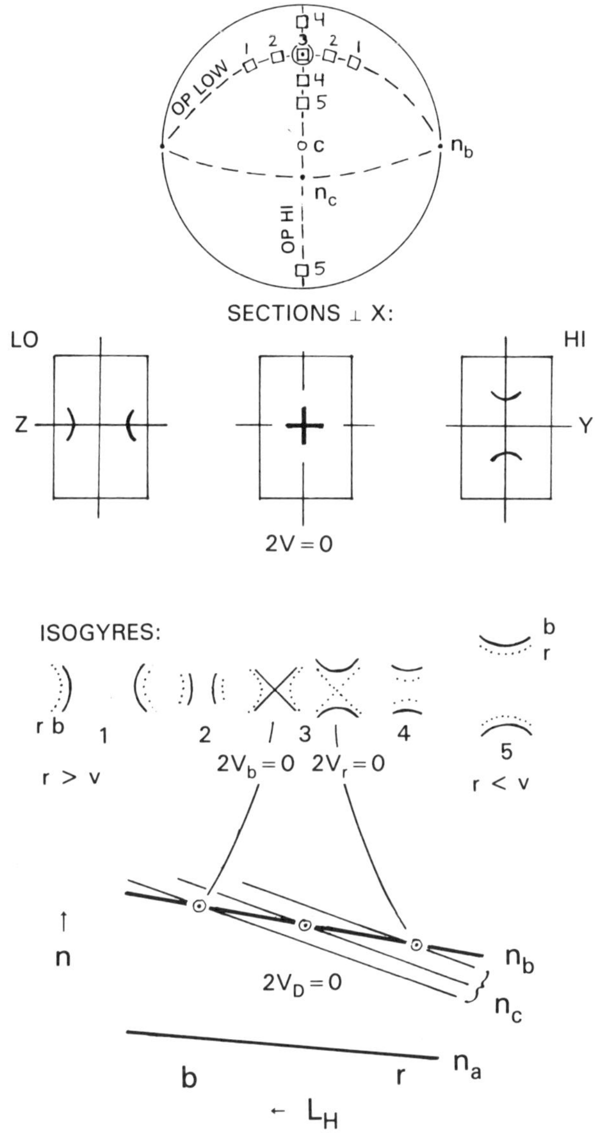

Fig. 14-23. The optic plane switch as a result of ordering in K feldspar.

Effect on rhombic dispersion. In the bottom part of Fig. 14-23 are shown the effects of dispersion seen when the crystals are illuminated with white light. As usual, the white light is represented by only two sample colors, red and blue. The single isogyres shown in the sections above for the D line are now double isogyres, one each for red and blue light (not the complementary colors seen, but the isogyres [dark places] for the actual colors considered). They do the same thing as the D-line isogyres, but now we see that the isogyres for blue light, closer together for low sanidine ($r > v$), come together before the isogyres for red light. The sequence is shown for all steps 1–5 above, with two representations of $2V = 0$, one each for red and blue light. The isogyres for blue light get to zero first, and then expand along the HI optic plane before those for red light, and in consequence, the dispersion formula changes to $r < v$.

The Hartmann dispersion diagram at the very bottom of Fig. 14-23 shows why the optic angle for blue light always stays ahead of that for red light. The refractive indices n_a and n_b are considered fixed for all wavelengths. The refractive index n_c has a steeper dispersion curve than n_b, and moves up past n_b so as to cross it first for blue light, then for yellow, then for red. At each of these three intersections the optic angle is zero for a specific wavelength only. When the optic angle is zero for the D line, it is past zero for blue light and yet to become zero for red light. As n_c sweeps past n_b with increasing disorder, the fortuitously uniaxial condition of the crystal sweeps through all the wavelengths of visible light from blue to red. In principle, such a transition could be directly observed and monitored by watching a Bxa figure of a crystal on a heating stage; in practice, the rate of disordering at ordinary pressures is too slow to make this a practicable idea.

In sum, the dispersion formula can distinguish between low and high sanidine.

Albite

The names of most natural disordered and ordered albite are simply high and low albite. (Analbite and monalbite are special names for high albite that can invert from or to monoclinic symmetry without diffusion of Al and Si; see, for example, Ribbe, 1983). The ordering of Na feldspar has exactly the same effect on crystal optics as that in K feldspar, except that in this case the two higher refractive indices do not cross over one another. Instead (Fig. 14-24), n_c approaches n_b with disorder, but does not reach it at high albite. On the other hand, n_c approaches n_a (α) in low albite, and the optic angle therefore passes through 90°, becoming positive (but still large) for low albite. At high albite, the 2V is only moderate in size, and negative.

Alkali Feldspar

The similar behavior of K and Na feldspar with ordering means that solid solutions of the two end members can be interpreted through

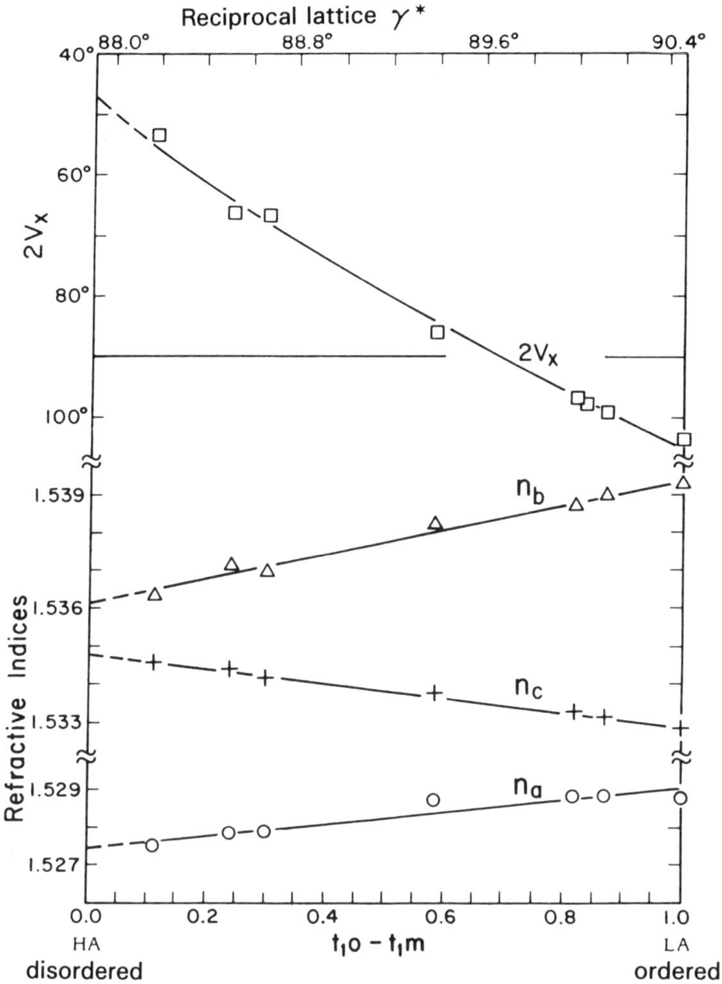

Fig. 14-24. Determinative diagram for ordering in albite, after Su et al. (1986b).

optical measurements as well as the end members. This is an important result because the pure end members are rare compared to intermediate compositions. Figure 14-25 shows the effect of order on 2V for the entire composition range from Ab to Or. Several generalizations can be made from this diagram. If the 2V is very large, i.e., 90±10°, the crystals are ordered (low-temperature forms) no matter what the composition. In this range, both composition and 2V must be known precisely to obtain a quantitative estimate of the order parameter. If 2V is in the range 0°–50°, the crystals are more disordered than ordered, and a good estimate of composition and 2V will yield a very good estimate of the order parameter. If the optic plane is (010), the crystals are Or-rich and disordered.

Because the α refractive index changes very little with order in both Ab and Or, it is a sensitive guide to composition. By reference to Figs. 14-22 and 14-24, one can see that the lowest refractive index is below 1.518 for all Or, and between 1.527 and 1.529 for all Ab. The following

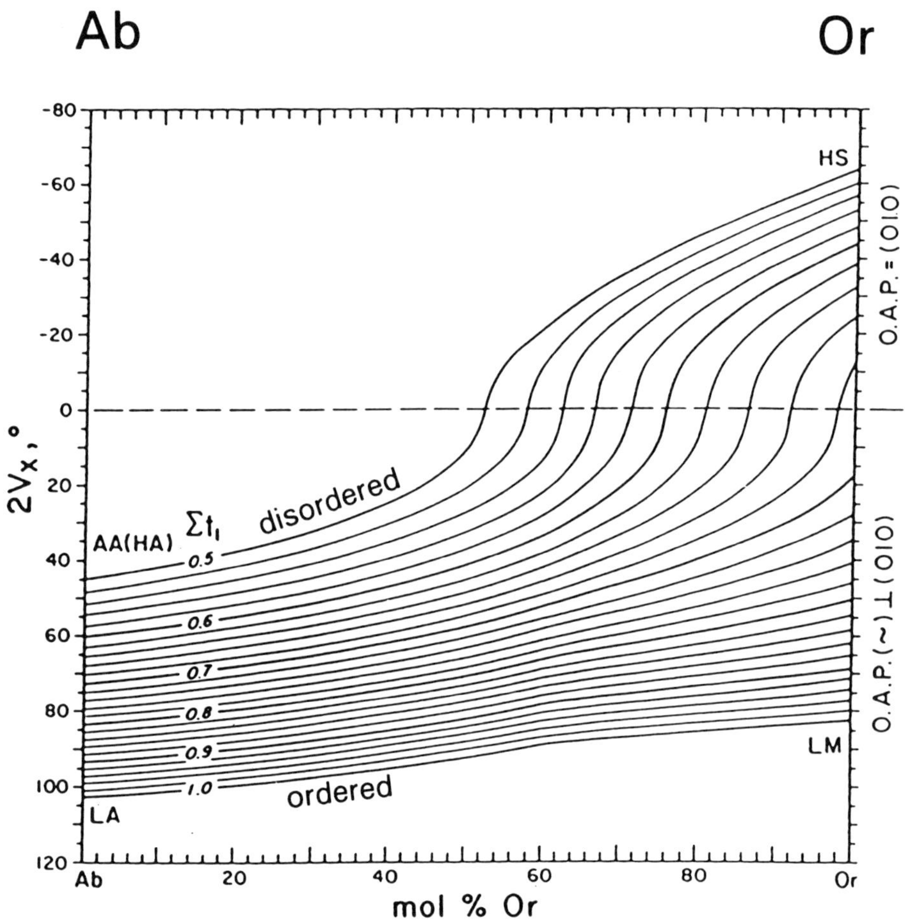

Fig. 14-25. Determinative diagram for alkali feldspar, after Su *et al.* (1986a).

determinative equations are derived, where X_{Or} is the mole fraction of Or, i.e., molar Or/(Ab + Or), and n_l is the lowest refractive index:

For disordered alkali feldspar:

$$X_{Or} = (1.5274 - n_l) / 0.0099 \qquad (14\text{-}2)$$

For ordered alkali feldspar:

$$X_{Or} = (1.529 - n_1) / 0.011 \quad (14\text{-}3)$$

If the state of order is unknown:

$$X_{Or} = (1.528 - n_1) / 0.01 \quad (14\text{-}4)$$

The last of these determinative relations will give compositions to within about 5 mole percent for intermediate and Or-rich feldspar. If 2V is then determined and a measure of disorder assigned, the composition can be refined by assigning appropriately different coefficients in the above equations. Specifically, using the order parameter in Fig. 14-25 but simplifying it to t_1,

$$n_o = 1.5274 + 0.0032(t_1 - 0.5) \quad (14\text{-}5)$$

$$\text{slope} = 0.0099 + 0.0022(t_1 - 0.5), \quad (14\text{-}6)$$

where n_o is the refractive index from which n_1 is subtracted in the above equations.

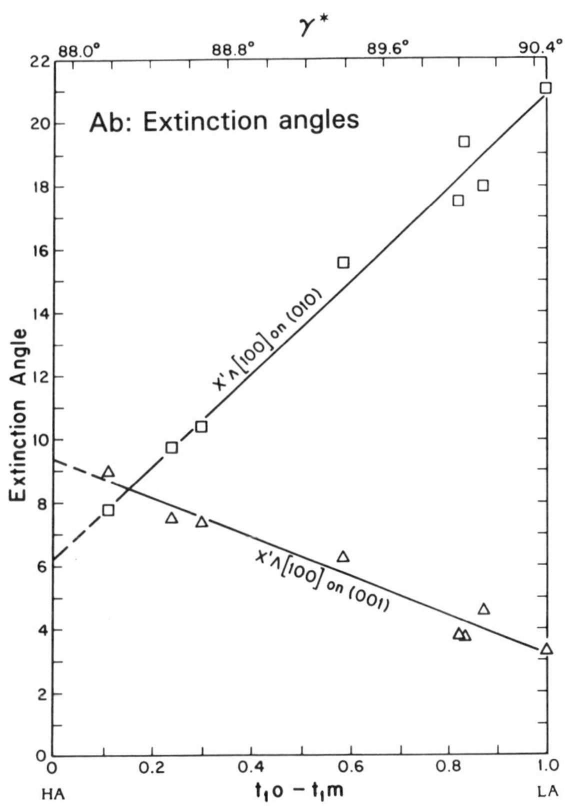

Fig. 14-26. Effect of ordering on extinction angles in albite (Su et al., 1986b).

Optic Orientation of Plagioclase

The natural plagioclase feldspars are all triclinic, so the orientation of the indicatrix is free to move with respect to the crystal axes. Not surprisingly, it does so in response to both composition and ordering. A simple and readily observable change of extinction angles with order is shown for the albite composition in Fig. 14-26. A complete statement of the change in optic orientation is shown in the stereogram of Fig. 14-27. The indicatrix axes X, Y, and Z all shift in concert to the right in the stereogram. A_1 and A_2 are the two optic axes, shown to shift position from low albite (LA) to high albite (HA) or analbite (AA). The track of A_1 is straightforward, but the track of A_2 encounters the equator of the net and re-emerges 180° away. An understanding of these migration curves is enhanced by reference to the perspective sketches.

Plagioclase composition varies away from pure albite by the substitution $CaAlNa_{-1}Si_{-1}$. The density and refractive indices increase in response to the addition of the heavy Ca in place of the light Na. Because the local changes in density and volume are themselves anisotropic, the orientation of the indicatrix changes as well as its shape and size. The shift of the indicatrix with composition is shown in Fig. 14-28 for six compositions of plagioclase. The nature of the shift is perhaps best seen by tracking the optic normal, Y, from its position near *c* in albite to a position nearly normal to *c* in anorthite. Stereograms that show this compositional migration, as well as the effects of ordering, are published in many compilations such as Tröger (1979). Although these migration curves were once used as determinative curves, they are laborious to apply, requiring careful measurements on individual grains in thin section with the universal stage. The dispersion method provides a much faster and more precise estimate of composition, from which estimates of ordering can be made by the various extinction angles plotted in standard compilations.

When only a thin section is available and an estimate of plagioclase composition is desired, the most reliable result from extinction angles is obtained by the *a*-normal method described in Chapter 13. In this method, the extinction angle is measured against the trace of the albite twin plane (010) in crystals whose *a* axis is normal to the section. This condition is recognized when two sets of cleavages [(010) and (001)] are sharp and do not move with changing focus. If such sections are very scarce (as may happen in coarse-grained rocks), the maximum symmetrical extinction method of Michel-Lévy is serviceable. This method uses the extinction angles in adjacent twins of the albite law, which are symmetrical if the (010) twin plane is normal to the section.

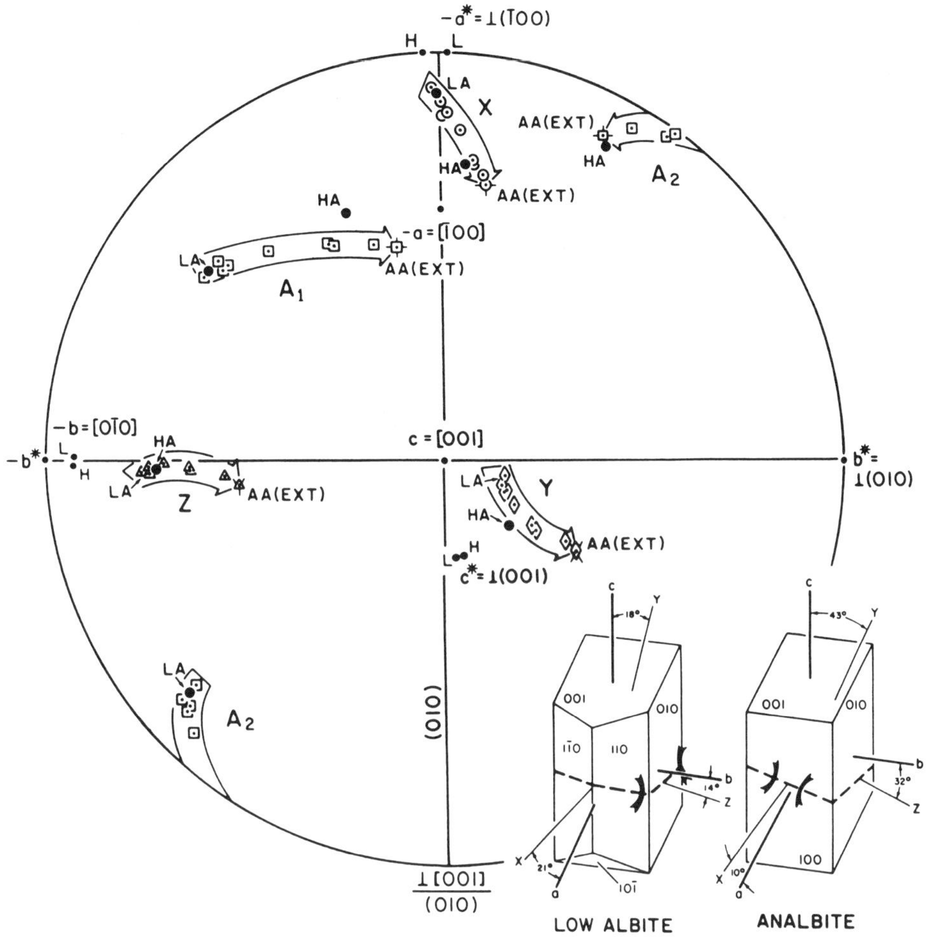

Fig. 14-27. Change of optic orientation with ordering in albite (Su et al., 1986b).

An especially clear description of these and other methods, with determinative diagrams, is given by Jones and Bloss (1980).

Although the ordering of feldspars is of some petrologic interest, its determination by optical or X-ray methods is not always of great importance. If the rock is volcanic, for example, the feldspars were almost surely disordered at the time of eruption. If the rock is a well-crystallized greenschist, the feldspars were and are almost surely well ordered. In such cases an independent determination of order is redundant or of trivial importance. In other cases, especially where one or more arrests have occurred in the cooling history, an independent determination of order may be fruitful. In short, the labor expended

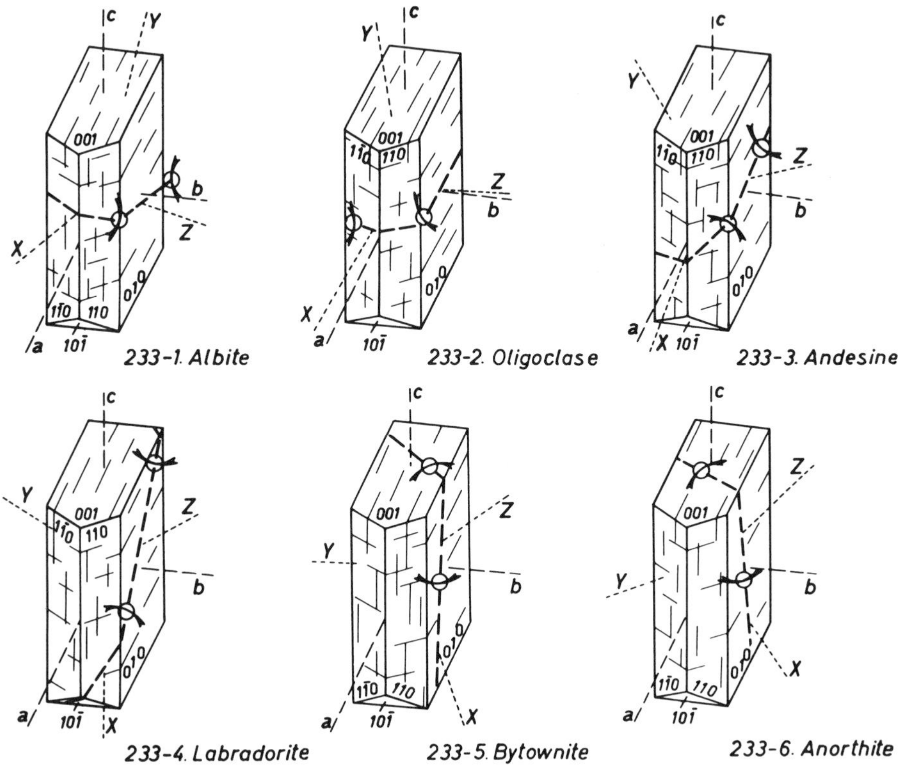

Fig. 14-28. Variation in optic orientation of plagioclase with composition, from Tröger (1979).

should be tailored to the potential information to be gained; the experiment should be well designed.

FORTUITOUS UNIAXIALITY AND DISPERSION

The ordering of sanidine produces a uniaxial condition where one refractive index migrates through another. This uniaxiality is fortuitous in the sense that it is not required by the high crystal symmetry (tetragonal or hexagonal) usually associated with uniaxiality. The crossover of two refractive indices requires a revised nomenclature for clarity, as discussed by Bloss (1985). When this occurs, the formula for the rhombic dispersion inevitably changes, as the uniaxial condition sweeps along the wavelength scale. The orientation of the optic plane may also change, and usually does because β is inevitably involved in the crossover. The case seen in sanidine (Figs. 14-22 and 14-23) is illustrative of the class in which uniaxiality and a shift in the dispersion formula result from a crossover, although it is unusual in being caused by ordering rather than composition.

Crossover behavior as a function of mineral composition is found in a number of important rock-forming minerals, where dispersion can be a useful guide to composition. Examples discussed by Bloss (1985) include Mn-bearing andalusite, ferrian and Fe-free zoisite (zoisite and pseudozoisite of Tröger), and Al- and Cr-chlorite. In zoisite, the crossover occurs for the lower two refractive indices and the crystal remains positive. In both the andalusite-kanonaite series and in the chlorites, all three refractive indices cross over at a single composition, so the mineral becomes fortuitously isotropic (but not, of course, morphologically isometric).

A further example of crossover is given by the alkali amphibole crossite in the glaucophane-magnesioriebeckite series. The mineral is negative and the crossover occurs between the two high refractive indices: n_b overtakes n_c as ferric iron replaces aluminum, and the optic plane shifts from (010) to an orientation normal to (010). The refractive index n_a remains alpha through the crossite composition, but as greater amounts of ferric iron are added, the Y indicatrix axis migrates from c toward a, and n_c therefore becomes interchanged with n_a. This mineral series behaves in such a complex manner that even the relabelling convention of Bloss (1985) is defeated. Zoned crystals of calcic amphibole, in which the crossover composition occurs at a specific zone, can be found in some alkalic igneous rocks. In such crystals, a Bxa figure can be made to close to uniaxial and reopen in the other plane by simply translating the section past the crossover composition. The concomitant change from $r<v$ to $r>>v$ can also be seen, and the dispersion is an important diagnostic tool.

15
Special Orientation Methods

INTRODUCTION

This chapter deals briefly with the universal stage and the spindle stage. As noted in Chapter 2, both these instruments have specialized applications that go far beyond routine crystal identification. Detailed discussions of their use may be found in the various books and papers referenced, particularly Muir (1981) for the universal stage and Bloss (1981) for the spindle stage. This chapter will also serve to suggest the pertinence of these instruments to problems of crystal identification, and to describe their manipulation for a few routine procedures.

Universal stages can be furnished by most manufacturers of petrographic microscopes. Many types are made; their variety is best appreciated by consulting the catalogs of the various manufacturers.[1]

UNIVERSAL STAGE

The universal stage (U-stage, U-Tisch, Fedorow stage) is a device that permits rotation of a selected optical or crystallographic direction or plane into a desired position, often parallel or perpendicular to the microscope axis. It is mounted with two screws upon the rotating stage of a petrographic microscope.

Universal stages are made in 2-, 3-, 4-, and 5-axis varieties, the most

[1] Certain "student model" petrographic microscopes are not designed to accept a universal stage, whereas almost all "research models" are. Critical requirements are the space available between stage and objective when the stage is fully lowered (or the tube raised), the positioning and threading of holes provided in the microscope stage, and whether the stage has a locking device. It may not be possible to match two generations of equipment without special adaptation of threads and holes. The U-stage is generally used with special objectives of long working distance.

common probably being the 4-axis. The axes are commonly named as in Table 15-1 for a 4-axis stage. Note that the microscope axis itself counts as a third vertical axis of rotation. There is no unanimity of designations. For the 5-axis stage, the use of A_0 for the additional E–W axis has been advocated by Hartshorne and Stuart (1970).

The universal stage is commonly used with thin sections. These are

Table 15-1
Nomenclature of Universal Stage Axes

Numerical Designation	Synonyms	Rest Position (Leitz)
A_1	Inner vertical, IV	0°
A_2	N–S horizontal, N–S	0°
A_3	Outer vertical, OV	90°
A_4	E–W horizontal, OEW, control axis	0°
A_5	Microscope axis, stage	360° = 0°

sandwiched with oil between glass hemispheres that permit angles of tilt up to 45° or 50°. Without hemispheres, the critical angle of refraction is exceeded and reflection occurs even at low angles of tilt, preventing observation of the crystal. The hemispheres also reduce the refraction correction needed to find the true orientation of a desired optical or crystallographic direction. For this reason, they are made in a variety of refractive indices, usually 1.515, 1.550, and 1.649, to match approximately the refractive indices of minerals commonly examined, such as calcite, quartz, or pyroxene. Refraction corrections may be determined graphically by means of a nomogram.[2] That the usual corrections based on Snell's law may not always be sufficient has been pointed out by Kamb (1962).

Manipulation

The fundamental operation of the universal stage is to bring a specific vibration direction of the crystal parallel either to A_4 (the control axis) or A_5 (the microscope axis). When this is done, the position of the vibration direction relative to the plane of the thin section can be plotted on a stereographic projection. This can be done in turn for other vibration directions or other directions such as optic axes until the orientation of all the optical elements of a crystal is completely known. Stereographic projections and their use are effectively summarized in Ford (1932) and in many elementary mineralogy and structural

[2] The nomogram of von Fedorow (1896) and other useful nomograms are contained in Tröger (1979).

geology textbooks. Examples of the operation of the U-stage are given by Haff (1942), Zussman (1977), Hutchison (1974), and Muir (1981).

The simplest example is the determination of quartz orientation in a thin section, a procedure common in petrofabric study. The c axis is brought either vertical (to A_5) or horizontal (to A_4). When the *c* axis is vertical, the circular section of the indicatrix is parallel to the microscope stage, and the grain therefore remains dark upon rotation about A_5, with crossed polars. When the *c* axis is horizontal, parallel to A_4, the circular section lies in the N–S vertical plane, and the grain remains dark under crossed polars upon rotation about A_4. In this event, if A_4 is placed NE–SW by rotating 45° on A_5, the interference colors will rise upon insertion of an accessory plate, since quartz is (+) and E lies along A_4.

The following procedure will illustrate the manipulation required to find the orientation of a quartz grain.

1. Lock all axes in rest position. Release locks as necessary in the following operations.
2. Orient the thin section in some convenient position, as with long edge N–S, select a grain, and *do not move the section with respect to the hemispheres* during observation of that grain. (A "Schmidt sledge" may be used with some U-stages to maintain the orientation of the section. This device is a simple 90° jig, one leg of which fits in a groove in the mount for the upper hemisphere.)
3. Turn grain to extinction on A_1.
4. Rotate on A_4:
   a. If grain becomes bright, go to step 5.
   b. If grain remains extinct, *c* is by chance either parallel to A_4 or normal to A_4. To determine if *c* is parallel to A_4, see step 11b. If *c* is normal to A_4, see step 10 or, to gain practice, rotate 90° on A_1 and go to step 5.
5. Leaving A_4 off zero, rotate to extinction on A_2. If falling interference colors occur on rotation, one is making *c* vertical. If rising colors occur, one is making *c* horizontal, hence obtaining maximum birefringence.
6. Set A_4 to zero. Refine extinction on A_1.
7. Rotate the other way on A_4.
8. Refine extinction on A_2.
9. Repeat steps 6–8 as necessary, until extinction persists on rotation of A_4.
10. If *c* is to be set vertical, a symmetry plane containing *c* is now normal to A_4. Rotate on A_5 until extinction vanishes, then rotate on A_4 until extinction is obtained. This can only mean that the

circular section is parallel to the stage, hence c = the optic axis is parallel to the microscope axis.

11. Test to confirm position of c. Rotate on A_5:
    a. If extinction persists, c is vertical.
    b. If extinction vanishes, place A_4 NE–SW and confirm that c is parallel to A_4 by inserting a gypsum plate; colors should rise.
12. Read arcs (A_1, A_2, and A_4) and plot c on a stereographic or equal area net. Lower hemisphere plots are usually used for petrofabrics. Set the zero (or 180°) mark on the net to the angle read on A_1. If the tilt on A_2 is to the right, count off the angle read on A_2 from the left side of the net along the equator. (Do the opposite if the tilt on A_2 is to the left.) If c is parallel to A_4, this is the point to plot for c, i.e., the position c would have in space if all axes were restored to their rest positions. If c is parallel to the microscope axis, the angle read on A_2 locates a great circle (a symmetry plane) containing c. If the tilt on A_4 is to the north, count off the angle read on A_4 northward from the equator and plot c.

The above procedure contains, in steps 6–9, the fundamental operations used to bring any symmetry plane of an indicatrix vertical and N–S, hence any vibration direction parallel to A_4. This is the desired result for most procedures with biaxial crystals. In most cases, two vibration directions can be brought parallel to A_4 in turn, and their positions plotted on a stereo net, making due corrections for the tilt of the inner stage. The third vibration direction can then be located by construction, since the vibration directions lie 90° from one another. The two known vibration directions are placed on the same great circle, and the pole of that great circle (i.e., a point on the equator 90° from that great circle) is the derived location of the third vibration direction. Rarely, if all three vibration directions are approximately equidistant from the microscope axis, all three can be located directly.

Optic Angle Measurement. In the determination of 2V, Y is set parallel to A_4 to make the optic plane vertical. That Y is parallel to A_4 is determined by rotating to the 45° position about A_5 and rotating on A_4 in a direction that causes interference colors to fall. This will produce extinction if an optic axis is accessible. It will produce extinction at two rotation positions on A_4 if 2V is sufficiently small that both optic axes are accessible. In this case, 2V is measured directly by rotation from one optic axis to the other and application of the appropriate refraction correction. If only one optic axis is accessible, V is measured by plotting the refraction-corrected position of the optic axis and reading off its distance from the plotted position of the Bxa along the great circle that represents the optic plane. In this case the Bxa will have been (or will need to be) plotted by bringing it parallel to A_5. The error in 2V so estimated will be greater than if measured directly from one optic axis to the other.

Extinction Angle Measurement. The measurement of extinction angles is accomplished by setting the desired reference plane [such as (010) in a monoclinic crystal for which $Z{:}c$ is sought] horizontal and then rotating about A_5, first to an extinction position and then to a recognizable crystallographic direction. This may be the trace of a cleavage or a twin plane, or rarely a crystal face or a train of inclusions. To set a reference crystallographic plane horizontal when it coincides with a symmetry plane of the indicatrix involves the location in stereographic projection of the three symmetry planes. One of them may then be set horizontal by reference to the angles on the stereo net. If, as in triclinic crystals, the desired crystallographic plane is not parallel to a symmetry plane of the indicatrix, it may sometimes be set horizontal by setting two cleavages (hence a zone axis) vertical. When a cleavage is vertical, its trace is a thin, sharp line, and as it is rotated away from a vertical orientation, the thin line broadens to a fuzzy band.

Conoscopic Observation

Observations with the universal stage need not be confined to the orthoscopic procedures described above. With special equipment, including a substage condensing lens, small hemispheres, and special objectives, conoscopic observations are possible. With the U-stage conoscope, an isogyre may be set coincident with a crosshair, and the symmetry plane thus oriented somewhat more sensitively than with orthoscopic observation.

Special Uses

The U-stage is extensively used to study the fabric of rocks, determined by the orientation of their constituent mineral grains. It is also extensively used in the study of twinning, particularly in the plagioclase feldspars. When the optical elements of two individuals in a twin are plotted, the operation needed to produce the twin can be deduced from their relationships on the stereo net. Since twin planes furnish identifiable crystallographic directions, the optic orientation of twinned crystals can be determined completely by means of the U-stage. The manipulations required for the determination of plagioclases with the

U-stage are conveniently presented by Slemmons (1962), using a method modified from Turner (1947). The Turner method minimizes the use of the stereographic projection. The plotting method making full use of the stereographic projection is described by Haff (1942), who also reviews in convenient form other useful U-stage procedures including preliminary adjustments.

Other special uses of the U-stage include studies of the lamellae of one mineral exsolved from another, as in many pyroxenes and amphiboles.

SPINDLE STAGE

The spindle stage is basically a very simple instrument consisting of a needle with a mineral grain stuck to its end with carpenter's glue and molasses. The needle lies on a flat stage and is restrained by clips or other means so that it maintains any given rotation position and an orientation perpendicular to the microscope axis. The business end of the needle and its adherent grain are immersed in a liquid of chosen refractive index contained in a small cell capped by a cover glass. The liquid may be changed by blotting it up, flushing the cell with a new liquid, and then adding a new liquid. The fundamental operation of the instrument is to bring an axis of the indicatrix parallel to the stage, after which it can be brought parallel to the polarizer for refractive index tests.

The immense utility of the spindle stage for refractive index determinations results from the fact that all three refractive indices can be determined on the same grain, no matter in what orientation it is mounted on the spindle. The basis of this fact can be appreciated if one imagines or looks at a model of an indicatrix (or any configuration of three mutually perpendicular axes). Imagine a rotation axis in any random position through the indicatrix. It will become apparent that upon rotation about this axis, any of the indicatrix axes can be made horizontal, i.e., parallel to the microscope stage. Once horizontal, the axis (a vibration direction) can be placed parallel to the polarizer by turning the microscope stage about its vertical axis to the appropriate extinction position.

In addition, all the indicatrix directions and the positions of the optic axes can be located and plotted by means of extinction curves combined with computations. Procedures for manipulating grains and obtaining all the necessary information from them with the spindle stage are described in detail by Bloss (1981) and more briefly on pp. 11–15 of

Fleischer et al. (1984). Only the bare outline of the approach is given here.

Manipulation

In practice, an indicatrix axis may easily be set horizontal by setting a symmetry plane of the indicatrix vertical. This is done by observing the interference figure and rotating the spindle and microscope stage until an isogyre bisects the field coincident with a crosshair. Then by rotation of the microscope stage, the isogyre can be brought coincident with the E–W crosshair, at which position the vibration direction normal to that symmetry plane lies N–S, in position for refractive index comparison with the immersion liquid, if the polarizer is oriented N–S.

In order to use the spindle stage efficiently, one should be able to identify the vibration direction that is to be placed N–S. If the isogyre is a CR, i.e., rotates counter to the direction of stage rotation, the symmetry plane that is vertical is the optic plane, and when the isogyre lies E–W the N–S vibration direction will be Y. The β index may then be measured if the polarizer is N–S. If the isogyre is an SR, rotating the same way as the stage rotates, the vertical symmetry plane is not the optic plane, and the normal to the symmetry plane is either X or Z, furnishing α or γ. Whether the horizontal vibration direction is X or Z is easily determined: set the isogyre E–W, view the grain orthoscopically, rotate the microscope stage 45° to the right, and insert an accessory plate. If the colors rise, Z is now NE; if colors fall, X is now NE. Restore the vibration direction to the polarizer direction before making refractive index tests.

The successive selection of vibration directions is made by successive rotations of the spindle axis followed by rotations of the microscope stage sufficient to restore extinction (or coincidence of an isogyre with a crosshair).

Other Procedures

There are many alternative procedures to the simple one described here for orienting a grain for observation. Determination of 2V is of course feasible by direct rotation from one optic axis to the other if Y is parallel to the spindle axis. With other orientations of the crystal on the spindle, 2V may be determined by construction.

The spindle stage is particularly useful with crystals whose habit or cleavage prevents measurement of one or more of the principal refractive indices by ordinary methods. Among minerals, micas and amphiboles are notorious in this respect. The spindle stage is also advantageous when the amount of material available for study is small.

16 Use of the Dispersion Method

INTRODUCTION

The task of crystal identification may be eased considerably by routine use of the dispersion method mentioned in Chapter 3, at page 73. In this procedure, the wavelength of light is varied to find a match between crystal and liquid. The chief practical advantage is that a critically oriented grain can be matched without having to disturb it by changing oils. Moreover, the matching criterion is more objective than the estimation of color fringes, requiring less experience in interpretation. The dispersion method (and its variant the double variation method, in which temperature is also varied in order to increase the range of a given liquid and to provide temperature control) has recently been brought to a high level of sophistication and accuracy by F.D. Bloss and his colleagues. In this chapter, we will concentrate on the simpler uses of the method for everyday crystal identification, and on applications to certain mineral groups in which a single reading can be translated to composition within a solid solution series.

OILS

Temperature Correction

In general, the refractive index of the oil will need to be corrected for the temperature of the immersion mount as measured or estimated on the stage of the microscope. The temperature coefficient *dn/dt* of the oil is usually printed on the bottle; its sign is always negative. (The symbol *d* here signifies a fixed increment delta, and is not meant as the differential.) The reference temperature is commonly taken as 25° C, which is conveniently close to the stage temperature of many microscopes in a room at 20° C after some period of use. If the refractive index of an oil is being checked at some temperature t, then its standard index is given by

$$n_{25} = n_t - \left[(t - 25)\left(\frac{dn}{dt}\right)\right] \tag{16-1}$$

For the more usual practical case of adjusting to the observed stage temperature t,

$$n_t = n_{25} + \left[(t - 25)\left(\frac{dn}{dt}\right)\right] \tag{16-2}$$

Oils can be calibrated in the immersion mounts themselves by use of standard glass powders (Sueno, 1933; Su, 1992).

Dispersion

New bottle labels generally give n_F and n_C directly, as well as n_D. Old bottles may give reciprocal dispersion as Abbe $V = (n_D - 1)/(n_F - n_C)$. In this case, the dispersion is given by

$$n_F - n_c = \frac{(n_D - 1)}{V} \tag{16-3}$$

The variation of dispersion with refractive index for old and new Cargille oils is shown, with equations, in Fig. 16-1.

Individual values of n_F and n_C may be found from the following rule:

$$n_F = M(n_F - n_c) + n_D \tag{16-4}$$

$$n_c = n_F - (n_F - n_C) \tag{16-5}$$

where M = the fractional dispersion, $(n_F - n_D)/(n_F - n_C)$,
= 0.7239 for special plagioclase liquids, Cargille Set 547
= 0.72 for $n_D \leq 1.78$, errors < 0.00008,
= 0.71 for n_D 1.79–1.80,
= 0.70 for n_D 1.81–1.90.

Indices for many lines including F and C for Cargille oils 1.670–11 are given in tables published by Cargille.

MONOCHROMATORS

The wavelength of light entering the crystal is conveniently varied by use of a substage Schott-wedge interference filter. This instrument may

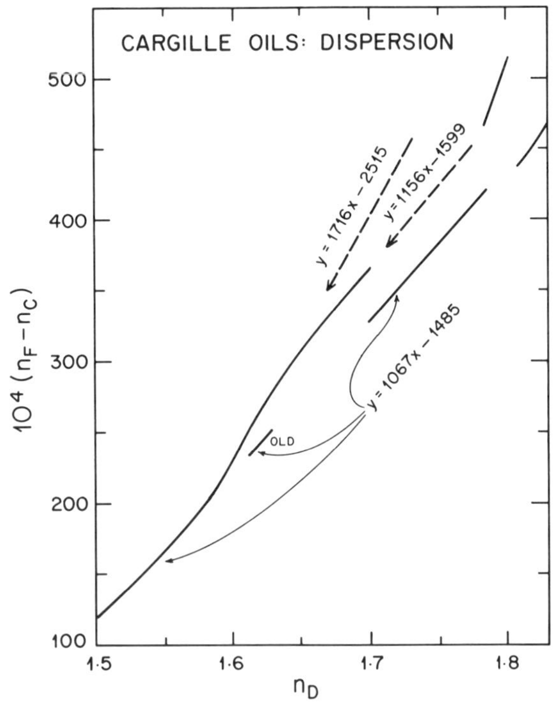

Fig. 16-1. Dispersion as a function of refractive index in Cargille index of refraction oils. The equations of various segments are given to facilitate direct calculation. Precise values should be obtained from the bottle or the manufacturer.

only loosely be called a monochromator, because it transmits a band of colors centered about a peak, rather than a single wavelength. The bandwidth may be controlled by a variable slit. The filter is best mounted in a shuttle that can be moved freely by hand so that the wavelength can be varied rapidly and the eye can easily detect a null point where oil and crystal are matched. The shuttle furnished by Leitz satisfies this requirement and mounts directly over the illuminator on both old and new models of their petrographic microscope.

The shuttle for the Leitz VERIL-S200 interference filter monochromator is a metal frame with a spring-loaded detent at one end to hold the filter in place. A window about 2 × 4 mm in dimensions should be cut or filed in the edge of the detent so that observations can be made in white light when the shuttle is mounted on the light source. It is then a simple matter to switch from color fringe to shadow-hopping in order for the individual observer to calibrate the interpretation of color fringes according to what s/he sees.

The frame of the interference filter is marked off in millimeters and these must be converted to wavelength using an individual calibration furnished by the manufacturer for each filter. This calibration is furnished as a list of conversion factors for selected wavelengths. These factors can either be fitted to a polynomial equation

(e. g. Bloss, 1981, p. 165) or interpolated into a conversion table that allows wavelength to be read for each millimeter on the scale.

For ordinary instruction or research purposes with small working groups it suffices to have one microscope in the laboratory dedicated to the dispersion method and provided with an interference filter and a thermometer. Observations can then be made in white light at individual stations, and the critical grains later examined in monochromatic light when the microscope slide is taken to the monochromator work station. For larger laboratory groups it is helpful and often adequate to have a monochromator work station set up on a trolley with a video camera and screen, so that critical observations can be made for the whole group to see.

REPRESENTATION OF WAVELENGTH VARIATION

Three of the many formulas for the variation of refractive index with wavelength are in common use, each for its own precision or suitability. The **Sellmeier equation** is discussed by Louisnathan et al. (1978) and by Su et al. (1987). In its linearized version, refractive index is represented by y:

$$y = a_0 + x a_1 \qquad (16\text{-}6)$$

where y equals $(n_\lambda^2 - 1)^{-1}$, x equals λ^{-2}, and n_λ equals the refractive index at wavelength λ. The variation of constants a_0 and a_1 is given by Louisnathan et al. (1978); it is smooth over the range of standard refractive index liquids for a_0, but jumpy for a_1.

The **Cauchy equation,**

$$n = c_1 + \frac{c_2}{\lambda_2} + \frac{c_3}{\lambda_4} \qquad (16\text{-}7)$$

where the c's are constants derived by fit and the wavelengths are in nm, is discussed by Su et al. (1987). This equation is particularly useful for precise representation of the refractive index of oils.

Here we shall concentrate on the simpler and more approximate **Hartmann equation** involving the empirical wavelength transform

$$L = \frac{10^5}{(\lambda - 200)}, \text{ for } \lambda \text{ in nm} . \qquad (16\text{-}8)$$

Several special values of L are often used:

$$\begin{aligned} L_F &= 349.528, \\ L_D &= 256.871, \\ L_C &= 219.154, \text{ and the difference,} \\ \Delta L_{(F-C)} &= 130.374. \end{aligned}$$

A plot of refractive index n against L is nearly straight, but for standard oils the observed index is higher than the Hartmann index at blue wavelengths. The practical cure for this is to work as close to the D line as feasible, or to check measurements in the blue against those made in another oil at more central wavelengths. Accepting these approximations as reasonable but always to be kept in mind, the Hartmann equation can be used easily as follows. If a graphical plot is desired, and only the D-line index and the dispersion are known from the bottle, then the index at the F line can be found from equations 16-4 and 16-5, above. For the more generally convenient algebraic use of the Hartmann equation, the refractive index can be expressed as a function of L by

$$n_L = aL + b \tag{16-9}$$

where a is the slope,

$$a = \frac{\Delta n_{(F-C)}}{\Delta L_{(F-C)}} = \frac{\Delta n_{(F-C)}}{130.374} \tag{16-10}$$

and b is the intercept,

$$b = n_D - (aL_D) = n_D - 256.871a \tag{16-11}$$

The Hartmann equation can be used both for oils and crystals, as illustrated in Fig. 16-2, where refractive indices are labelled at the F, D, and C lines for the oil and at D for the crystal. The circle indicates a match of crystal and liquid at L_m (bottom scale) or λ_m (top scale). The problem in crystal identification is to find n_D for the crystal, knowing L_m and the equation for the oil. This involves either assuming a slope for the crystal or measuring it by finding a match in another oil.

CRYSTALS

If high precision is desired, the dispersion curve of the crystal is best determined directly by making observations of the wavelength of match in several oils, and performing a linear regression on the resulting values of n_λ (Huntington, 1979). However, for routine purposes, a first approximation to the dispersion slope will usually suffice. Dispersion rises with n, in general. For an unknown mineral, the dispersion can be estimated from n for the purpose of solving for n_D when a match has been made at some other wavelength. Obviously, if the match has been made near D, the error in extrapolation will be less. Using data tabulated in Tröger (1979), the approximate correlation is

$$\log(n_F - n_C) = (5.9\log n_D) - 3.2 \tag{16-12}$$

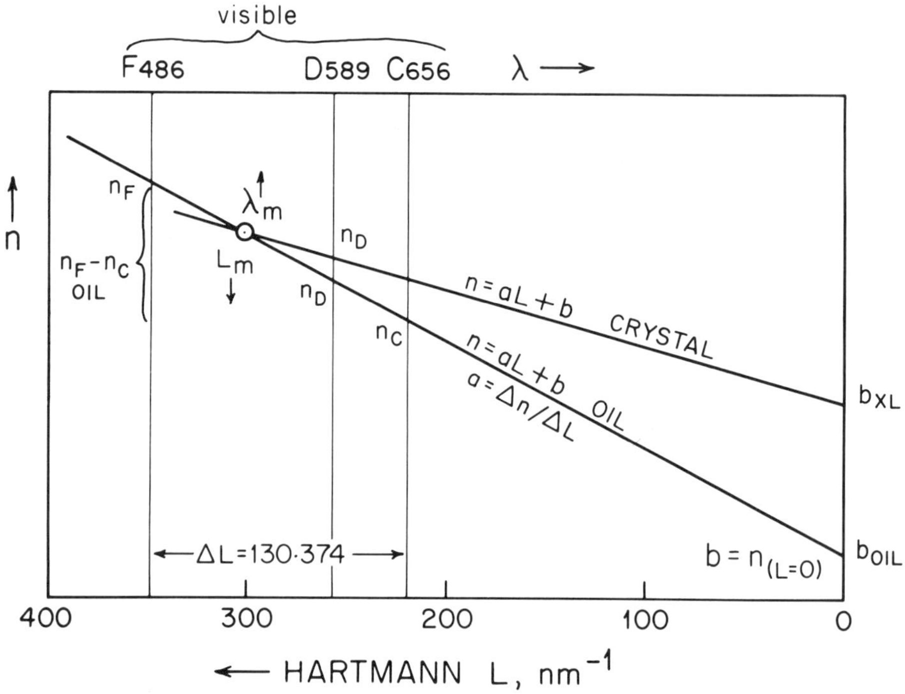

Fig. 16-2. Dispersion curves for oil and crystal in Hartmann space, showing the derivation of the slope *a*, the intercept *b*, and the dispersion equation 16-9.

Hence the expected dispersion is, with estimated uncertainties:

n_D	$n_F - n_C$
1.4	0.005±0.0025
1.5	0.007
1.55	0.008
1.6	0.010±0.004
1.65	0.012
1.7	0.014
1.75	0.017
1.8	0.020±0.007.

SU'S METHOD FOR FINDING *n* AT ANY WAVELENGTH

The preceding discussion concerned a method for finding the refractive index at any wavelength for either the oil or the liquid taken singly. This method can also be used to find n_D for the crystal by combining the equations shown in Fig. 16-2. A more direct method due to Su

(1993) yields n_D (or any n_λ) for the crystal directly from the wavelength of match and the known dispersion slopes of the oil and the crystal. The solution of the problem is accomplished through the use of a fractional difference, in Hartmann space, between the wavelength of match and the wavelength of the D line. In Su's notation this fractional difference is denoted k_D, defined as

$$k_D = \frac{X_0 - X_D}{X_F - X_C} \tag{16-13}$$

where $X = (\lambda - 200)^{-1}$, similar to L, the Hartmann wavelength scale in equation 16-8. The equation for the refractive index of the crystal at the D line is then derived as follows.

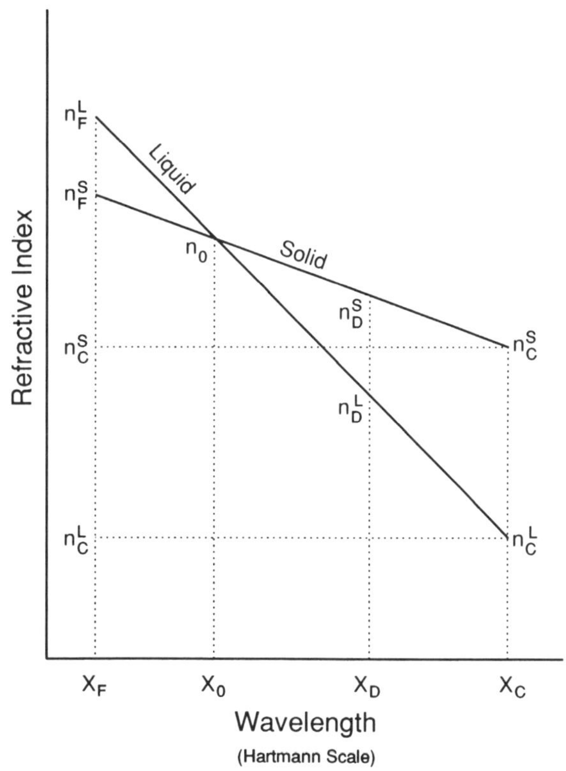

Fig. 16-3. Derivation of Su's method for finding the crystal index n_D^S from the wavelength of match in Hartmann space.

For straight Hartmann dispersion lines intersecting as in Fig. 16-3, we have from similar triangles

$$\frac{n_F^L - n_F^S}{n_D^S - n_D^L} = \frac{X_F - X_O}{X_O - X_D} \tag{16-14}$$

and

$$\frac{n_C^S - n_C^L}{n_D^S - n_D^L} = \frac{X_O - X_C}{X_O - D_D}\,. \tag{16-15}$$

Combining these, we have

$$\frac{(n_F^L - n_C^L) - (n_F^S - n_C^S)}{(n_D^S - n_D^L)} = \frac{X_F - X_C}{X_O - X_D} . \tag{16-16}$$

Let the dispersions of liquid and solid be

$$\Delta^L = n_F^L - n_C^L, \quad \Delta^S = n_F^S - n_C^S . \tag{16-17}$$

Using equation 16-13 we have

$$n_D^S = n_D^L + (\Delta^L - \Delta^S) \cdot k_D . \tag{16-18}$$

Because all the wavelengths but λ_0, the matching wavelength, are known in advance, the fractional Hartmann wavelength reduces to

$$k_D = \frac{(\lambda_0 - 200)^{-1} - (589.3 - 200)^{-1}}{(486.1 - 200)^{-1} - (656.3 - 200)^{-1}} \tag{16-19}$$

and hence

$$= \frac{(\lambda_0 - 200)^{-1} - 0.002569}{0.001304} , \tag{16-20}$$

which can be calculated or tabulated for all λ_0 (Table 16-1).

The final result is the determinative equation

$$n_D^S = n_D^L + (\Delta^L - \Delta^S) \cdot k_D \tag{16-21}$$

which indicates that the crystal index at the D line is simply that of the liquid at the D line plus the difference in the two dispersions ($n_F - n_C$) times the fractional wavelength. The same equation with subscripts i can be used to solve for n at any wavelength i using the generalized fractional wavelength $k_i = [(\lambda_0 - 200)^{-1} - (i - 200)^{-1}]/0.001304$ (Su, 1993).

If the dispersion of the crystal is in question, it can be determined directly by finding a matching wavelength in another oil and applying the **double liquid method** (Su, 1993). Assuming that the refractive index of liquid 1 is lower than that of liquid 2, and the matching wavelengths are determined to be λ_0^1 and λ_0^2, respectively, then the refractive index of the crystal can be calculated from the following equation:

$$n_D^S = n_D^1 + (n_D^2 - n_D^1) \cdot k_{D2} , \tag{16-22}$$

where

$$k_{D2} = \frac{(X_0^1 - X_D)}{(X_0^1 - X_0^2)} . \tag{16-23}$$

The Hartmann wavelength values X are as defined below equation 16-13, and here refer to the matching wavelengths in liquids 1 and 2, respectively. As an example of how the double liquid method is used, a test calculation can be made on sample TS-1, using the refractive index worksheet (Fig. 4-1) at the end of Chapter 4. Because in this case the crystal matched liquid 2 very near to the D line, the result with the double liquid method is essentially the same as for liquid 2, i.e., $n = 1.5447$.

PROCEDURE FOR FINDING n_D

The recipe for determining a refractive index is, then, as follows. First find an oil that matches the index of the crystal at some wavelength. (If the crystal has a low birefringence, this condition may obtain for all the principal refractive indices.)

For the oil:

Find and record the wavelength of match λ_0 (in nm) and the temperature of match t_m in degrees C.

Transform the wavelength to k_D from Table 16-1. Correct the D-line index of the oil for temperature.

Find the dispersion ($n_F - n_C$) of the oil from the label or a table or graph.

For the crystal:

Estimate the provisional dispersion from equation 16-12.

Find n_D of the crystal from equation 16-21.

If the crystal is now identified and its actual dispersion is known from the literature, recalculate the index with the corrected dispersion instead of the provisional dispersion. Otherwise evaluate the probable error in the result from the expected errors listed under equation 16-12 and the wavelength of match; if the error exceeds the desired tolerance, then a new mount should be made in another oil so that the dispersion can be determined directly by the double liquid method. This will rarely need doing in routine work unless the match has been made very far from the D line. Note that in crystals of high birefringence, equation 16-12 should be used for each index separately, because the dispersion may well differ between the low and high limits of the index range.

The procedure is rendered simple by use of a worksheet, here reproduced as Fig. 16-4. The worksheet provides all the necessary steps and constants, and allows the worker to concentrate on collecting reproducible data. The sheet also provides for comparison of calculated and observed 2V, which should be made in order to monitor the quality of the recorded data, and contains spaces for recording a few other

Table 16-1
Conversion of λ_0 to k_D

λ_0	0	1	2	3	4	5	6	7	8	9
400	1.8649	1.8458	1.8269	1.8082	1.7897	1.7713	1.7532	1.7352	1.7174	1.6997
410	1.6822	1.6649	1.6478	1.6308	1.6140	1.5973	1.5808	1.5644	1.5482	1.5321
420	1.5162	1.5004	1.4848	1.4693	1.4539	1.4387	1.4236	1.4087	1.3939	1.3792
430	1.3646	1.3502	1.3359	1.3217	1.3076	1.2937	1.2798	1.2661	1.2525	1.2390
440	1.2257	1.2124	1.1993	1.1862	1.1733	1.1604	1.1477	1.1351	1.1226	1.1102
450	1.0978	1.0856	1.0735	1.0615	1.0495	1.0377	1.0259	1.0143	1.0027	.9912
460	.9798	.9685	.9573	.9462	.9351	.9242	.9133	.9025	.8918	.8811
470	.8706	.8601	.8497	.8393	.8291	.8189	.8088	.7988	.7888	.7789
480	.7691	.7594	.7497	.7401	.7305	.7210	.7116	.7023	.6930	.6838
490	.6746	.6656	.6565	.6476	.6387	.6298	.6210	.6123	.6036	.5950
500	.5865	.5780	.5695	.5612	.5528	.5446	.5363	.5282	.5201	.5120
510	.5040	.4961	.4881	.4803	.4725	.4647	.4570	.4494	.4418	.4342
520	.4267	.4192	.4118	.4044	.3971	.3898	.3826	.3754	.3682	.3611
530	.3541	.3470	.3400	.3331	.3262	.3194	.3125	.3058	.2990	.2923
540	.2857	.2791	.2725	.2660	.2595	.2530	.2466	.2402	.2338	.2275
550	.2212	.2150	.2088	.2026	.1965	.1904	.1843	.1783	.1723	.1663
560	.1604	.1545	.1496	.1427	.1369	.1312	.1254	.1197	.1140	.1084
570	.1028	.0972	.0916	.0861	.0806	.0751	.0697	.0643	.0589	.0535
580	.0482	.0429	.0377	.0324	.0272	.0220	.0168	.0117	.0066	.0015
590	−.0035	−.0086	−.0136	−.0185	−.0235	−.0284	−.0333	−.0382	−.0431	−.0479
600	−.0527	−.0575	−.0622	−.0670	−.0717	−.0764	−.0810	−.0857	−.0903	−.0949
610	−.0995	−.1040	−.1086	−.1131	−.1175	−.1220	−.1265	−.1309	−.1353	−.1397
620	−.1440	−.1484	−.1527	−.1570	−.1612	−.1655	−.1697	−.1740	−.1782	−.1823
630	−.1865	−.1906	−.1947	−.1988	−.2029	−.2070	−.2110	−.2151	−.2191	−.2231
640	−.2270	−.2310	−.2349	−.2388	−.2427	−.2466	−.2505	−.2543	−.2582	−.2620
650	−.2658	−.2695	−.2733	−.2771	−.2808	−.2845	−.2882	−.2919	−.2955	−.2992
660	−.3028	−.3064	−.3100	−.3136	−.3172	−.3208	−.3243	−.3278	−.3313	−.3348
670	−.3383	−.3418	−.3452	−.3486	−.3521	−.3555	−.3589	−.3622	−.3656	−.3690
680	−.3723	−.3756	−.3789	−.3822	−.3855	−.3888	−.3920	−.3953	−.3985	−.4017
690	−.4049	−.4081	−.4113	−.4144	−.4176	−.4207	−.4238	−.4270	−.4301	−.4331
700	−.4362	−.4393	−.4423	−.4454	−.4484	−.4514	−.4544	−.4574	−.4604	−.4633

REFRACTIVE INDEX WORKSHEET

SAMPLE NO.__________

	ALPHA α (NLOW)	BETA β (OMEGA)	GAMMA γ (NHI)
Oil:	______	______	______
Shuttle reading	______ mm	______ mm	______ mm
Lambda match = λ_0	______ nm	______ nm	______ nm
k_D, from table	______	______	______
t match	______ °C	______ °C	______ °C
dn/dt (bottle)	— ______E-4	— ______E-4	— ______E-4
$n_D{}^L = n_{25} + [(t\text{-}25)(dn/dt)]$	______	______	______
Δ^L(F-C) (bottle or graph)	______	______	______

Crystal:

Calculate Δ^S(F-C) = 10↑(5.9 log n - 3.2) or = exp (5.9 ln n -7.368):

Provisional Dispersion, Δ^S(F-C)	______	______	______
$n_D{}^S = n_D{}^L + k_D(\Delta^L - \Delta^S)$	______	______	______

(If warranted, recycle with refined values of the crystal dispersion)

$\gamma - \beta$ = ______ B = ______ $Vx \approx \arcsin \sqrt{(\gamma - \beta) / B}$ = ____

$2V_{calc}$ ______ $2V_{obs}$ ______ from figure ____________________

Sign ____ Cleavages ______ ______ ______ Color, etc. ____________________

Optic orientation ________________________ Crystal system ____________

Distinguishing features ____

Mineral Name __________________ Composition __________________

Observer __________________ Date__________________

Fig. 16-4. Worksheet for the determination of refractive index by the dispersion method.

features of interest. Worked examples based on the worksheet shown in Fig. 16-4 are shown in Figures 4-1, 9-7, and 12-12.

APPLICATION TO MAFIC MINERALS

The substitution of ferrous iron for magnesium ($MgFe_{-1}$) is important in many groups of so-called mafic minerals, and the composition in terms of this cation exchange function is often wanted, whether for routine petrographic description or for calculations. From the rule of Gladstone and Dale (Chapter 1), it will be anticipated that the denser Fe end members will have the higher refractive index. If the dispersion of the mineral is known as a function of composition (or directly as a function of refractive index), a single observation of match between a known oil and the crystal whose species is known will furnish the composition in terms of the atom fraction of the Mg end member:

$$X_{Mg} = \frac{Mg}{(Mg + Fe^{2+})} \tag{16-24}$$

For such mafic minerals, the refractive index commonly varies linearly with X_{Mg},

$$n_D = n_{D,0} - (\Delta n_D X_{Mg}) \tag{16-25}$$

where $n_{D,0} = n_D$ for the Fe end member at $X_{Mg} = 0$ and Δn_D is the difference in D-line refractive index between the pure Mg and Fe end members.

The resulting **determinative equation**, valid for all wavelengths when n_0 is suitably defined, is

$$X_{Mg} = \frac{(n_0 - n)}{\Delta n_\lambda} \tag{16-26}$$

where Δn_λ is the refractive index difference between the end members at the wavelength λ. Because composition is normalized to $X_{Mg} = 1.0$, the quantity Δn_λ is simply the slope of the determinative curve of the mineral at a given wavelength. The derivation applies in principle to any binary series when composition is normalized to 1.0. If the composition is given in terms of the high-index end member (as for plagioclase in the next section), n_1 and n_0 will be reversed and the numerator in the determinative equation will be $(n - n_0)$.

For the three common mafic mineral species olivine, orthopyroxene, and augite, the dispersion values have been determined by Berg and Morse (1981) as a function of the D-line refractive index. For use in Su's method, we define, using Su's (1994) notation,

$$\Delta^S = n_F^S - n_C^S = z + w \cdot n_D^S \,, \qquad (16\text{-}27)$$

where z and w are constants determined by Berg and Morse (1981) and tabulated on the worksheet shown in Fig. 16-5. Substituting equation 16-21 into 16-27, we derive (Su, 1994)

$$n_D^S = \frac{[n_D^L + (\Delta^L - z) \cdot k_D]}{(1 + w \cdot k_D)} \,. \qquad (16\text{-}28)$$

The resulting value of the D-line refractive index is then used in a determinative equation of the form of equation 16-26, using constants n_0 and Δn as found by Berg and Morse (1981) and also tabulated here in the worksheet, Fig. 16-5.

For olivine and augite, the β refractive index is used and a counter-rotating isogyre is verified to lie on or near the crosshair that is normal to the direction of the lower polar. Suitable candidate grains for olivine show low interference colors or, if they are very near an optic axis, do not go completely to extinction. Suitable grains for augite are those on the (100) parting, showing parallel extinction along the crosshair normal to the lower polar. A worked example using augite is shown in Fig. 16-6. For orthopyroxene, the γ refractive index is used and a same-rotating isogyre is verified to lie on or near the crosshair that is normal to the direction of the lower polar. Nearly every grain is a candidate because of the excellent prismatic cleavage in this orthorhombic mineral; the γ index is observable when the grain shows a faint green pleochroic color rather than faint pink.

The precise determination of mafic minerals by the dispersion method is enhanced if observations are made on a number of grains. Berg and Morse (1981) note that a precision of 1–2 mole percent is attainable on good material in which the natural variation is not extreme when 10 grains per sample are measured. If such abundant data are to be taken for more than a few samples, it will be helpful to reduce the data by computer. In any case, a statement about the composition of each grain and the standard (or mean) deviation and range in compositions will afford potentially important information on homogeneity or compositional zoning within the sample. The results have a precision and utility comparable to that of the electron microprobe for the Mg ratio.

The dispersion method can be adapted to any binary solid solution for which the dispersion is known or can be compiled from the literature or determined by the double liquid method. Among common mafic minerals that should lend themselves to such routine determination are the amphibole series such as anthophyllite-gedrite, cummingtonite, actinolite, glaucophane, arfvedsonite, and pargasite. The Tschermaki-

REFRACTIVE INDEX WORKSHEET -- MAFIC MINERALS

SAMPLE NO.__________

	BETA β (OL,AUG)	GAMMA γ (OPX)
Oil:	______	______
Shuttle reading	______ mm	______ mm
Lambda match = λ_0	______ nm	______ nm
k_D, from table	______	______
t match	______ °C	______ °C
dn/dt (bottle)	–______E-4	–______E-4
$n_D^L = n_{25} + [(t-25)(dn/dt)]$	______	______
Δ^L(F-C) (bottle or graph)	______	______

Crystal:

Choose one set:

CONSTANTS:

	OL beta	OPX gamma	AUG beta
z	-0.12265	-0.07526	-0.09479
w	0.07911	0.05047	0.06236
n_0	1.8650	1.7886	1.7447
Δn	0.2140	0.1236	0.0686

(Berg & Morse '81 Am. Min.)

Find $n_D^S = [n_D^L + (\Delta^L - z)\cdot k_D] / (1 + w\cdot k_D)$ = 1.______

Find $X_{Mg} = (n_0 - n_D^S) / \Delta n$ = 0.______

Mineral Group ______________ Composition: (Fo)(En)__________

Observer ______________ Date ______________

Fig. 16-5. Worksheet for the determination of mafic minerals.

tic substitution $[AlAl][MgSi]_{-1}$ can be determined for common chlorites and the $MgFe_{-1}$ substitution for common biotites. In many mafic minerals such as garnet too many components vary to be determinable by refractive index alone, but within a given setting of common bulk composition, much of the variation of interest is the $MgFe_{-1}$ substitu-

REFRACTIVE INDEX WORKSHEET -- MAFIC MINERALS

SAMPLE NO. KI-4 Aug.

	BETA β (OL,AUG)	GAMMA γ (OPX)
Oil:	1.692	______
Shuttle reading	97 mm	______ mm
Lambda match = λ_0	548 nm	______ nm
k_D, from table	0.2338	______
t match	22.5 °C	______ °C
dn/dt (bottle)	−4.5 E-4	−____ E-4
$n_D^L = n_{25} + [(t-25)(dn/dt)]$	1.6931_3	______
Δ^L(F-C) (bottle or graph)	0.0356	______

Crystal:

Choose one set:

CONSTANTS:

	OL beta	OPX gamma	AUG beta
z	-0.12265	-0.07526	-0.09479
w	0.07911	0.05047	0.06236
n_0	1.8650	1.7886	1.7447
Δn	0.2140	0.1236	0.0686

(Berg & Morse '81 Am. Min.)

Find $n_D^S = [n_D^L + (\Delta^L - z) \cdot k_D] / (1 + w \cdot k_D)$ = 1.6988

Find $X_{Mg} = (n_0 - n_D^S) / \Delta n$ = 0.668

Mineral Group augite

Composition: (~~Fo~~)(En) 66.8

Observer Morse

Date 14 Nov. 83

Fig. 16-6. Worked example of the determination of a mafic mineral using the worksheet of Fig. 16-5.

tion, for which the dispersion method could furnish a rapid relative determination.

Precision

A 1983 test at the University of Massachusetts involved eight students, each of whom determined an olivine and an orthopyroxene by the dispersion method. The consensus for the olivine was $\beta = 1.6674$ with a standard deviation of 0.0010; the range was -0.0008 to $+0.0018$. The mineral composition was determined to be Fo 92.3±0.5. Clearly, the material is quite homogeneous and most individual workers succeeded in reproducing the consensus refractive index within 0.001. For the orthopyroxene, the consensus for gamma was $\gamma = 1.7157$, with a standard deviation of 0.002 and a range of -0.0035 to $+0.0022$. The composition was determined to be En 59.0±1.5. It appears likely that there is some real grain-to-grain variation in this sample at the mole percent level.

APPLICATION TO PLAGIOCLASE

The plagioclase feldspars display the substitution $[CaAl][NaSi]_{-1}$ and the conventional measure of composition is the mole fraction of anorthite, conveniently abbreviated

$$X_{An} = \frac{Ca}{(Ca+Na)} \tag{16-29}$$

The value of the low (α') refractive index on (001) cleavage flakes (Tsuboi, 1923) provides the best measure of composition for low (i.e., plutonic, slowly cooled) plagioclase. Revised and tested refractive index curves are summarized by Morse (1978) as follows:

For		Δn_D	n_D^{Ab}
X_{An} 0.0–0.24	$n_D =$	0.05165 X_{An} +	1.5287
0.24–0.31		0.05587	1.5277
0.31–0.84		0.05144	1.5290
0.84–1.0		0.04688	1.5328

(16-30)

A plagioclase dispersion chart was constructed by Tsuboi (1934) and later by Morse (1968; see also Tröger, 1979, p. 128). Dispersion curves of calibrated oils can be drawn on an overlay and shifted up and down to adjust for temperature. When the intersection of the oil and matching wavelength is found, the plagioclase composition can be read off the chart directly.

An algebraic solution to the problem will often be preferable. A review of Tsuboi's (1923) dispersion determinations suggests that the dispersion rises with An as

$$10^4\,(n_F - n_C) = 83 + 21X_{An}\,, \tag{16-31}$$

or

$$\Delta^S = (n_F - n_C) = 0.0083 + 0.0021 \cdot X_{An}\,. \tag{16-32}$$

Anorthite is the high-index end member of the plagioclase series, and the determinative equation has the form

$$X_{An} = \frac{(n_D^S - n_D^{Ab})}{\Delta n_D} \tag{16-33}$$

Substituting equation 16-32 into equation 16-21, we have

$$n_D^S = n_D^L + (\Delta^L - 0.0083 - 0.0021 \cdot X_{An}) \cdot k_D\,. \tag{16-34}$$

Substituting equation 16-34 into 16-33 and rearranging, we have the complete determinative equation using the dispersion method (Su, 1994):

$$X_{An} = \frac{[n_D^L + (\Delta^L - 0.0083) \cdot k_D] - n_D^{Ab}}{(\Delta n_D + 0.0021 \cdot k_D)}\,. \tag{16-35}$$

Procedure

A well-calibrated set of plagioclase dispersion oils in the range 1.530–1.578 in steps of 0.006 is effective for the determination of plagioclase by the dispersion method. These can be calibrated individually or obtained commercially as Cargille Plagioclase Dispersion Set 547. If nothing is known about the mineral composition, an oil near the middle of the set should be chosen at the start. As soon as a nearly matching oil is found, a preliminary estimate of the plagioclase composition can be made from equation 16-30. The wavelength of match and the temperature are recorded, and the calculation of the result is then straightforward using the worksheet, Fig. 16-7.

A worked example is shown in Fig. 16-8. The calculation is made using the mean result from the observation of 17 grains. The D-line value of the refractive index can also be calculated using the equation near the bottom of the worksheet, as shown in the example. The use of 10 or more grains per mount not only increases the precision of the mean, but also provides useful information on the range of composition (Morse, 1978).

REFRACTIVE INDEX WORKSHEET -- PLAGIOCLASE

SAMPLE NO.__________

	ALPHA PRIME (001)	
Oil:	_______	
Shuttle reading	_______	mm
Lambda match $= \lambda_0$	_______	nm
k_D, from table	_______	
t match	_______	°C
dn/dt (bottle)	−_____	E-4
$n_D^L = n_{25}+[(t\text{-}25)(dn/dt)]$	_______	
Δ^L(F-C) (bottle or graph)	_______	

CONSTANTS

	An 0-24	An 24-31	An 31-84	An 84-100
n_D^{Ab}	1.5287	1.5277	1.5290	1.5328
Δn_D	0.05165	0.05587	0.05144	0.04688

Crystal:

Find $X_{An} = \{[n_D^L + (\Delta^L - 0.0083)\cdot k_D] - n_D^{Ab}\} / (\Delta n_D + 0.0021\cdot k_D)$

X_{An} = 0.______

And if desired find $n_D^S = n_D^L + [\Delta^L - 0.0083 - (0.0021\cdot X_{An})]\cdot k_D$ = 1.______

Composition, An____________

Observer ________________ Date____________________

Fig. 16-7. Worksheet for the determination of plagioclase by the dispersion method.

REFRACTIVE INDEX WORKSHEET -- PLAGIOCLASE

SAMPLE NO. PG-721

	ALPHA PRIME (001)
Oil:	1.554
Shuttle reading	93.3 mm (N=17)
Lambda match = λ_0	539 nm
k_D, from table	0.2923
t match	24.5 °C
dn/dt (bottle)	– 4.1 E-4
$n_D^L = n_{25}+[(t\text{-}25)(dn/dt)]$	1.55421
Δ^L(F-C) (bottle or graph)	0.01764

CONSTANTS

	An 0-24	An 24-31	An 31-84	An 84-100
n_D^{Ab}	1.5287	1.5277	1.5290	1.5328
Δn_D	0.05165	0.05587	0.05144	0.04688

Crystal:

Find $X_{An} = \{[n_D^L + (\Delta^L - 0.0083)\cdot k_D] - n_D^{Ab}\} / (\Delta n_D + 0.0021\cdot k_D)$

X_{An} = 0.537

And if desired find $n_D^S = n_D^L + [\Delta^L - 0.0083 - (0.0021\cdot X_{An})]\cdot k_D$ = 1.5566

Composition, An 53.7

Observer Morse Date 5 Dec. '83 / 19 Jan '88

Fig. 16-8. Worked example of the determination of plagioclase using the worksheet of Fig. 16-7.

Precision

A test with six students in 1983 at the University of Massachusetts on a labradorite sample gave a consensus composition of An 54.7±0.2, with a range from −0.2 to +0.3. The implied standard deviation of the α' refractive index is 0.0001. One would not expect such precision among learning practitioners every week in the year, but it evidently can occur.

17 Crystal Identification Recipes

INTRODUCTION

We treat first the identification of crystals in grain mounts by the immersion method, then of whole rocks crushed in grain mounts, and then an overview of thin section procedures.

IMMERSION METHOD

Principles

The name of the game with the immersion method is to find a match (Fig. 17-1) between the crystal and a liquid of known refractive index. The criterion of match with white light is an orange-red color fringe on one side of the grain that resembles the fluorescent orange-red of certain traffic pylons (Fig. 17-2), here called pylon color. The desired color is not as red as a ripe cherry tomato. The refractive index on the label of the bottle is taken as the index of the matching crystal. (It is expected that the oils rise with step 0.002 until 1.7, and after that step 0.005.) Other optical properties are also noted. The crystal is identified

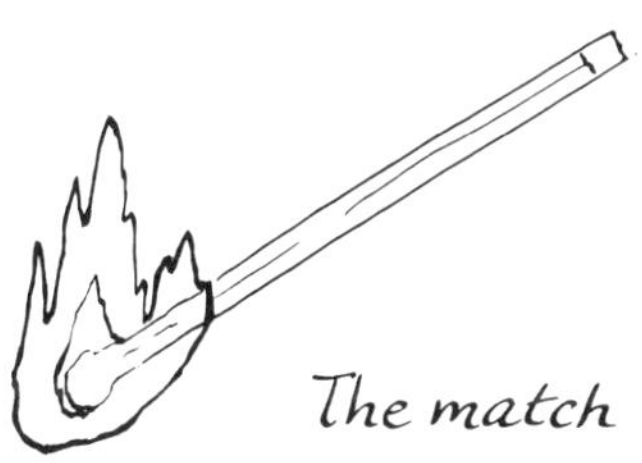

Fig. 17-1. Match.

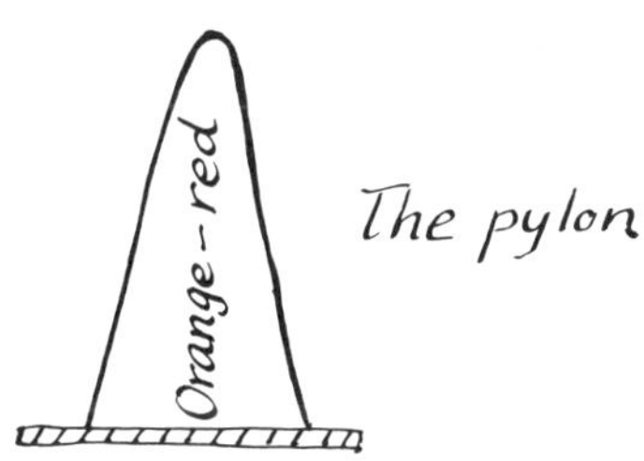

Fig. 17-2. Pylon.

by reference to a table, such as Tröger (1979) or USGS 1627 (1984), organized by the value of the refractive index.

Illumination by white light is assumed. At any stage of the procedure, a transfer to monochromatic light may be worthwhile in order to determine a refractive index that does not match the oil in the mount at the D line, but does so for some other wavelength of light. Worksheets for such investigations are given in Chapter 16. It is particularly valuable to use the monochromator when a limiting refractive index is sought and few suitably oriented grains are available, but one is discovered.

Procedure

1. *Sample Preparation.* If the sample is already prepared, go to step 2.

   Sample Preparation. Crush the sample in a pin mortar or between two steel plates until enough fine material is gained. Pour the crush onto a piece of paper and start shaking it off onto another piece. When the right grain size (fine sand to silt) starts coming off, collect it on a clean piece of paper. Make a funnel of this and pour the sample into a labelled vial or other container for safe keeping. Merge the fines with the coarse crush and save until the work is finished. If time permits and a nice range of sizes is wanted, sieve the working fraction to 80–200 mesh (180–75 micrometers). Better yet, wash in a beaker of clean water, decant fines, flush with alcohol or acetone, and dry the sample on an absorbent filter paper in the air stream of a fume hood. Such washing will remove unwanted dust.

2. *Slide Preparation.* Tap a few grains from the vial onto a clean glass slide (27 × 46 mm recommended). With tweezers, place a small cover slip (8 × 8 mm recommended) over the grains.
3. *Oil Immersion.* Choose an oil near n = 1.55 and introduce a small drop near the edge of the cover glass so that the drop will run under by capillary attraction. DO NOT OVERFILL. One may stir the mount gently with the edge of a cork or pencil eraser pressed very lightly ON TOP of the cover glass.
4. *Test for Anisotropy.* Examine with low power objective lens to focus and test for anisotropy.
   A. Focus grains (if you can't see them do [B] first)
   B. Cross the polars.
      1. If grains are dark record ISO and go to step 5.
      2. If you see bright or colored grains record ANISO and go to step 7.

5. It is an **ISOTROPIC CRYSTAL.** Examine grain with 10× objective. **GO TO TABLE 17-1.** Make as many of the tests as appropriate to be sure of the result.
   A. If the crystal is HIGH, choose a higher oil and go to (C).
   B. If the crystal is LOW, choose a lower oil.
   C. Write down the name (index value) of the oil decided upon
   D. Make a new mount with that oil.
   E. Repeat step 5 until a match is found.

Table 17-1
Refractive Index Tests by the Immersion Method

1. Focus grain with 10× objective.
2. Stop down illumination with substage diaphragm.
3. Determine by one or more tests whether the crystal is high or low with respect to the oil.

FOR BECKE LINE:

4. RAISE the focus (lower the stage:)
   Becke line **OUT**, crystal **LOW**
   Becke line **IN**, crystal **HI**

FOR OBLIQUE ILLUMINATION:

5. SHADOW the field with frame of accessory plate or analyzer
   Shadow **OPP**osite, crystal **LOW**
   Shadow **NEAR**, crystal **HIGH**

FOR COLOR FRINGES:

6. SHADOW the field with frame of accessory plate or analyzer. Observe the color fringe that is not blue.
   Color fringe **RED,** crystal **LOW**
   Color fringe **OR**ange or
   YEellow, crystal **HI**
   Color fringe **PYLON COLOR**, crystal **ON**.

For more details refer to Fig. 17-3, which shows the estimation of match from the color fringe.

Strategy. The method of the exercise is illustrated in Fig. 17-4. The actual refractive index of the crystal falls at the heavy line. Lighter lines represent oils. The first oil tried was (1). The result was XL LOW. Another, much lower, oil was chosen (2). For oil (2), it is important to try going beyond the index of the crystal so as to avoid false steps. For (2), the result was XL HI. Oil (3) was chosen HALFWAY BETWEEN oils (1) and (2), for maximum efficiency in searching for a match. The result was XL LO. Oil (4) was then chosen halfway between oils (2) and (3). The result was XL HI. Oil (5) was chosen halfway between (3) and (4). The result was STILL XL HI. The only place left to go was oil (6), halfway between (3) and (5). In this a match was found (Fig. 17-4).

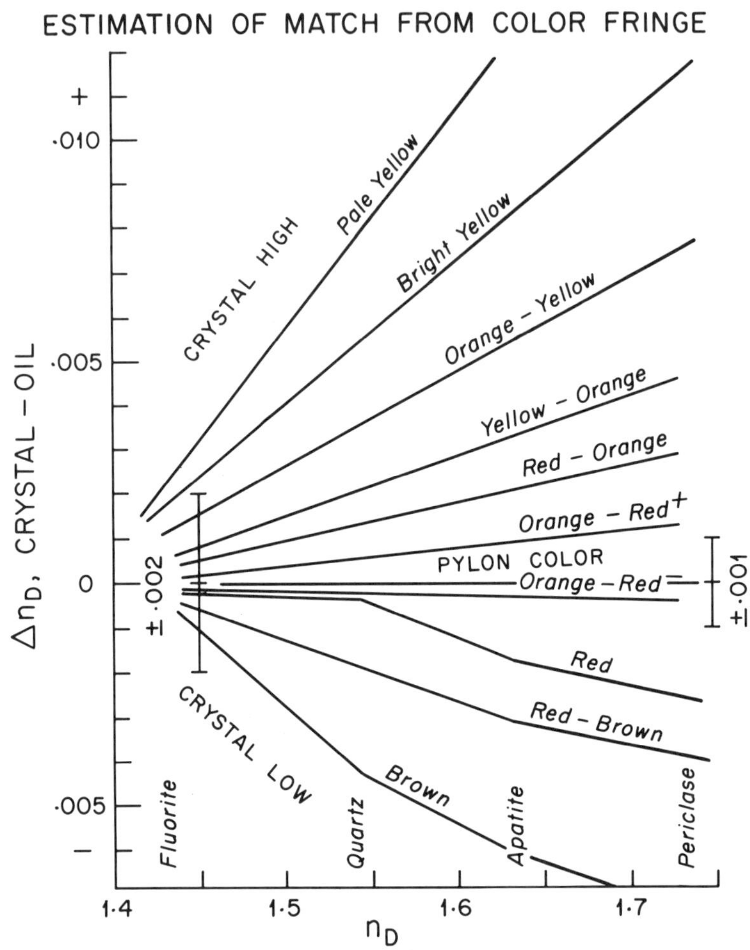

Fig. 17-3. Variation of color fringe with refractive index.

With greater experience, some shortcuts are possible. In oil (3), the color fringe was red-brown and the index was ~1.56. From Fig. 17-3, this result suggests that the crystal is low by only about 0.002. An experienced operator would therefore go directly to oil (6), assuming that the oils in Fig. 17-4 are 0.002 apart. A novice operator may get burned trying this direct route.

It is important to make a new mount as soon as a decision has been reached about the old one. GET DATA. AVOID RAPTURES OF THE DEEP.

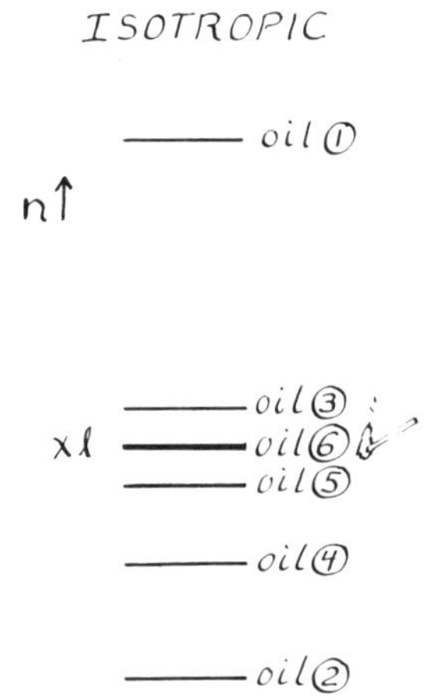

Fig. 17-4. Stepwise approach of oils used to the actual refractive index of the crystal. Vertical scale is refractive index.

6. Use refractive index to enter a table such as the endpapers of Tröger to identify the crystal. Use any other properties noted to exclude other candidates and to confirm the identification. Write down the name of the crystal or mineral, and its chemical formula.
7. It is an **ANISOTROPIC CRYSTAL.** A range of refractive index exists, and the upper and lower limits must be determined. These limits are specific refractive indices related to the direction of vibration of polarized light in the crystal. If the crystal is uniaxial, no special intermediate values exist and all intermediate values are called epsilon prime (ϵ'). If the crystal is biaxial, a special intermediate value exists and is called beta (β). One proceeds on the assumption that the crystal is uniaxial until more evidence arrives. For this reason, one determines first the limiting values of refractive index, leaving β (if there is one) till last. THE CRYSTAL IS UNIAXIAL UNTIL PROVEN OTHERWISE. REPEAT STEP 5 UNTIL IN SOME OIL CONFLICTING RESULTS ARE OBTAINED: SOME GRAINS ARE XL LO, OTHERS ARE XL HI. THEN:

A. Pick either the high or low limit and determine it separately as though it were the only value to determine.

At the high limit will be seen:
XL ON AND OTHERS XL LOW; NONE XL HI. Value is n_h.
At the low limit will be seen:

XL ON AND OTHERS XL HI; NONE XL LOW. Value is n_1

THE GRAIN MUST BE AT EXTINCTION DURING ALL SUCH OBSERVATIONS

B. If either limiting index is seen in every grain at some position of stage rotation, call it omega (ω) and assume that the mineral is uniaxial. Assume that the other index is ϵ. When its limit is found, take an interference figure. The figure should be a uniaxial flash figure. If it is not, the limiting value of the refractive index has not been reached, or the crystal is biaxial. If one limiting refractive index does not appear in every grain, the crystal is biaxial: **GO TO STEP 8.**

C. Determine the uniaxial sign from the identity of the refractive indices as determined in (B):
   (+) if $\epsilon > \omega$,
   (−) if $\epsilon < \omega$.
   OR determine the sign using a gypsum (GIPS; 1° red) plate on an interference figure, either the
   flash figure (see Fig. 9-2) or the
   optic axis figure (see Fig. 9-3).

D. Note any other properties and enter the tables to identify the crystal. Write down the name and formula.

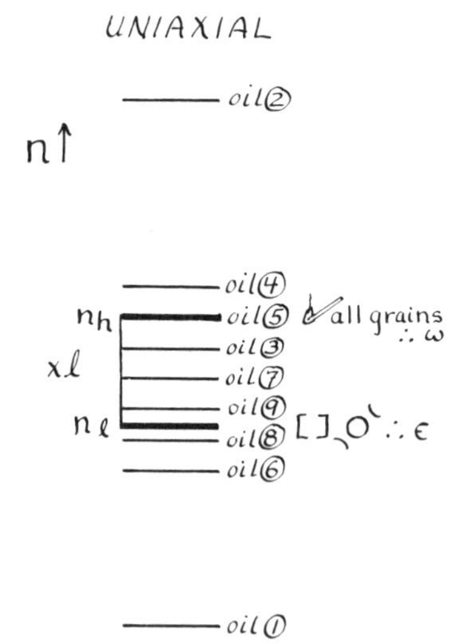

Fig. 17-5. Stepwise approach to the refractive indices (heavy lines) of a uniaxial crystal.

Strategy. An example of **UNIAXIAL** identification is shown in Fig. 17-5. The principles are the same as with isotropic crystals, Fig. 17-4. The limiting refractive indices, n_h and n_l, are shown as thick bars. The first oil showed XL HI and oil (2) showed XL LOW. Oil (3), halfway between the other two, landed by chance within the range between the limiting values. Both red and yellow color fringes were seen. The yellow ones seemed more common, so the decision was made to find n_h first, especially as it was nearby. Oil (4) was chosen (based on Fig. 17-3) on purpose to overshoot the upper refractive index, in order to bracket the value as closely as possible. Oil (5) then landed right on the upper limit and showed a pylon color fringe showing a match **IN A GRAIN AT EXTINCTION**. On further examination, it turned out that every grain matched oil (5) at some rotation position of the stage (**AT ONE EXTINCTION POSITION**). Of the two uniaxial indices, ϵ and ω, this could only be true for the ordinary index ω.

Oil (6) was chosen halfway between (5) and (1) in a search for n_l. The remaining distances were halved, as nearly as permitted by the fixed interval of the oils, by oils (7), (8), and (9). None yielded a match, but the limiting index was XL HI in oil (8) and XL LOW in oil (9), so its value lay between the two, which formed a **bracket [ ]** instead of a match. In each of the two oils, the grain giving the critical signal was found to show a flash figure, so the limiting value of ϵ was indeed found, not some intermediate ϵ'.

In oil (8), the result was: ALL XL HI, SOME VERY NEAR (orange color fringe) and SOME FAR (yellow fringe). In oil (9), the result was: MOST XL HI; A FEW XL LOW, RED COLOR FRINGE.

8. It is a **BIAXIAL CRYSTAL.** You have reached step 8 after determining the high and low limits of the range of refractive index. **ALL INDICES MUST BE FOUND AT AN EXTINCTION POSITION.**

   A. Now name these limits:

   $n_h = \gamma$
   $n_l = \alpha.$

   B. For each limit, confirm that a same-rotating one-symmetry plane (SR1SP) interference figure occurs in the grain showing a match. This figure shows an isogyre coincident with the analyzer direction at the stage rotation position where the desired index is being observed in plane light. When the stage is then rotated a little bit to the right, some part of the isogyre rotates a little bit to the right. Such an isogyre may well be part of a figure of higher symmetry such as a Bxa or Bxo, but the only requirement is that a single symmetry plane be parallel to the analyzer, so that the desired indicatrix axis (normal to the symmetry plane) is parallel to the polarizer.

9. Determine the optic sign and estimate the optic angle from one or more interference figures.
   For the **optic sign** refer to Figs. 12-2 or 12-3.
   For the **optic angle** (2V) refer to Fig. 11-15 or 12-4.
10. Estimate the value of beta from the following considerations.
    If the 2V is large, near 90°, β lies about halfway between α and γ.
    If the 2V is small, β lies near α if XL (+) and near γ if XL (−)
    If the 2V is intermediate, β lies correspondingly closer to the midpoint between the limiting indices.
    More precisely,
    Let $B = \gamma - \alpha$
    Let V_X be the half optic angle about X

$$\text{Then } \beta \approx \gamma - (B \sin^2 V_X) \tag{17-1}$$

IT IS VITAL TO ESTIMATE BETA BECAUSE OTHERWISE EFFORT WILL BE WASTED IN OILS THAT DO NOT MATCH BETA.

11. Seek β in oils near its estimated value.
    A. LOOK FOR THE PYLON COLOR FRINGE.
    B. TEST THE GRAIN FOR EXTINCTION.
       If not at extinction, put it there and return to (A)
       If grain remains at extinction on rotation of the stage, the index is β at all rotation positions
    C. TEST THE GRAIN FOR CR1SP, the counter-rotating one-symmetry plane figure. Only grains showing this figure have the β index; the figure, however, may be part of another figure (Bxa or Bxo) of higher symmetry.
    D. Repeat as necessary until a β is confirmed.

12. Calculate 2V from the refractive indices from equation 10-9 and see that it is consistent with observed 2V and sign.
13. Record data on cleavage forms (flat tops, straight edges) and extinction angles. State probable crystal system and optic orientation if possible.
14. Enter tables with value of β and identify crystal.

Strategy. An example of **BIAXIAL** identification is shown in Fig. 17-6. The heavy bars are the three principal refractive indices.

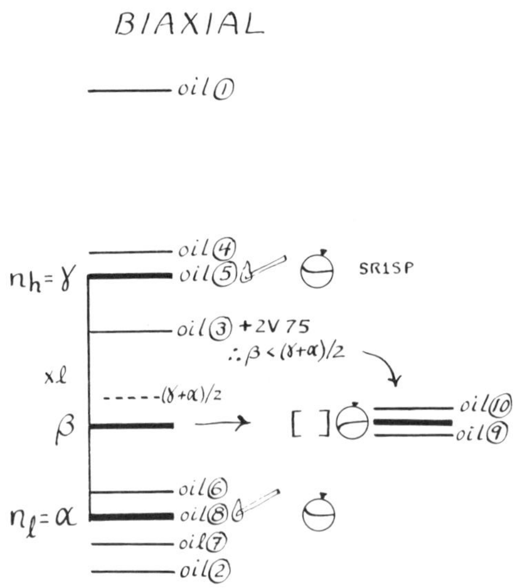

Fig. 17-6. Stepwise approach to refractive indices of a biaxial crystal.

ALL OBSERVATIONS ARE TO BE MADE WITH THE GRAIN AT EXTINCTION. AVOID REFLEXIVE ROTATION OF THE STAGE; THINK FIRST; YOU MAY BE STOPPING THE EXPERIMENT.

Oil (1) showed XL LOW. Oil (2) showed XL HI but near for some grains. Oil (3), halfway between the latter two, showed some yellow color fringes (XL HI) and many opposite shadows (XL LOW). A Bxa figure showed a rotation angle of 12°, suggesting 2V ≈ 75° from Fig. 11-15. When the isogyres had left the field in the NE–SW quadrants, it was known that the Bxo lay in these quadrants. A GIPS showed colors falling in the interference field and in the grain viewed orthoscopically, therefore Bxo = X, therefore Bxa = Z, crystal (+). This information will be needed later. Oil (4) showed some red color fringes (XL LOW but near) and many opposite shadows. The quality of the color fringe suggested that the high index of the crystal was only low by about 0.002, so oil (5) was chosen and a match was found. The matching grain was found to have a SR1SP interference figure, confirming that the index matching the oil with a pylon color fringe was γ and not γ'.

A search was then made for α, as recorded by oils (6) through (8).

Again, a pylon match was found and confirmed as α by means of a SR1SP figure.

Beta was estimated from the sign, 2V, and B = ($\gamma - \alpha$). From inspection, it could be stipulated that β lay not far below the midpoint of the range between γ and α, as the 2V was not far from 90°. The β index was then soon bracketed between oils (9) and (10). CR1SP figures were seen in both oils, verifying that the β index was being observed.

DISPERSION METHOD

Often a grain is found with a special refractive index indicated by a 1-SP interference figure, but the oil does not exactly match the crystal at the D line. If it shows a color fringe, however, the chances are good that a match can be found for some wavelength of visible light. In that case, a monochromator should be introduced over the white light source, and the wavelength of match found, without disturbing the crystal. Only two measurements are needed, the wavelength of match and the stage temperature. Some further data will be needed from the bottle or from tables. The data reduction can be made with the aid of a worksheet, Fig. 16-4. Worked examples using the dispersion method are shown at the ends of Chapters 4, 9, and 12, and in Chapter 16.

CRUSHED ROCKS

Introduction

Crystal identification in samples of crushed rock is a speedy and precise method of determining the mineralogy and mode (percentage of minerals) of the rock, and it can even be done in the field (Jaffe, 1993). It is a valuable supplement to examination in thin section; first because it gives better estimates of mineral composition, and second, because it often reveals trace minerals that may escape notice in, or be absent from, a thin section. The thin section, on the other hand, reveals textures and orientations not available in grain mounts. Each method serves its special purpose. Speed is the essence of the immersion method. A skilled operator can determine the essential mineralogy of a rock in 10–15 minutes, about the time it takes to saw, clean, and begin to dry the chip for a thin section.

Often a suite of rocks is to be sectioned, or crushed for chemical analysis. In such a case, it is well worthwhile to grab an aliquant, such as the first rough slice of the rock or a suitably fine crush, for examination with the immersion method. The effective volume thus sampled may be far greater than that sampled by the thin section, and comparable to that sampled for analysis. Trace minerals may thereby be captured.

For rough and ready examination, it suffices to knock a small piece off the rock and process it by hand, as for minerals in step 1 of the immersion method. For more elaborate studies, separation of felsic and mafic minerals in heavy liquids, followed by magnetic separation, may be helpful.

Procedure

The examination of rocks proceeds in a very different way from routine crystal identification of a complete unknown. The general nature of a rock is usually known from the hand specimen, and certain minerals are regularly expected. Accordingly, the strategy is to set a series of **traps** for these common minerals, and to detect their presence or absence in the trap by the presence or absence of color fringes. Exhaustive documentation is seldom necessary; one or two additional tests will usually suffice to confirm the identity of an anticipated mineral caught in the trap.

The Traps

For **felsic rocks** the traps and intended captives are:

1.544 oil—ω of quartz
—α'_{001} composition of plagioclase, An25–40

1.524 oil—discriminates between lowest index of plagioclase (positive relief) and highest index of alkali feldspar (negative relief)

1.630 oil—catches apatite, nearly ubiquitous in rocks. Also catches biotite (Mg>Fe), tourmaline, anthophyllite, andalusite, and some chlorite.

Then if the need seems indicated from earlier observations:

1.720 oil—catches kyanite, reveals staurolite and other high index minerals

1.8 oil—if garnet suspected

1.92 oil—catches zircon, sphene (titanite).

For **mafic rocks** omit the 1.524 oil unless orthoclase is suspected and substitute:

1.554 oil—gives plagioclase composition in range An40–60, hence allows classification of rock.

1.690 oil—catches intermediate to magnesian olivine, orthopyroxene, augite, and some amphiboles.

The indicated traps are not expected to catch all the minerals in the rock, but unexpected ones will be revealed during examination in these oils because they will not match the oil, or will look different from the common ones seen. Then they can be chased down individually.

Experience and common sense will suggest variations on the procedure and the specific traps chosen, depending on the rocks being studied and their setting. The main idea is to tailor the search to the purpose, and not to start from scratch on an exhaustive quest. Another main principle is to work from known associations: if alkalic minerals are found, look for nepheline. If metamorphic minerals are found, look for the common index minerals for that grade of metamorphism.

Examples

A sample of *Kinsman granite* from New Hampshire turned up the following minerals:

Quartz, $\omega = 1.544$, conchoidal fracture, uniaxial (+) figure.

Microcline, α 1.519, near Or 90 (equation 14-3), plaid twinning.

Plagioclase, α'_{001} 1.548, An 36, albite twinning.

Biotite, $\gamma \approx \beta$ 1.622, near Mg 70, pleochroic in browns.

Sillimanite, $n \approx 1.66$, parallel extinction, (+) elongation.

Muscovite, $n \approx 1.65$–1.658, (−) 2V ~ 40°. Looks nothing like mica seen in thin section because the perfect cleavage flakes have low birefringence.

Garnet, $n \approx 1.8$, conchoidal fracture, faint but distinctive pink color.

Opaque, bladed: probably ilmenite.

Apatite, color fringes seen on a single grain too small to see, in 1.630 oil.

Unknown, possibly Mn-andalusite. Needs more work.

A sample of *Gassetts schist* from Vermont in immersion mounts showed the following:

Quartz
Muscovite
Staurolite (distinctive yellow pleochroism)
Garnet
Kyanite, n ≈ 1.72, inclined extinction
Chlorite (pale green)
Biotite (brown)
Tourmaline
Bladed opaque, ilmenite assumed.

However, the survey missed two known minerals in the rock. One is chloritoid, seen in thin section only as rare, minute inclusions in garnet. The other is paragonite, the sodium white mica occurring on the other side of the solvus from muscovite. The two white micas in this rock are normally distinguished only with the help of X-ray diffraction or electron microprobe analysis.

These two examples contain more minerals than most rocks. One would usually expect to find a somewhat simpler mineralogy. The examples do indicate the kind of success that can be achieved with crushed rocks.

An *orthopyroxene megacryst* from Labrador, with abundant, oriented exsolution lamellae of plagioclase, was separated in heavy liquids (bromoform, $\rho = 2.8$, and methylene iodide, $\rho = 3.2$) and a magnetic (Frantz Isodynamic) separator; the fractions were examined under the binocular microscope and in oil immersion. The complete list of minerals seen in this previously homogeneous but now exsolved single crystal is as follows:

Hypersthene, ≈ En 50
Plagioclase, bytownite, ≈ An 75
Augite
Olivine
Apatite
Biotite
Magnetite (hand magnet)
Ilmenite (magnetic separator)
Pyrite.

This list of exsolution products vividly justifies the frequent allegation that pyroxenes are really just rocks in disguise.

THIN SECTIONS

Introduction

Crystal identification in thin section proceeds very differently from the immersion procedure. The starting point, nevertheless, is refractive index. Because the only reference points are the mounting medium, usually near 1.54, and the other minerals (once they are known), the characterization of refractive index is centered on **relief**. Minerals are characterized as having high, intermediate, or low relief. If the relief is caused by indices above the mounting medium, it is called positive, and if below, negative. Relative relief may also be a useful property, as for example in distinguishing olivine by its higher relief relative to pyroxene, or orthoclase by its lower relief relative to albite.

Cleavages normal to the thin section appear as fine cracks, and those inclined to the section as broader or fuzzy cracks. Thin sections strained during grinding may have excessive cleavage cracks. Some cleavages are distinctive: the single good cleavage seen in end sections of prismatic sillimanite serves as a good example (Fig. 11-2).

Practice

Certain optical effects are unique to thin sections. Among these is the **shagreen surface** seen especially on olivine and apatite. It is a finely pock-marked effect (Fig. 17-7) caused by plucking during grinding and revealed because of the high refractive index of the material. It is absent in nicely polished thin sections, which makes them difficult to interpret at first (Fig. 17-8). The shagreen surface is so distinctive that it is one of the major clues to these minerals. Another distinctive effect of grinding is seen in micas, where it is called the **birdseye maple effect**. People unfortunate enough not to have seen birdseye maple gunstocks or furniture are not stirred by this description, so the well-equipped laboratory needs an example. The effect is a telltale for the presence of micas. Another useful characteristic of minerals with high birefringence such as micas and particularly carbonates is **differential relief** caused by the great range in refractive index as the stage is turned.

Recognition of alkali feldspar without grid twinning is a learned trick in thin section. Staining (Hutchison, 1974) may be done to aid in point counting, but with practice, K feldspar can be recognized if the substage diaphragm is partly closed and the focus slightly raised. The low relief of the K feldspar, combined with a faint pinkish tinge, suffice to make it easily seen once the eye becomes adjusted to subtleties in color.

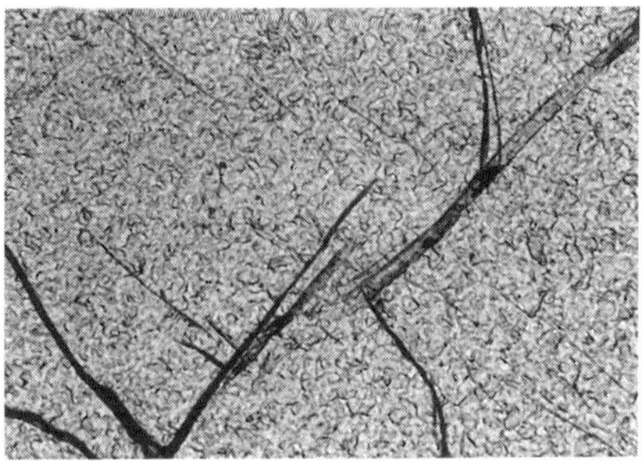

Fig. 17-7. Shagreen surface on olivine.

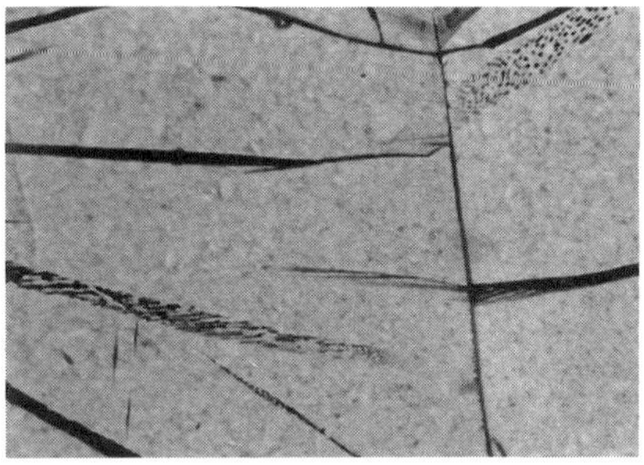

Fig. 17-8. Doubly polished thin section of olivine showing the lack of a shagreen surface.

When the famous analytical geochemist E.G. Zies was 80, wearing glasses as thick as bottle bases, he called a younger associate over to a lab bench where there were two Erlenmeyer flasks sitting on white paper. He asked the observer to describe the color of the liquid in each flask. The answer was "clear" for both. A second opinion was requested. Longer study was urged. At last, a difference was hesitantly proclaimed. But then, the question was, what was the difference? One appeared a little gray, perhaps. "Oh, no," was the answer. "That is a faint pinkish lavender. That is the end point of a permanganate titration for ferrous iron. If it looks more colored than that, the end point has been passed." The moral of this story is that most of us can see subtler colors than we think we can, through practice.

Extinction angles are often diagnostic in thin section. For example, the calcic pyroxenes commonly have extinction angles greater than 40°, whereas amphiboles rarely have extinction angles greater than 20°. Likewise, the prismatic cleavage angles in sections normal to *c* are diagnostic between pyroxenes (88°–92°) and amphiboles (124°–56°). These are often not seen in immersion because of the poor basal cleavage.

Color and pleochroism play an important role in the identification of mafic minerals. These vary with the orientation of the cut, and may be misleading until one learns to associate the various colors and shades seen in a single mineral species.

Birefringence and interference colors may also be confusing at first in thin section because of varied orientation. Olivine grains cut normal to an optic axis are almost dark under crossed polars, but never go quite to extinction in white light. Grains cut parallel to the optic plane show bright interference colors. The clue that links these grains with very unlike interference colors is the shagreen surface and the common appearance in plane light. **The foremost rule of thin section study is: In plane light, grains that look alike probably are alike.** They are probably the same species if they look similar, have similar pleochroic colors at one rotation position or another, and have the same relief. This rule notwithstanding, different grains of the same mineral can look different, for example when one is cut normal to an absorption axis and fails to show the range of pleochroic colors that occur for other orientations.

As an observer becomes practiced, more and more minerals are recognized by sight and confirmed by tests. Sight recognition is a desirable goal, as long as it is backed up by sufficiently rigorous tests.

The more common rock-forming minerals are listed in Appendix A according to their optical properties in thin section. The table is arranged according to relief. For more information and visual aids, various compilations are available. Among the most useful are Philpotts (1989) and the atlases by MacKenzie and Guilford (1980), and MacKenzie et al. (1982). The latter contain stunning and instructive color photographs.

Appendix A

Optical Properties of Common Rock-Forming Minerals

Optical Properties of Common Rock-Forming Minerals

J. B. Lyons, S. A. Morse, and R. E. Stoiber

Mineral	Chemical Composition	Xl. System and Best Cleavage	Sign, 2V	Indices and Relief	Birefringence and Color	Distinguishing Characteristics [a]Characteristically parallel, but see Fig. 13-3.
			A. High Positive Relief			
Zircon	$ZrSiO_4$	Tet.	(+)	ω=1.940	High biref. (.055)	Small euhedral grains show parallel[a] extinction; may cause pleochroic haloes if enclosed in other minerals
Sphene (Titanite)	$CaTiSiO_5$	Mon. (110)	(+) 30–50	β=1.895 to 1.935	High biref. (0.108–.135) Often brownish in color	Wedge-shaped grains; may show (110) cleavage or (100) or (221) parting; $Z \wedge c = 51°$; very high relief; $r > v$ extreme.
Garnet	$A_3B_2(SiO_4)_3$ where $A = R^{2+}$ and $B = R^{3+}$	Iso.		High 1.7–1.9	Grandite often weakly birefracting.	Very pale pink commonest color; inclusions common. Indices vary widely with composition. Crystals often euhedral. Uvarovite green, very rare.
Staurolite	$H_2FeAl_4Si_2O_{12}$ (approximately)	Orth. (010)	(+) 2V = 87	β=1.750	Low biref. (.012)	Pleochroic colorless to golden yellow; one good cleavage; twins cruciform or oblique; metamorphic.
Olivine Series	Mg_2SiO_4 to Fe_2SiO_4	Orth. Orth.	(+) 2V=85 (−) 2V=47 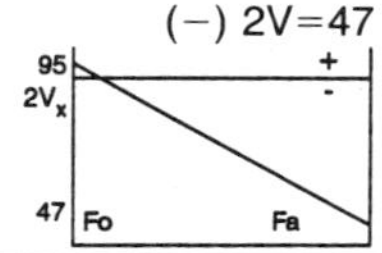	β=1.651 to β=1.865	High biref. (.035) High biref. (.051)	Colorless (Fo) to yellow or pale brown (Fa); high relief. Shagreen (mottled) surface; often cracked and altered to serpentine. Poor (010) and (100) cleavages. Extinction parallel.[a]

Tourmaline	$Na(Mg,Fe,Mn,Li,Al)_3$-$[Si_6O_{18}](BO_3)_3$-$(OH,F)_4$	Hex.	(−)	ω=1.636 to 1.698	Mod. biref. (.020)	Generally pleochroic; *maximum absorption across the length*. Colors chiefly brown, green, blue. Triangular outline on basal section.
Amphibole		(110)				Characteristic 124°–56° cleavage. Generally pleochroic.
Anthophyllite-Gedrite Cummingtonite	$(MgFe)_7(SiAl)_8O_{22}(OH)_2$ + Na in gedrite Same but no Na	Orth. Mon.	(+) and (−) 2V=80± same	β=1.605–1.689 1.65–71	Biref. mod. to low (.013 to .026)	Z=*c*; Y=*b*; may be pleochroic in tans; found only in metamorphic rocks; parallel[a] extinction.
Tremolite	$Ca_2Mg_5Si_8O_{22}(OH)_2$	Mon.	(−) 2V = 85	β=1.615	Biref. mod. to low (.026)	Z∧*c*=20°; colorless; a metamorphic mineral.
Actinolite	$Ca_2(MgFe)_5Si_8O_{22}(OH)_2$	Mon.	(−) 2V = 75	β=1.642	Mod. (.020)	Z∧*c*=15°; pleochroic yellow to green.
Hornblende	Complex Al amphibole	Mon.	(−) 2V near 80	β=1.63–1.7	Mod. (.020)	Z∧*c*=15°–25°; pleochroic yellow, green, brown.
Arfvedsonite	Soda-rich amphibole	Mon.	2V large, (−)	β=1.69–1.70	Low (.005)	X∧*c*=20°; strong pleochroism in blues and greens; found chiefly in sodic igneous rocks.
Glaucophane	$Na_2Mg_3Al_2Si_8O_{22}(OH)_2$	Mon.	(−) 2V = 45	β=1.638	Mod. (.018)	Z∧*c*=5°; distinct lavender blue to blue-black pleochroism. A metamorphic mineral.
Sillimanite	Al_2SiO_5	Orth. (010)	(+) 2V = 30 r>v	β=1.658–1.670	Mod. (.020)	Y ⊥ single perfect cleavage; Z=c; may be in interlacing needles (fibrolite). Metamorphic rocks. End section:
Calcite	$CaCO_3$	Hex. $(10\bar{1}1)$	(−)	ω=1.658 ϵ=1.486	Extreme (.172)	Differential relief on stage rotation; one index far above 1.54, other below; rhombohedral cleavage; *often twinned*; told from dolomite etc. only by index.

Mineral	Chemical Composition	Xl. System and Best Cleavage	Sign, 2V	Indices and Relief	Birefringence and Color	Distinguishing Characteristics
Chloritoid	$H_2FeAl_2SiO_7$	Mon. (110)	(+) 2V 36–68	β=1.720–1.770	Low (.007–.016)	Pleochroic in greenish-black to gray. Epidote has better cleavage. X parallel to (010) parting.
Zoisite	$HCa_2Al_3Si_3O_{13}$	Orth. (010)	(+) 2V = 30±, r<v	β=1.700	Low (.006)	Abnormal blue interference colors; one cleavage; parallel[a] extinction; colorless.
Epidote	$H(CaFe)_2Al_3Si_3O_{13}$	Mon. (001)	(−) 2V = 75	β=1.753–1.763	High (.033–to .051)	Abnormal interference colors; Z∧cleavage= 14°–24°; weakly pleochroic in pistachio greens; usually in fine aggregates.
Kyanite	Al_2SiO_5	Tri.	(−) 2V = 83	β=1.722	Low (.012)	High relief; 2 good cleavages, best is (100); Z∧*c*=30°; alters to sericite; nice colorless blades.
Pyroxene		(110) 93–87°				Cleavage characteristic; cf. amphibole.
Enstatite	$MgSiO_3$	Orth. (210)	(+) 2V = 55	β=1.652–1.670	Low biref.	Parallel[a] extinction. Colorless.
Hypersthene	$(MgFe)SiO_3$	Orth. (210)	(−) 2V = 75	β=1.680–1.702	Low biref.	Parallel[a] extinction; faint pink to green pleochroism distinctive.
Pigeonite	$(MgFe)SiO_3 + CaSiO_3$	Mon. (110)	(+) 2V = 0 − 20	β=1.654–1.660	Low biref.	Colorless; Z∧*c*=22°–45°; low 2V diagnostic.
Diopside	$CaMgSi_2O_6$	Mon. (110)	(+) 2V = 60	β=1.672	Mod. biref.	Z∧*c*=38°; colorless; pure form common chiefly in contact metamorphic rocks.

Continued

(*Pyroxene*, cont.)

Augite	$Ca(MgFe)Si_2O_6$	Mon. (110)	(+) 2V = 60	β=1.676–1.745	Mod. biref.	$Z \wedge c$=48°–54°; may twin on (100) or (001); light colored in thin section, but reddish or violet when Ti is abundant.
Aegirinaugite	$Ca(MgFe)Si_2O_6$ – $NaFeSi_2O_6$	Mon. (110)	(+) 2V = 65	β=1.680–1.720	Mod. biref. Green	$Z \wedge c$=65°–75°; pleochroic in greens; may show inclined dispersion; common in alkalic rocks.
Apatite	$Ca_5(F,Cl)(PO_4)_3$	Hex.	(−)	ω=1.632–1.648	Low (.004)	Small hexagonal prisms; shagreen surface; colorless. Small inclusions common.
Andalusite	Al_2SiO_5	Orth. (110)	(−) 2V = 85	β=1.640	Low (.011)	*Faint pink pleochroism*; X=c; carbonaceous inclusions; alters to sericite.
			B. *Intermediate Positive Relief*			
Micas		(001)				Perfect basal cleavage
Muscovite	$KAl_2[AlSi_3]O_{10}(OH)_2$	Mon.	(−) 2V = 45±	β=1.582–1.610	High biref.	High index parallels cleavage; colorless.
Sericite						Fine grained muscovite.
Biotite	$K(Mg,Fe)_3[AlSi_3]O_{10}(OH)_2$	Mon.	(−) 2V = 0 − 10	β=1.564–1.690	High biref. Green, brown	Pleochroic; max. absorption and high index parallel cleavage; pleochroic haloes around inclusions; green ferric, brown titanian.
Paragonite	$NaAl_2[AlSi_3]O_{10}(OH)_2$	Mon.	Na analog of muscovite, with which it may be intergrown in schists; X-ray or probe needed to identify.			
Chlorite	$(Mg,Al,Fe)_6[Al,Si]_4O_{10}(OH)_8$	Mon. (001)	(+) or (−) small-zero	β=1.570–1.65	Very low B; pleochroic in greens	Perf. (001) cleavage; may show abnormal interference colors; max. absorption parallel to cleavage; high index parallel or normal to cleavage.

Mineral	Chemical Composition	Xl. System and Best Cleavage	Sign, 2V	Indices and Relief	Birefringence and Color	Distinguishing Characteristics
Serpentine	$Mg_6Si_4O_{10}(OH)_8$	Mon.	(−)	n=1.55±	Very low B	Colorless to green; fibrous or radial; often alters from olivine.
Talc	$Mg_3Si_4O_{10}(OH)_2$	Mon. (001)	(−)2V= 0–30	β=1.59	High biref. (.03–.05)	Difficult to distinguish from muscovite except by its association and slightly lower indices and 2V.
		C. *Low Relief (Positive and Negative)*				
Feldspar		(001) (010) (100)			Colorless	Variously twinned.
Orthoclase	$KAlSi_3O_8$	Mon. (001) (010)	(−) 2V = 30–60	β=1.523	Low (.006)	Index lower than 1.54; 2 good cleavages; no polysynthetic twinning.
Sanidine	$KAlSi_3O_8$		(−)2V 0–40	β=1.523	Low	High temperature polymorph of orthoclase.
Microcline	$KAlSi_3O_8$	Tri. (001) (010)	(−) 2V=83	β=1.522	Low (.007)	Cross-hatch (plaid) twinning distinctive.
Anorthoclase	$(Na,K)AlSi_3O_8$	Tri.	(−) 2V=50	β=1.528	Low (.006)	Cross-hatch twinning; uncommon.
Plagioclase Albite	$NaAlSi_3O_8$, An = 0–10 Plagioclase end member	Tri. (001) (010) (100)	(+) 2V=70	β=1.529	Low (.010)	Albite twinning characteristic. All other plagioclase has *n* > 1.54.
Oligoclase	An 10–30	-do-	(−) 2V=86			May be antiperthitic (with exsolution of Or). May be iridescent (peristerite).
Andesine	An 30-50	-do-	(+) 2V=87			May be antiperthitic up to about An 40.
Labradorite	An 50-70	-do-	(+) 2V=78			May be iridescent when slowly cooled.

Continued

Bytownite	An 70-90	-do-	(−) 2V=83			Fine lamellar exsolution common.
Anorthite	$CaAl_2Si_2O_8$, An 90–100 Plagioclase end member	-do-	(−) 2V=77	β=1.583		Metamorphic and arc volcanic rocks.
Quartz	SiO_2	Hex.	(+)	ϵ=1.553 ω=1.544	Low (.009)	No color, cleavage, or alteration. Often has inclusions; undulatory extinction common.
Nepheline	$NaAlSiO_4$	Hex.	(−)	ω=1.536–1.549	Low (.005)	Resembles quartz, but sign negative. Told from orthoclase by higher index and uniaxial character; often altered to cancrinite or sodalite (blue).
Cordierite	$Mg_2Al_4Si_5O_{18}$	Orth.	(−) 2V 40–80	β=1.543	Low (.011)	Often confused with quartz but biaxial; also may show inclusions of sillimanite, yellow pleochroic haloes, and characteristic twins.
Scapolite	Marialite ~Ab + NaCl Meionite ~An + $CaCO_3$	Tet.	(−)	ω=1.539–1.596	(.002–.039) often *zoned* int. colors	Alteration of plagioclase, mostly contact-metamorphic. Zoned interference colors characteristic.
			D. *Moderate Negative Relief*			
Leucite	$KAlSi_2O_6$	Iso.		n=1.508		Generally weakly birefracting; inclusions common; trapezohedral crystals.
Analcite	$NaAlSi_2O_6 \cdot H_2O$	Iso.		n=1.489		Trapezohedral form; generally lacks inclusions.

Mineral	Chemical Composition	Xl. System and Best Cleavage	Sign, 2V	Indices and Relief	Birefringence and Color	Distinguishing Characteristics
Calcite	$CaCO_3$	Hex. $(10\bar{1}1)$	(−)	$\omega=1.658$ $\epsilon=1.484$	Extreme biref. (.172)	Differential relief on rotation of stage; one index below 1.54, other far above; rhombohedral cleavage; often twinned; told from other rhombohedral carbonates only by index.
Cristobalite	SiO_2	Tet.	(−)	$\omega=1.487$ $\epsilon=1.484$	Low	Often looks isotropic; multiple twinning; generally in fibrous balls in cavities; negative relief.
Tridymite	SiO_2	Orth.	(+) 2V=35	$\beta=1.469$	Low	Lath-shaped forms; negative elongation; wedge-shaped grains; slight pinkish color; negative relief.
Zeolites	Complex hydrous Na,Ca,Al silicate	Orth.	(+) or rarely (−)		Low-Mod.	Generally low index, weak biref.; rarely colored; often fibrous. See tables for more data.
Fluorite	CaF_2	Iso. (111)		n=1.4339		Often in euhedral octahedra. Common accessory in some granites.

[a]Characteristically parallel, but see Fig. 13-3.

Appendix B
Identification of Fibrous Asbestos

The most common asbestos minerals are two types of amphibole (riebeckite and grunerite) and the serpentine mineral chrysotile. The amphibole fibers are particularly hazardous to health, and the list of fibrous minerals regulated by the U.S. Environmental Protection Agency also includes the amphiboles tremolite, actinolite, and anthophyllite. Details of the mineralogy and health effects of asbestos minerals are given by Skinner et al. (1988).

Blue amphibole asbestos (riebeckite, also called crocidolite) was heavily used in coating pipes and bulkheads in warships during World War II. Its use in later applications was more limited, but for a long time amphibole fibers were routinely mixed in with chrysotile, so many fiber-bearing materials contain some of both minerals. Typical applications during the mid-century are found in insulation, and the manufacture of floor and ceiling tiles used in public buildings. "Management in place" is often recommended where the material involved is unlikely to release fibers in everyday situations. In other situations, removal is often practiced whether or not it might be warranted by the mineralogy and etiology of the fibers.

It is of clear importance to public health and risk management to identify correctly the minerals present in any fibrous product to which people may be routinely exposed. Optical examination offers a low-cost means of crystal identification and fiber counting, especially as a precursor to more specialized methods of examination such as electron microscopy.

The names and properties of the three most common asbestos minerals are listed in Table B-1. All are monoclinic. Amosite is another name for grunerite. The amphiboles are chain silicates with fiber axes parallel to the *c* crystallographic axis. Chrysotile is a septechlorite sheet silicate in which fibers are rolled-up sheets elongate parallel to the *a* crystallographic axis, but along which the indicatrix axis may be either Z or X. The so-called Z fibers may be the more common.

Measurement of refractive index furnishes definitive information leading to the identification of asbestos minerals. This is done by the immersion method using standard refractive index liquids. If white light is used for illumination, the criterion of match in oblique illumination is the orange-red ("pylon") color described in Chapter 3 (see Fig. 3-15 or Fig. 17-3). Dispersion staining is used in many industrial laboratories to enhance the appearance of asbestos fibers and determine their refractive index. If dispersion staining is used with central screening (Chapter 3; McCrone, 1987; McCrone et al. 1977), the color indicating the condition of match is deep violet (Fleischer et al., 1984, Table 1). Other colors may be interpreted, according to the method used, by reference to either of these figures or the table. Either method leads to the correct result.

The two amphiboles have high refractive index, near 1.7, whereas chrysotile has a moderate refractive index near 1.5; this distinction provides a fundamental and rapid basis for identification. If the unknown fiber sample is immersed in a liquid of n = 1.56 and shows low relief or color fringes it cannot be one of the amphiboles. Conversely, if the sample is immersed in a liquid of n = 1.7 any amphiboles present will have low relief or show color fringes, whereas any chrysotile present will have high negative relief.

The sign of the elongation is diagnostic between the amphiboles, being positive in grunerite and negative in riebeckite (the terms length-slow and length-fast, respectively, are also in common use). Birefringence, hence interference color, is also diagnostic, being much larger in grunerite. The most spectacular diagnostic feature is the blue color of riebeckite, easily recognized in fibers only a few micrometers thick. The woolly material is pale blue to the naked eye, and unprocessed fibers are dark blue.

The high refractive index along the length of grunerite is about the same as the low refractive index along the length of riebeckite. A 1.698 oil is effective in seeking a color fringe to identify these minerals because it is more stable than the common 1.7+ oils and hence needs less frequent calibration.

The dispersion method affords a rapid means of establishing quantitatively the diagnostic refractive index of fibrous asbestos. It is used according to the methods described in Chapter 16.

Values of the dispersion, $n_F - n_C$, were determined at the University of Massachusetts from standard samples of the three mineral groups, with the results shown in Table B-1. Fibers were matched at varying wavelengths of light in different oils, and the results regressed in Hartmann space. A binder present in the grunerite sample was removed by dissolution in water, followed by rinsing in acetone and drying in air. Results were easily obtained for grunerite and chrysotile, but the high color of riebeckite makes the recognition of color fringes and shadows difficult. To enhance the use of the oblique illumination stop inserted from the NW, the polarizer

Table B-1
Properties of Common Asbestos Minerals

	Amphibole		Serpentine
	Grunerite (Amosite)	Riebeckite (Crocidolite)	Chrysotile
Approximate chemical composition	$Fe_7Si_8O_{22}(OH)_2$	$Na_2Fe_3(Al,Fe)_2$ $Si_8O_{22}(OH)_2$	$Mg_6Si_4O_{10}(OH)_8$
Crystal System	Monoclinic	Monoclinic	Monoclinic
Fiber axis	*c*	*c*	*a*
Elongation	+ (length slow)	− (length fast)	+ (Z fiber) or − (X fiber)
n along fiber	γ' ≈ 1.70	α' ≈ 1.70	γ' (Z); α' (X) ≈ 1.56; ≈ 1.55
Reported range	1.679–1.719	1.685–1.700	1.545–61 (Z) 1.538–60 (X)
Standard sample (UMass)	1.6995	1.6995	1.559 (Z)
Birefringence	strong, 0.04	weak, 0.004	weak, < 0.008
$n_F - n_C$	0.012	0.019	0.010
Color of fibers	Colorless to pale yellow	Dark blue to greenish yellow	Colorless
At matching refractive index:			
Color fringe in oblique illumination	Orange-red	Orange-red (may be obscured)	Orange-red
Dispersion color (central screen)	Deep violet	Deep violet (may be obscured)	Deep violet
Diagnostic features	+ elong, B, γ' near 1.7	Blue color, α' near 1.7	low n, ± elong colorless

and the fiber were oriented NE, so the shadow appeared evenly along the length of the fiber. Slight misorientation is of no concern because of the low birefringence. The refractive indices n_D found for γ' of grunerite and α' of riebeckite were both 1.6995. The chrysotile sample was determined to have $\gamma' = 1.559 \pm 0.001$, the relatively large error bracket being due mainly to demonstrably real variation among the fibers rather than uncertainty in the wavelength of match.

Other possibly fibrous minerals that may occur in asbestos samples are talc (*n* along fiber 1.59), brucite (1.56), tremolite (1.63), actinolite

(1.67), anthophyllite (1.63), and wollastonite (1.63). Of these, only brucite is likely to be confused with any of the major three asbestos minerals, and it can be distinguished by its high birefringence and low hardness. Talc is also distinguished by its very high birefringence and very low hardness. The other minerals in the "1.6" group can be identified by the immersion method using all relevant optical properties and reference to standard tables. A single mount will often serve to determine the refractive index of tremolite, actinolite, or anthophyllite by the dispersion method (Su, 1993).

Synthetic and natural organic fibers may also occur in asbestos samples, and most can be distinguished by their low refractive indices. A comprehensive treatment of these is given by McCrone (1987).

References

ADAMS, L.H. (1954) Fred E. Wright: 1878–1953. *Science* **120**, 241–242. **112***

ALLEN, R.D. (1956) A new equation relating index of refraction and specific gravity. *Am. Mineralogist* **41**, 245–57. **16**

BECKE, F.J. (1893) Über die Bestimmbarkeit der Gesteinsgemengtheile, besonders der Plagioclase auf Grund ihres Lichtbrechungsvermögens. *Sitz. Akad. Wien* **52**, Abt. 1, 359. **54, 55**

BECKE, F.J. (1896) Unterschiedung von optisch + und -zweiaxigen Mineralien mit dem Mikroskop. *Tschermaks Min. Petrog. Mitt.* **XVI**, 181. **185**

BERG, J.H., and MORSE, S.A. (1981) Dispersion method for olivine, orthopyroxene, and augite. *Am. Mineralogist* **66**, 985–89. **300, 301**

BETHKE, C.M., and BIRNIE, R.W. (1980) Computer synthesis of optical interference figures. *Am. Mineralogist* **65**, 1294–1301. **178**

BIOT, J.B. (1820) Mémoir sur les lois genérales de la double réfraction et de la polarisation, dans les corps regulièrement cristallisés. *Mém. Acad. France* **III**, 177–384. **106, 165**

BLOSS, F.D. (1961) *An Introduction to the Methods of Optical Crystallography* (New York: Holt, Rinehart, & Winston, Inc.). **82**

BLOSS, F.D. (1981) *The Spindle Stage: Principles and Practice* (Cambridge: Cambridge University Press). **ix, 40, 70, 282, 287**

BLOSS, F.D. (1985) Labelling refractive index curves for mineral series. *Am. Mineralogist* **70**, 428–32. **280, 281**

BORN, MAX, and WOLF, EMIL (1964) *Principles of Optics*, 2nd (revised) Ed. (Oxford: Pergamon Press). **1, 17, 30**

BRAGG, W.L. (1913) The diffraction of short electromagnetic waves by a crystal. *Proc. Camb. Phil. Soc.* **17**, 43–57. **98**

BRAGG, W.L. (1937) *Atomic Structure of Minerals* (Ithaca, NY: Cornell University Press). **120**

BRAGG, W.L. (1975) *The Development of X-ray Analysis* (London: G. Bell and Sons Ltd.). **98**

BUERGER, M.J. (1933) *Optical Identification of Crystalline Compounds* (Cambridge, MA: M.I.T. Letter Shop; copyright by John Wiley & Sons, Inc., New York). **ix**

BUERGER, M.J. (1933) Optical properties of ideal solution immersion liquids. *Am. Mineralogist* **18**, 325–34. **64**

BURNS, R.G. (1966) Apparatus for measuring polarized absorption spectra of small crystals. *Jour. Sci. Instrum.* **43**, 58–60. **138**

BURNS, R.G. (1970) Crystal field spectra and evidence of cation ordering in olivine minerals. *Am. Mineralogist* **55**, 1608–32. **138**

BURRI, CONRAD (1950) *Das Polarisationsmikroskop* (Basel: Birkhäuser). **113**

*Page numbers where citation occurs.

BUTLER, M.H. (1971) Application of the polarizing microscope in the conservation of paintings and other works of art. *Bull. Internat. Inst. Conservators, Am. Group.* **May,** 113. **2**

BUTLER, R.D. (1933) Immersion liquids of intermediate refraction. *Am. Mineralogist* **18,** 386–401. **28, 65**

CAMERON, E.N. (1961) *Ore Microscopy* (New York: John Wiley & Sons, Inc.). **83**

CHAPMAN, C.A. (1940) A model to illustrate isogyres in interference figures. *Am. Jour. Sci.* **238,** 805–10. **178**

CHAPMAN, C.A. (1943) Large magnesia-rich triphylite crystals in pegmatite. *Am. Mineralogist* **28,** 90–98. **200**

CHERKASOV, YU.A. (1960) Application of "focal screening" to measurement of indices of refraction by the immersion method. *Internat. Geol. Rev.* **2,** 218–35. **70**

CHRISTIANSEN, C. (1884) Untersuchungen über die optischen Eigenschaften von fein vertheilten Körpern. *Wiedemanns Ann. d. Physik u. Chemie (Neue Folge)* **23,** 298–306; *ibid.* (1885) **24,** 439. **63**

CRAIG, J.R., and VAUGHAN, D.J. (1981) *Ore Microscopy and Ore Petrography* (New York: John Wiley and Sons). **48**

DALY, R.A. (1899) On the optical characters of the vertical zone of amphiboles and pyroxenes. *Proc. Am. Acad. Arts and Sciences* 311–23. **238**

DEER, W.A.; HOWIE, R.A.; and ZUSSMAN, JACK (1962–1986) *Rock Forming Minerals,* (London: Longmans) as follows: **viii, xii**
(1962) **1**: *Ortho- and Ring Silicates.*
(1962) **5**: *Non-silicates.*
(1963) **2**: *Chain Silicates.*
(1967) **3**: *Sheet Silicates.*
(1967) **4**: *Framework Silicates.*
(1978) 2nd Ed. **2A**: *Single-chain Silicates.*
(1982) 2nd Ed. **1A**: *Orthosilicates.*
(1986) 2nd Ed. **1B**: *Disilicates and Ring Silicates.*

DEER, W.A.; HOWIE, R.A.; and ZUSSMAN, JACK (1992) *An Introduction to the Rock Forming Minerals,* 2nd Ed. (London: Longmans; New York: John Wiley & Sons, Inc.) **viii, xii**

DODGE, N.B. (1948) The dark-field color immersion method. *Am. Mineralogist* **33,** 541–49. **70**

EMMONS, R.C. (1943) *The Universal Stage.* Geol. Soc. Am. Mem. **8.** **75**

EMMONS, R.C., and GATES, R.M. (1957) Misconception of the intermediate refractive index. *Bull. Geol, Soc. Am.* **68,** 1413. **145**

FAIRBAIRN, H.W. (1943) Gelatin coated slides for refractive index immersion mounts. *Am. Mineralogist* **28,** 396. **45, 214**

FAIRBAIRN, H.W. (1943) Packing in ionic minerals, *Bull. Geol, Soc. Am.* **54,** 1320. **15**

FAIRBAIRN, H.W., and SHEPPARD, C.W. (1945) Maximum error in some mineralogic computations. *Am. Mineralogist* **30,** 673–703. **209**

FAUST, G.T., and GABRIEL, ALTON (1940) Petrographic methods and their application to the examination of non-metallic minerals. *U.S. Bur. Mines Inform. Circ.* **7129,** 3–5. **46**

VON FEDOROW, E.S. (1896) Universalmethode und Feldspathstudien. *Z. Krist.* **26,** 246. **283**

FLEISCHER, MICHAEL; WILCOX, R.E.; and MATZKO, J.J. (1984) *Microscopic Determination of the Nonopaque Minerals, 3rd Ed.* U.S. Geol. Surv. Bull. 1627 (revision of Bull 848). Includes data for many opaque minerals as well. **viii, 14, 70, 71, 77, 82, 136, 142, 167, 199, 288**

FLETCHER, L. (1891), The optical indicatrix and the transmission of light in crystals. *Mineralogical Mag.* **9,** 278–388. **117**

FLETCHER, L. (1892) *The Optical Indicatrix and the Transmission of Light in Crystals* (annotated reprint of 1891 article; London: H. Frowde). **117**

FORD, W.E. (1932) *Dana's Textbook of Mineralogy*, 4th Ed. (New York: John Wiley & Sons, Inc.). **42, 284**

FRESNEL, A.J. (1827) Memóir sur la double réfraction. *Mém. Acad. France* **VII,** 45–176. **165**

GALOPIN, R., and HENRY, N.F.M. (1972) *Microscopic Study of Opaque Minerals* (Cambridge: W. Heffer and Sons, Ltd.). **83**

GORANSEN, R.W. (1926) The determination of plagioclase feldspars. *Am. Mineralogist* **11,** 139–54. **234**

HAFF, J.C. (1942) Fedorow method (universal stage) of indicatrix orientation. *Colorado School of Mines Quarterly*, **37,** No. 3. **284, 287**

HARKER, ALFRED (1901) On a question relative to extinction in rockslices. *Mineralogical Mag.* **13,** 66–8. **233**

HARTSHORNE, N.H., and STUART, A. (1970) *Crystals and the Polarising Microscope*, 4th Ed. (London: Edward Arnold). **2, 7, 40, 113, 283**

HAUSWALDT, HANS (1902) *Interferenzerscheinungen an doppeltbrechenden Kristallplatten in konvergenten polarisierten Licht* (Magdeburg: Hausdruckerei J. G. Hauswaldt.) **127, 134, 185**

HESS, H.H. (1960) *Stillwater Igneous Complex*. Geol. Soc. Am. Mem. **80.** **90, 145**

HINTZE, C. (1915) *Handbuch der Mineralogie* (Leipzig: von Veit) Vol. 1, p. 2237. **65**

HUMPHRIES, D.W. (1992) *The Preparation of Thin Sections of Rocks, Minerals, and Ceramics* (Oxford: Oxford University Press). **47**

HUNTINGTON, H.D. (1979) Kiglapait mineralogy I: Apatite, biotite, and volatiles. *J. Petrology* **20,** 625–52. **293**

HUTCHISON, C.S. (1974) *Laboratory Handbook of Petrographic Techniques* (London: John Wiley and Sons). **284, 322**

HUYGENS, CHRISTIAAN (1690) *Treatise on Light*. (1912 English translation reprinted 1962. New York: Dover Publications, Inc.) **5, 10–13, 99, 100**

JAFFE, H.W. (1956) Application of the rule of Gladstone and Dale to minerals. *Am. Mineralogist* **41,** 757–77. **15, 16**

JAFFE, H.W. (1988) *Crystal Chemistry and Refractivity* (Cambridge: Cambridge University Press.) **15, 16**

JAFFE, H.W.; ROBINSON, PETER; TRACY, R.J.; and ROSS, MALCOLM (1975) Orientation of pigeonite exsolution lamellae in metamorphic augite: Correlation with composition and calculated optimal phase boundaries. *Am. Mineralogist* **60,** 9–28. **264**

JELLEY, E.E. (1934) A microrefractometer and its use in chemical microscopy. *J. Royal Microscop. Soc. London* **54,** 234–45. **40, 41**

JOHANNSEN, ALBERT (1918, reprinted 1968), *Manual of Petrographic Methods*, 2nd Ed. (New York: McGraw-Hill Book Co.) **93, 107, 110, 113, 124, 165**

JONES, N.W., and BLOSS, F.D. (1980) *Laboratory Manual for Optical Mineralogy* (Minneapolis: Burgess Publishing Co.) **viii, 234. 279**

JUDD, J.W. (1908) Henry Clifton Sorby, and the birth of microscopical petrology. *Geol. Mag.* **V,** 192–204. **47**

KAMB, W.B. (1958) Isogyres in interference figures. *Am. Mineralogist* **43,** 1029–67. **176, 188, 189, 208, 209**

KAMB, W.B. (1962) Refraction correction for universal stage measurements: I. Uniaxial crystals. *Am. Mineralogist* **47,** 227–45. **207, 283**

KEMP, D.R.C., and HARVEY, A.D., Eds. (1983) *The Petrology of Archaeological Artefacts* (Oxford: The Clarendon Press). **2**

KORDES, E. (1960) *Optische Daten zur Bestimmung Anorganischer Substanzen mit dem Polarisationsmikroskop* (Weinheim: Verlag Chemie GMBH). **77**

KRUMBEIN, W.C., and PETTIJOHN, F.J. (1938) *Manual of Sedimentary Petrography* (New York: D. Appleton Century Co.). **45, 80**

LARSEN, E.S. JR., and BERMAN, HARRY (1934) *The Microscopic Determination of the Non-opaque Minerals*, 2nd Ed., U.S. Geol. Surv. Bull. **848**. **viii, 14, 15, 65, 136, 140, 199**

LIPSON, H.S. (1970) *Crystals and X-rays* (London: Wykeham Publications Ltd.). **98, 100**

LONSDALE, KATHLEEN (1968) Human stones, *Science* **159,** 1199–1207. **2**

LOUISNATHAN, S.J.; BLOSS, F.D.; and KORDA, E.J. (1978) Measurement of refractive indices and their dispersion. *Am. Mineralogist* **63,** 394–400. **292**

LOUPEKINE, I.S. (1947) Graphical derivation of refractive index ε for the trigonal carbonates. *Am. Mineralogist* **32,** 502–7. **149**

MACKENZIE, W.S., and GUILFORD, C. (1980) *Atlas of Rock-Forming Minerals in Thin Section* (New York: John Wiley & Sons, Inc.) **324**

MACKENZIE, W.S.; DONALDSON, C.H.; and GUILFORD, C. (1982) *Atlas of Igneous Rocks and their Textures* (New York: John Wiley & Sons, Inc.) **324**

MALLARD, E. (1882) Sur la mesure de l'angle des axes optiques, *Bull. Soc. Mineral. France* **V,** 77–87. **209, 210**

MANDARINO, J.A. (1959) Absorption and pleochroism: two much-neglected optical properties of crystals. *Am. Mineralogist* **44,** 65–77. **138, 202**

MANDARINO, J.A. (1976) The Gladstone-Dale relationship. Part I. Derivation of new constants. *Can. Mineralogist* **14,** 498–502. (Also Parts II–IV, 1978, 1979, 1981: see Jaffe [1988]). **15**

MANGE, M.A., and MAURER, H.F.W. (1991) *Heavy Minerals in Colour* (London: Chapman and Hall). **43, 45, 46, 80**

MCCRONE, W.C. (1987) *Asbestos Identification* (Chicago: McCrone Research Institute). **334, 336**

MCCRONE, W.C.; MCCRONE, L.B.; and DELLY, J.G. (1977) Dispersion staining. Chapter X in *Polarized Light Microscopy*, 8th Ed. (Chicago: McCrone Research Institute), 169–96. **70, 334**

MERWIN, H.E., and LARSEN, E.S. JR. (1912) Mixtures of amorphous sulphur and selenium as immersion media for the determination of high refractive indices with the microscope. *Am. Jour. Sci.* **34,** 42–7. **73**

MICHEL-LÉVY, A. (1888) *Les minéraux des roches* (Paris: Baudry). **114, 278**

MILNER, H.B. (1962) *Sedimentary Petrography*, 4th Rev. Ed., 2 vols. (London: Murby & Co.). **45, 80**

MORSE, S.A. (1968) Revised dispersion method for low plagioclase, *Am. Mineralogist* **53,** 105–16. **74, 214**

MORSE, S.A. (1978) Test of plagioclase dispersion method and rapid probe analysis. *Am. Mineralogist* **63,** 768–70. **304**

MUIR, I.D. (1981) *The 4-axis Universal Stage* (Chicago: Microscope Publications Ltd.). **282, 284**

MUNRO, M. (1963) Errors in the measurement of 2V with the universal stage. *Am. Mineralogist* **48,** 308–23. **207**

PALACHE, CHARLES; BERMAN, HARRY; and FRONDEL, CLIFFORD (1944, 1951, 1962) *Dana's System of Mineralogy*, 7th Ed., 3 vols. (New York: John Wiley & Sons, Inc.). **86, 159**

PECKETT, A(NDREW) (1992) *The Colours of Opaque Minerals* (Chichester: John Wiley and Sons). **3, 48, 78**

PHILPOTTS, A.R. (1989) *Petrography of Igneous and Metamorphic Rocks* (Englewood Cliffs: Prentice Hall). **viii, 324**

POSNJAK, E., and MERWIN, H.E. (1922) The system ferric oxide-sulfur trioxide-water. *Jour. Am. Chem. Soc.* **44,** 1965–94. **73**

RIBBE, P.H. (1983a) Chemistry, structure and nomenclature of feldspars; 1–19 in Ribbe, 1983c. **270**

RIBBE, P.H. (1983b) Interference colors; 266–70 in Ribbe, 1983c. **101**

RIBBE, P.H., Ed. (1983c) *Feldspar Mineralogy*. Mineralogical Soc. Am. Rev. Mineral. **2,** 2nd Ed. **274**

RIBBE, P.H., and ROSENBERG, P.E. (1971) Optical and X-ray determinative methods for fluorine in topaz. *Am. Mineralogist* **56,** 1812–21. **225, 226**

ROBINSON, PETER; JAFFE, H.W.; ROSS, MALCOLM; and KLEIN, CORNELIS JR. (1971) Orientation of exsolution lamellae in clinopyroxenes and clinoamphiboles: consideration of optimal phase boundaries. *Am. Mineralogist* **56,** 909–39. **263**

ROSENFELD, J.L., and CHASE, A.B. (1961) Pressure and temperature of crystallization from elastic effects around solid inclusions in minerals? *Am. J. Sci.* **259,** 519–41. **47**

SAYLOR, C.P. (1935) Accuracy of microscopical methods for determining refractive index by immersion. *Jour. Res. Natl. Bur. Stds.* **15,** 277–94. **55, 71, 72**

SCHUSTER, M. (1880) Über die optische Orientierung der Plagioklase. *Tschermaks Min. Petrog. Mitt.* **3,** 117–284. **230**

SHORT, M.N. (1940) Microscopic determination of the ore minerals, *U.S. Geol. Surv. Bull.* **914.** **viii, 12, 13, 33, 46, 83, 120, 160, 207, 279**

SHUBNIKOV, A.V. (1960) *Principles of Optical Crystallography* (New York: Consultants Bureau). **118**

SHURCLIFF, W.A. (1962) *Polarized Light* (Cambridge: Harvard University Press). **36**

SKINNER, H.C.W.; ROSS, MALCOLM; and FRONDEL, CLIFFORD (1988) *Asbestos and Other Fibrous Minerals: Mineralogy, Crystal Chemistry, and Health Effects.* (Oxford University Press) **333**

SLEMMONS, D.B. (1962) Determination of volcanic and plutonic plagoclases using a three- or four-axis universal stage. *Geol. Soc. Am. Special Paper* **69.** **287**

SMITH, H.T.U. (1938) Model to aid in visualizing the optical properties of crystals. *Am. Mineralogist* **23,** 629–43. **160**

STAPLES, L.W. (1936) Mineral determination by microchemical methods. *Am. Mineralogist* **21,** 613–34. **46**

STEVENSON, J.S., and STEVENSON, L.S. (1973) Well developed growth zoning in a struvite bladder stone. *Canad. Mineralogist* **11,** 985–90. **2**

STOIBER, R.E., and MORSE, S.A. (1972) *Microscopic Identification of Crystals* (New York: Ronald Press). Reprinted 1979, 1981, 1986 by R.E. Krieger, Melbourne, FL. **ix**

STRENS, R.G.J., and FREER, ROBERT (1978) The physical basis of mineral optics, I. Classical theory. *Mineral. Mag.* **42,** 19–30. **30**

SU, S.C. (1992) Calibration of refractive index liquids by using optical glass standards with dispersion staining. *Microscope* **40,** 95–108. **73, 290**

SU, S.C. (1993) Determination of refractive index of solids by dispersion staining method: An analytical approach. *Proc. Ann. Mtg. Microscopy Soc. Am. Abstr.* **51,** 456–457. **294, 296, 333**

SU, S.C. (1994) A revised dispersion method for determining the composition of olivine, orthopyroxene, augite, and plagioclase. *Am. Mineralogist,* submitted. **300, 301, 305**

SU, S.C.; BLOSS, F.D.; and GUNTER, MICKEY (1987) Procedures and computer program to refine the double variation method. *Am. Mineralogist* **72,** 1011–13. **40, 72, 75, 292**

SU, S.C.; BLOSS, F.D.; RIBBE, P.H.; and STEWART, D.B. (1984) Optic axial angle, a precise measure of Al,Si ordering in T_1 tetrahedral sites of K-rich alkali feldspars. *Am. Mineralogist* **69,** 440–48. **272**

SU, S.C.; RIBBE, P.H.; and BLOSS, F.D. (1986a) Alkali feldspars: Structural state determination from composition and optic axial angle, 2V. *Am. Mineralogist* **71,** 1285–96. **276**

SU, S.C.; RIBBE, P.H.; BLOSS, F.D.; and GOLDSMITH, J.R. (1986b) Optical properties of single crystals in the order-disorder series low albite-high albite. *Am. Mineralogist* **71,** 1384–92. **275, 277, 279**

SUENO, T. (1933) On the use of standard glass powders in refractive index determination, *Am. Mineralogist* **18,** 421–30. **73, 290**

TAYLOR, E.C. SR. (1959) Peristome teeth in polarized light. *The Bryologist,* **62,** 149–55. **2**

TERTSCH, H. (1938) Zwei graphische Methoden zur Bestimmung optische Werter an bestimmten Kristallflächen. *Tschermaks Min. Petrog. Mitt.* **51,** 166. **238**

TILTON, L.W. (1942) Testing of Abbe refractometers. *J. Opt. Soc. Am.* **32,** 371–81. **40**

TOBI, A.C. (1956) A chart for measurement of optic axial angles. *Am. Mineralogist* **41,** 516–18. **209**

TRÖGER, W.E. (1979) *Optical Determination of Rock-Forming Minerals,* Part 1, Determinative Tables. Revised by H. U. Bambauer, F. Taborsky, and H.D. Trochim; English edition of the 4th German edition (Stuttgart: E. Schweizerbart'sche Verlagsbuchhandlung). **viii, ix, xii, 77, 82, 97, 207, 209, 214, 222, 224, 226, 234, 238, 278, 280, 281, 283, 293, 304, 310, 313**

TSUBOI, SEITARÔ (1923) A dispersion method of determining the plagioclases in cleavage flakes. *Mineralogical Mag.* **20,** 108–22. **74, 214, 305**

TSUBOI, SEITARÔ (1934) A straight-line diagram for determining plagioclase by the dispersion method. *Japanese Jour. Geol. Geog.* **11,** 325–26. **74, 304**

TURNER, F.J. (1947) Determination of plagioclases with the 4-axis universal stage. *Am. Mineralogist* **32,** 389–410. **287**

USGS 848 See LARSEN and BERMAN (1934).

USGS 1627 See FLEISCHER et al. (1984).

WILCOX, R.E. (1983) Refractive index determination using the central focal masking technique with dispersion colors. *Am. Mineralogist* **68,** 1226–36. **70**

WINCHELL, A.N., and WINCHELL, HORACE (1951) *Elements of Optical Mineralogy,* 4th Ed., Pt. II, Descriptions of Minerals (New York: John Wiley & Sons, Inc.). **76, 77, 136, 199**

WINCHELL, HORACE (1946) A chart for measurement of interference figures. *Am. Mineralogist* **31,** 43–50. **209**

WINCHELL, HORACE (1965) *Optical Properties of Minerals; a Determinative Table* (New York: Academic Press). **203**

WINCHELL, N.H., and WINCHELL, A.N. (1929) *Elements of Optical Mineralogy,* 2nd Ed., Pt. III, Determinative Tables (New York: John Wiley & Sons, Inc.). **97**

WRIGHT, F.E. (1905) The determination of the optical character of birefracting minerals. *Am. Jour. Sci.* (4), **20,** 285–96. **208**

WRIGHT, F.E. (1911) *Methods of Petrographic Microscopic Research.* Carnegie Inst. Wash. Publ. **158.** **63, 112**

WRIGHT, F.E. (1951) Computation of the optical angle from the three principal refractive indices. *Am. Mineralogist* **36,** 543–56. **166, 167**

WÜLFING, E.A. (1895) Beiträge zur Kenntnis der Pyroxenfamilie. *Tschermaks Min. Petrog. Mitt.* **15,** 38. **220**

WYLLIE, P.J. (1959) Discrepancies between optic axial angles of olivines measured over different bisectrices. *Am. Mineralogist* **44,** 49–64. **207**

ZUSSMAN, JACK, Ed. (1977) *Physical Methods in Determinative Mineralogy* 2nd Ed. (London: Academic Press). **284**

Index

a-normal method, 234, 278
Abbe refractometer, 23, 40, 72, 73
  and critical angle, 23
  high accuracy, 40, 73
Absorption, 6, 25–27, 29, 47, 82, 135, 137–139, 199, 201, 202, 324, 327, 329
  biaxial crystals, 201
  defined, 29
  determination, 138, 201–202
Absorption axes, 201
Absorption formula, 137, 138, 201, 202
  biaxial, 202
  uniaxial, 137
Accessory plate, 38, 57–59, 107–109, 142–145, 148, 151, 153, 182, 187, 204, 205, 212, 229, 237, 284, 288, 311
  colors rise or fall, 109
  effect of, 108
  principle, 107
  retardation in, 107
  three most common types, 107
  use of, 106, 109
  used to identify interference color, 112
Accuracy, 16, 35, 40, 42, 71–73, 86, 112, 149, 158, 210, 289
  defined, 72
  improving, 72
  in routine work, 71
  of dispersion method, 289, 293
  of immersion method, 71
  tested, 72
Acicular crystal habit, 235
Actinolite, 301, 327, 333, 335, 336
Activity, optical, 9
Acute bisectrix (Bxa), 125, 163, 164, 180, 220, 224
  defined, 163
Aegirinaugite, 329
Aenigmatite, 254, 255
Air, 3, 14, 17, 18, 23, 24, 34, 164, 165, 186, 310, 334
  color of light in, 24
  n taken as 1 in, 14
  refractive index of, 14
  wave velocity, 14
Albite, 215, 230, 234, 240–242, 266, 274, 275, 277–279, 320, 322, 330
  twin law, 240
Alcohol, 14, 44–46, 310
Alkali feldspar, 274, 276, 277, 319, 322
Allowed direction, 6
Alpha refractive index, 159–167
  confirmation, 212
Alpha prime (α') index, 161, 181, 214
Alpha-chloronaphthalene, 27
Alunite, 140, 145, 152
Amici, G.B., 40, 124
Ammonium chloride, 27, 28
Amosite, 333, 335
Amphibole, 101, 168, 202, 214, 227, 238, 263, 281, 287, 288, 301, 320, 323, 327, 328, 333–335
  calcic, 281
Amplitude, 3–7, 87–89, 91, 93, 97
  wave, 4
Analbite, 274, 278
Analcite, 331
Analyzer, 36, 57, 78, 91–93, 107, 109, 113, 156, 311, 315
  need to remove, 109, 156
  rotating, 113
Andalusite, 281, 319, 320, 329, 281, 329
Andesine, 330
Angle
  axial, 163, 216
  critical, 22
  extinction, 49
  glancing, 99
  minimum deviation, 41
  of incidence, 17, 18, 22, 23, 99
  of reflection, 22
  of refraction, 17, 23, 25, 41, 79, 283
  optic, 49, 163–167, 188, 189, 207–210
  silhouette, 151
Angles, stereo coordinate, 238
Anhydrite, 218

Anisotropic crystal plate, 20, 89, 90
Anisotropic crystals, 8, 14, 21, 25, 30, 47, 49, 76–78, 102, 104, 180
Anisotropic substances, 3, 6, 10, 14, 16, 23, 28, 139
Anisotropy, test for, 310
Annular screening, 70
Anorthite, vii, 208, 234, 266, 278, 304, 331, 208, 331
Anorthoclase, 330
Anthophyllite, 327, 333
Anthophyllite-gedrite, 301
Apatite, 69, 110, 136, 140, 152, 158, 200, 319–322, 329
Aperture diaphragm, 38
Aperture; *see* Numerical aperture, N.A.
Aqua regia, 82
Aragonite, 194, 209, 235, 239
Arfvedsonite, 301, 327
Asbestos, 333
Augite, viii, xi, 203, 208, 241, 259, 261–264, 300, 301, 320, 321, 329
Axial angle, 163 (*see also* optic angle)
Axial character, 124, 136, 199

Babingtonite, 196, 254–256
Bakelite, 90
Barite, 169, 184, 197
Bartholinus and calcite, 12
Basal section, 127, 143, 235, 327
Basalt, 241
Becke, F., 55
Becke line, 50, 52–56, 61, 69, 72, 74, 76, 79, 83, 149, 215, 311
  colored, 56,
  crystal high (ill.), 52
  crystal low (ill.), 52
  from lamellae, 262
  practice, 52
  principle, 54
  rule for, 54
  with colored crystals, 55
Berek compensator, 113
Bertrand lens, 35, 37, 38, 77, 123, 142
  described, 37
Bertrand, E., 40, 124
Beryl, 122, 199
Beta refractive index, 159–167
  confirmation, 212
  defined, 159
  estimation, 316
Beta prime, 162
Biaxial crystal identification, 225, 239
  properties used, 199
Biaxial crystals, 8, 9, 15, 30, 104, 106, 115, 118, 123, 136, 159, 163, 168, 199–202, 211, 215, 216, 227, 231, 232, 285
  identification, 200
Biaxial indicatrix, 126, 160–162, 167, 173, 246
Biaxial interference figures, 125, 172–175, 198
  naming, 173
Biaxial minerals, abundance, 199
Biographical comment
  Becke, F., 55
  Bertrand, E., 124
  Biot, J.B., 165
  Bragg, W.H., 100
  Bragg, W.L., 100
  Fermat, P. de, 90
  Fletcher, L., 117
  Fraunhofer, J., 27
  Fresnel, A.J., 165
  Huygens, C., 10
  Larsen, E.S. Jr., 73
  Merwin, H.E., 73
  Posnjak, E., 73
  Rutherford, E., 41
  Schairer, J.F., 75
  Snell, W., 17
  Sorby, H.C., 47
  Thomson, J.J., 100
  Wright, F.E., 112
  Zies, E.G., 323
Biot, J.B., 1, 165
Biot-Fresnel Law, 30, 165, 166, 176, 178, 179
  approximation, 178
Biot-Fresnel rule, 178, 255
  approximate, 255
Biotite, 136, 177, 202, 207, 209, 319–321, 329, 136, 302, 329
Birdseye maple effect, 79, 322
Birefringence, 28, 44, 71, 78, 103, 104, 109, 111, 113, 120, 124, 127, 129, 130, 132, 133, 136, 144–146, 150, 156, 166, 167, 182, 183, 185, 199, 214, 224, 226, 284, 297, 320, 322, 326, 334–336
  and retardation, 90
  biaxial crystals, 202
  biaxial, strength, 202
  defined, 2, 21
  dispersion of, 141
  estimated from interference figures, 140
  estimated in thin section, 141
  estimation, 203
  estimation in immersion, 139
  in thin section, 323
  of a section, 21, 114, 202
  of a substance, 21
  partial, 21, 202
  strength, 139
  uniaxial crystals, 139

Birnie, R.W., ix, 178
Bisectrix, 125, 163, 164, 176, 180, 215, 220, 224
acute (Bxa), 163
obtuse (Bxo), 163
Bloss, F.D., viii, ix, x, 40, 70, 73, 75, 82, 207, 214, 234, 271, 279, 280–282, 287, 289, 292
Borax, 222
Bracket, refractive index, 315, 335
Bragg, W. H., 100
Bragg, W. L., 100
Bragg's Law, 98–101
Brightness, 93, 94, 97, 98
British Museum, 117
Brookite, 219
Brucite, 153, 335, 336
Buerger, M.J., ix, 64
Buerger diagrams for color fringes, 64
Bunsen, R.W., 27
Butler, R.D., ix
Bxa, 125, 163, 164, 167, 173–193
Bxa direction, 176, 183, 187
Bxa figure, 175–184, 186, 187, 190–195, 194, 195, 197, 201, 204, 205, 206, 209, 216, 218, 222–224, 226, 237, 238, 257, 259, 274, 281, 317
aragonite, 194
color curves, 182
distinguished from other 2-SP, 185–189
example with color curves, 182, 193
muscovite, 182, 194
vibration directions, 176
vibration vectors, 179
Bxa′, 180, 182, 183, 204
Bxo, 125, 163, 164, 169, 173–176, 183–189
Bxo direction, 186, 187, 204–206
Bxo figure, 184, 204, 210, 257, 258, 261, 267
distinguished from other 2-SP, 185–189
examples, 184
vibration vectors, 184
Bytownite, 196, 203, 321, 331

c axis, 13, 29, 119, 120, 122, 126, 127, 131–133, 142, 153, 155, 177, 230, 235, 238, 247, 250, 251, 254, 271, 284
Calcite, ix, 5, 12, 13, 21, 29, 36, 95, 104, 105, 113, 117, 127, 132, 134, 140, 145, 148–150, 154, 155, 199, 204, 283, 327, 332
Calcite experiment, ix, 12, 21, 29
Cambridge, University of, ix, 99, 100
Cancrinite, 105, 133, 140, 145, 154, 331
Carbonate group, 122, 149
Carbonate minerals, 120, 149, 150, 153
Carbonates, 136, 139, 149, 200, 322, 332
Cargille, 45, 69, 290, 291, 305
Carnegie Institution of Washington, 73, 112
Cauchy equation, 292
Cauchy, A.L., 1
Cavendish Professor, 100
Cell edges, 271
in feldspar, 271
Cell mitosis, 2
Cellulose tape, 184
Central illumination test, 52
Central screening, 70, 71, 334
Chapman, C.A., ix
Chemistry, 2, 99, 207
Cherry tomato, 67, 68, 309
Chlorite, 136, 141, 153, 214, 224, 281, 319, 321, 329, 136, 281, 302, 329
Chloritoid, 321, 328
Chondrodite, 202, 240
Christiansen effect, 63
Chrysotile, 333
Circular section, 29, 117, 118, 127, 129, 160–164, 167, 180, 284, 285
and optic axes, 162
biaxial, 160
normal to, equals optic axis, 163
origin, 162
Class, xi, 25, 30, 72, 119, 136, 168, 199, 247, 248, 281
holosymmetric, 119
Classification, viii, 8, 125, 126, 172, 320
Cleavage
and crystal system, 241
basal, 153
effects of, 148
examples, 154
flat-topped fragments, 154
fragments, 81
in immersion mounts, 153
in isometric crystals, 80
in thin section, 80, 153
in uniaxial crystals, 152
pinacoidal, 236
prismatic, 153, 235
prismatic form, 243
prismatic versus pinacoidal, 243
recognition, 154, 242, 322
rhombohedral, 153
types, 242
Cleavage flakes, 95, 145, 151, 196, 197, 258, 265, 304, 320
epsilon prime on, 148
Cleavage forms, 207, 243–245, 316
recognition, 243
Cleavage fragments
platy or rod-like, 214
Clinohumite, 240
Coal, 2
Color, 2, 24–27

in air, 24
in mafic minerals, 323
isotropic crystals, 78
of crystals, 28, 29
of riebeckite, 334
Color bands, 124, 144 (*see also* color curves)
Color chips, 67, 72
for color fringe test, 67
Color curves, 124, 126–128, 130–134, 140, 141, 144–146, 176, 183–185, 193, 198, 203
and birefringence, 124, 127, 128–130, 203, 204
in Bxa figure, 182
in Bxo figure, 184
in flash figure, 133, 134, 185
in interference figures, 124
in optic axis figure, 128, 195
not curves of equal birefringence, 183
Color fringe, 56, 62, 64, 65, 67–69, 72, 73, 83, 147, 156, 211, 224, 291, 309, 311, 312, 315–318, 334, 335
and dispersion, 63, 65
calibration, 67
in oblique illumination, 62
intensity, 62
orange-red, 62, 65
principle, 63
pylon color, 65
reference standard, 65
rules, 62
Color of light, 24, 69, 220
and frequency, 24
Color-blind people, 63, 145
Compensation, 109, 110, 112
with accessory plate, 112
Compensator, 112, 113
Berek, 113
Ehringhaus, 113
elliptic, 113
Sénarmont, 113
Components
of vibration, 89
plane polarized, 102
Composition plane, 81, 154, 155, 228, 230, 239–241
Conchoidal fracture, 320
Condenser, 31, 32, 35–38, 54, 58, 123
diaphragm, 37
substage, 37
Confirmation, 2, 148, 212
Conoscope, 37, 124, 286
Conoscopic illumination, 123
Conoscopic observation, 37, 102, 286
Conventions, 103, 104, 109, 143, 170, 239, 250
for naming sections or planes, 104
Converging lens, 37, 38, 57, 58, 77, 102, 123
Cordierite, 215, 239, 240, 331, 240, 331
Corundum, 136
Counter-rotating figure (CR), 174
Cover glass, 34, 43–45, 51, 287, 310
CR and SR figures, 190–192, 194
CR and SR isogyres, 192
CR figure, 192, 194, 193, 195
CR1SP figure, 258, 254, 258, 259, 318
Cracks, cleavage, 80
Crest, wave, 4, 87
Cristobalite, 78, 332
Critical angle, 22, 23, 40, 49, 52, 54, 283
defined, 22
Crocidolite, 333, 335
Crossed axial plane dispersion, 170, 219, 235
Crossed nicols, 36
Crossed polars, 36
Crosshairs, 32, 33, 37, 62, 126, 133, 173, 175, 213
Crossite, 281
Crossover,
refractive index, 280, 281
Crushed grains, 43, 44, 76, 82, 140, 155
mounting, 44
preparation, 43
Crushed rock, crystal identification in xii, 318
Crystal,
aggregates, 46
biaxial, 315
cleavage, 80
defined, 8
form, 79, 150
size, 150
system, 8
uniaxial, 313
Crystal identification,
defined, 1
isotropic, 76
worked example, 83, 155, 223
Crystal structure analysis, 101
Crystal symmetry, 8, 30, 119, 215, 280
and vibration directions, 30
Crystal system, viii, ix, 8, 13, 30, 76, 119, 155, 168, 228, 231, 232–234, 239, 241–243, 245–247, 316, 335
determination, 155
examples of identification, 243
hexagonal, 118
identification from cleavage, 243
identification in thin section, 245
isometric (cubic), 8
monoclinic, 257
orthorhombic, 260
tetragonal, 118

triclinic, 254
Crystalline precipitate, 46, 82
Crystallization, 79
Crystallographic directions, 2, 8, 9, 35, 116, 118, 163, 168, 201, 228, 229, 233–236, 235, 236, 238, 245, 251, 286
identification, 235
Crystallographic properties, 77, 79, 136, 155, 168, 199, 224, 265
Crystallographic review, 247
Crystallography, 2, 8, 150, 247
operations, 247
Crystals
anisotropic, 49
euhedral, 80
grown on slides, 151
hexagonal, 151, 155
isotropic, 49
monoclinic, 168
orthorhombic, 168
tetragonal, 155
triclinic, 168
Cubic system, 8
Cummingtonite, 301, 327, 301, 327

D line, 51, 69, 83, 86, 272–274, 293, 295–297, 310, 318
standard wavelength, 28
Darkness, 89, 92
Daylight, 35, 51, 106
Density, 2, 15, 16, 35, 45, 81, 82, 86, 271, 278
of minerals, 81
Descartes, R., 17
Determinative equation, 86, 296, 300, 301, 305
alkali feldspar, 276
dispersion method, 296
mafic minerals, 300
plagioclase, 305
Detrital mineral grains, 45
Diaphragm, 35, 37, 38, 52, 57, 72, 83, 123, 311, 322
aperture, iris, 38
field, 38
pinhole, 123
substage, 38
Dichroism, 36, 137
Dickey, J.S. Jr., ix, 241
Differential relief, 322, 327, 332
Diffraction, 3, 98–101, 253, 321
and Bragg's Law, 98
optical, demonstrated, 98
related to wavelength, 98
X-ray, 99
Diopside, 208, 240, 259, 260, 262, 328, 208, 328
Dirac, P., 1
Directions image, 123
Dispersion, 24
and crystal symmetry, 215
and extinction, 220
and optic orientation, 237
axial angle, 216
crossed, 220
crossed axial plane, 170, 219, 235
determination of, 223
effect on extinction, 106
estimation of slope, 293
examples, 216, 222
fractional, 290
horizontal, 220
in interference figures, 198
inclined, 220
monoclinic, 220, 237
of crystals, 293, 297
of other optical properties, 28
of the birefringence, 141
of the optic angle (axes), 216
of the optic plane, 220
of the refractive index, 25, 27, 150, 216
of the vibration directions, 170, 219
reciprocal (Abbe V), 290
rhombic, 216
showing biaxial character, 200
strength of, 27, 220
triclinic crystals, 222
use in optic orientation, 237
Dispersion curves, 27, 28, 64, 66, 73, 74, 85, 294, 304
algebraic, 74
liquid and solid, 28
Dispersion formula, 274, 281
Dispersion method, vii, viii, xi, xii, 40, 51, 71–75, 83, 84, 150, 155, 156, 216, 226, 278, 289, 292, 299, 301, 304, 305, 306, 318, 334, 336
asbestos, 334
biaxial crystals, 226
described, 73
in worked example, 83, 156, 226
precision, 74, 301, 304, 308
procedure, 297
recipes, 318
use of, 289
Dispersion staining, 69–71, 334
asbestos minerals, 334
Dolomite, 149, 154, 327
Double diaphragm method, 72
Double liquid method, 296, 297, 301
Double refraction, 5, 21 (*see also* birefringence)

Double variation method, 75, 289
Dravite, 137

E direction, 133
E ray, 116, 117, 122
E vibration direction, 117, 119, 120, 125, 142, 153
E', 132, 134, 143–145, 148
Ehringhaus compensator, 113
Einstein, A., 1
Electric field, 29, 122
  in crystals, 29
Electron microprobe, xi, 3, 48, 82, 301, 321
Ellipsoid, 29, 117, 119, 120, 126, 149, 160, 162, 165, 167, 168
  of revolution, 117, 126, 149
  spin, 117
  triaxial, 160
Elliptic compensator, 113
Elongation, 235, 320, 332, 334, 335
  applied to cleavages, 228
  crystal, 227
  examples, 152
  limitations, 152
  negative, 152
  positive, 152
  rules, 152
  rules for, biaxial crystals, 227
  sign of, 151, 227
  uniaxial crystals, 151
Energy, 9, 11, 12
Enstatite, 328
Epidote, 141, 328
Epsilon, 147, 148, 156, 313
  confirmation of, 148
  ε', 119, 129, 132, 135, 137, 142, 143, 147–149, 161, 162, 180, 181, 214, 230, 304, 308, 313, 315, 317, 335
  defined, 117
Epsomite, 218
Equipment, 31, 71, 111, 140, 188, 209, 230, 282, 286
Error, 16, 56, 67, 71, 74, 86, 139, 151, 167, 209, 234, 286, 293, 297, 335
Ethyldiiodoarsene, 14
Euclase, 185
Euhedral crystals, 80, 81, 151, 152, 227
Extinction, 78, 97
  complete, 104
  defined, 97
  effect of dispersion on, 106, 171
  explained, 97
  imperfect, 104
  in biaxial crystals, 231
  inclined, 106, 232, 233
  inclined, in orthorhombic crystals, 231
  observed in mica, 97
  parallel, 105, 153, 231
  straight, 105
  symmetrical, 105, 154, 231
  table, 232
  types, 105, 106, 231–234
  undulatory, 105
Extinction angle, 49, 103, 106, 221, 228–234, 237, 238, 251, 254, 259, 260, 278, 286
  expression, 230
  in thin section, 323
  measurement, 228
  measurement with U-stage, 286
  special uses, 234
Extinction positions, 131, 132, 151, 153, 154, 173, 175, 228
Extraordinary ray, 116
Eyepiece, 31, 32, 123, 124

F line, 293
Fairbairn, H.W., ix
Falling colors, 144
Fast ray, 103
Fedorow stage, 282
Feldspar, viii, 55, 74, 203, 207, 215, 230, 234, 240, 265, 266, 268–274, 276–279, 286, 304, 319, 322, 330
  cell edges, 271
  composition, 266
  optic orientation, 265
  ordering, 268
  structure, 268
Feldspathoids, 76, 136, 200
Fermat, P. de, 1, 90
Fermat's principle, 89, 108
Ferroaugite, 220, 223
Fibers
  asbestiform, 333
  organic, 336
Fibrous minerals
  in asbestos, 335
Field diameter, 186
Field diaphragm, 38
Filter, 31, 35, 38, 44, 51, 85, 290–292, 310
  interference, 290
Flame tests, 26
Flash figure, 125, 126, 132–134, 138, 142, 143, 145, 147, 148, 156, 203–205, 212, 213, 259, 261, 314, 315
  biaxial (ON), 185
  color curves in, 133, 134, 185
  distinguished from other 2-SP, 185–189
  uniaxial, 132, 314
  used for pleochroism, 138
Flattening, sign of, 227, 228
Fletcher, L., 117

Fluorapatite, 156
Fluorite, 16, 29, 69, 76, 80, 332
Focal masking, 69
Focal plane, 37, 38, 54, 57, 69, 70, 77, 123
Focal screening, 69, 71
Focus, 31, 33, 34, 37, 38, 49, 52, 54–58, 60, 124, 228, 246, 253, 266, 278, 310, 311, 322
Form, 2, 8, 79, 150–152, 227
Fractional difference
  Hartmann wavelength, 295
Fractional dispersion, 290
Fraunhofer lines, 27
Fraunhofer spectrum, 27
Fraunhofer, J., 1, 27
Frequency of vibration, 5, 24, 95
Fresnel, A., 1, 165
Frontispiece, 62, 64, 97, 107, 110, 111, 114, 263

Gamma refractive index, 159–167
  confirmation, 212
Gamma prime (γ') index, 161, 181, 214
Garnet, 46, 76, 79, 302, 319–321, 326, 78, 302, 326
Gassetts schist, 320
Gedrite, 327
Gelatin mounts, 45, 214
Geometrical surfaces, 118
Geophysical Laboratory, 73, 75, 112
Gips, 107, 144, 314, 317 (*see also* gypsum plate)
Gladstone and Dale
  rule of, 15, 16, 271, 300
Glasses, 2, 32, 33, 43, 45, 73, 76, 323
Glauberite, 218
Glaucophane, 281, 301, 327, 301, 327
Glimmer, 107 (*see also* mica plate)
Goldsmith, J.R., x
Grating, diffraction, 100
Grunerite, 333–335
Gypsum plate, 107, 108, 110, 142, 144–146, 152, 187, 224, 285
  as generic name, 107
  at extinction position, 230
  in optic sign determination, 144, 205, 206

H_2O, 82, 331
Halite, 76, 80, 83, 86
  dispersion of, 86
Hardness, 82, 226, 336
Hartmann equation, vii, 292, 293
Hartmann net, 74
Hartmann space, 69, 85, 294, 295, 334
HCl, 82
Heavy minerals, 45, 46
Hemimorphite, 208
Hemispheres, 283, 284, 286
Heulandite, 106, 209, 221
Hexagonal crystals, 151, 155
Hexagonal system, 120, 154, 155
HNO_3, 82
Hornblende, 170, 203, 327
Human eye,
  color sensitivity, 24
Humite group, 240
Huygens, C., 1, 5, 10–13, 99, 100
Huygens' construction, 10, 11, 99, 100
  defined, 10
Hypersthene, xi, 168, 260, 262, 263, 321, 328, 260, 328

Iceland spar, 5
Illumination, 35
  adjustment, 35, 37
  conoscopic, 123
  daylight, 106
  Köhler, 38
  orthoscopic, 123
Ilmenite, 241, 320, 321
Image,
  real, 31
  virtual, 31
Immersion method, vii, xi, xii, 49–51, 55, 71, 72, 76, 210, 309, 311, 318, 319, 333, 336
  strategy, 50
In phase, 88
Index of refraction, 14, 16–18, 28, 49, 79, 133, 149, 150, 291 (*see also* refractive index)
Indicatrix, 117
  axes, 160
  biaxial, 160
  circular section, 117
  equations, 165
  models, 160
  orientation, 119, 161
  sections, 117
  symmetry planes, 118
  terminology, 163
  uniaxial, 116
Intensity, 5, 7, 24, 26, 29, 56, 62, 64, 81, 89, 93–95, 97, 112, 137, 138, 141, 202
Interference, 87
  constructive, 88
  destructive, 89
Interference color, 87
  abnormal, 141
  above fourth order, 95
  compensation, 112
  defined, 93
  fall, rise, 108, 109
  first-order red, 108
  graphical explanation, 96

gray, 78, 106, 108
origin, 94
seen in calcite, 95
seen in mica, 94
unique, 111
white, 106
Interference color chart, 114
Interference color scale, 95, 97, 128, 129
seen in quartz wedge, 95
Interference figure, 77, 118, 314
0-SP, 126
1-SP, 126
2-SP, 126
and cleavage type, 243
and elongation, 227
and flattening, 228
biaxial, 172
classification, 125, 172
CR1SP, 316
defined, 123
effect of crystal orientation, 125
effect of dispersion, 198
extinction in, 124
how to obtain, 123
n-SP, 126
photographs of, 127, 134, 185, 241
SR1SP, 315
types, 126
uniaxial, 126
uniaxial and biaxial, 123
use, uniaxial, 134
Intergrowths, 259, 261, 262, 264
of crystals, 261
Iridescence,
in minerals, 101
Iris diaphragm, 37, 38
Isochromatic curves (*see* color curves)
Isochromes (*see* color curves)
Isogyres, 124
behavior, 126
colored fringes on (dispersion), 216–223
CR and SR, 190–194
curvature, 208
defined, 124
leave the field, or not, 186
origin, 127
Isometric crystals, 80, 81, 151
Isometric system, 8, 78
Isomorph, 2
Isotropic,
inorganic crystals, 76
minerals, 76
Isotropic character
determination, 77
Isotropic crystal identification, 76, 84
Isotropic crystals, 8, 49, 66, 78, 81, 149, 154, 314
Isotropic medium, 9, 10, 17
Isotropic minerals
abundance, 76
Isotropic substances, 3, 20, 76, 77, 79, 104, 145

Jaffe, H.W., ix, x, 15, 16, 264

K feldspar, 215, 269, 271, 274, 322
Kamb, B., ix
Kanonaite, 281
kD, Hartmann wavelength, 85, 295, 297, 298
Kinsman granite, 320
Kirchhoff, G.R., 27
Kyanite, 106, 257, 258, 319, 321, 328

Labradorite, 101, 330
Lamellae, 101, 154, 240, 241, 261, 262, 287, 321
twin, 154, 240
Land, E.H., 36
Larsen, E.S., 73
Lasaulx method, 124
Laser, 98, 99
Lattice, 261, 268, 269
reciprocal, 268
Lemniscates, 183
color curves, 183
Length-fast, 152, 334
Length-slow, 152, 334
Lens, 7, 31–35, 37, 38, 57–60, 77, 81, 102, 123, 124, 142, 224, 257, 286, 310
Lensoid grains, 55
Leucite, 78, 331, 78, 331
Light
circularly polarized, 6
defined, 3
elliptically polarized, 6
monochromatic, 24
nature of, 1
plane polarized, 6
sky, 215
sodium, 70, 127, 134, 185
source, 4
theories of, 165
white, 24
Liquid, refractive index
Cargille, 290
dispersion, 39
properties, 39
refractive index, 39, 50, 290
special plagioclase, 290
Lithiophilite, 200

Lorentz, H.A., 16
Lorenz, L., 16
Lyons, J.B., viii, 326

Mafic minerals, 300–302, 319, 323
  dispersion method for, 300
Magnesite, 149
Magnification, 32–34, 43, 123
Mallard's constant, 209, 210
Mallard's method, 209
Marble, 153, 204
Maskelynite, 117
Match, vii, 51, 52, 54, 56, 63–65, 67–69, 71, 74, 76, 83, 85, 86, 147, 150, 156, 211, 212, 282, 283, 289, 293, 295, 297, 300, 305, 309–312, 315–318, 320, 333–335
Matching condition with dispersion method, 85
Matching wavelength, 70, 74, 296
Maxwell, J.C., 1, 27
McCrone, W.C., ix
Media, 16, 39, 44, 82, 149, 151
Melatope, 178, 181, 183, 192–195, 205, 213, 218, 251, 254
  defined, 172
Melilite, 154
Melts, 44
  refractive index, 39
Merwin, H.E., 40, 73
Metamorphic rocks, 46, 105, 269, 327, 328, 331
Methyl alcohol, 14
Mica, 94, 100, 107, 109, 110, 122, 182, 214, 320, 321, 122, 168
Mica plate, 107, 110, 107
Micas, 329
Michel-Lévy chart, 114
Michel-Lévy method, 278
Microchemical reactions, 46, 82
Microcline, 241, 242, 269, 271, 320, 330
  maximum, 269
Micrometer, 33, 209
Micrometer scale, 32
Microprobe, see electron, 3
Microscope,
  binocular, 321
  compound, 2, 31
  polarizing, petrographic, 1, 31
Miller index, 229, 249, 253, 261
Minimum deviation condition, 100
Minimum deviation method, 41, 73
Mixed species in a sample, 214
Modal analysis, 1, 34
Monalbite, 274
Monochromatic light, 87, 216, 310
Monochromator, xi, xii, 35, 69, 74, 156, 292, 310, 318
  interference filter, 85, 291
Monoclinic crystals, 168, 201, 214, 219, 221, 222, 230, 232, 269
  extinction in, 232
Monoclinic system, 219, 232, 237, 239, 243, 246, 248
Moon, 112
Morphological crystallography
  review, 247
Mosses, 2
Mounting medium, 43, 46, 46, 49, 50, 79, 80, 149, 150, 215, 322
  desirable refractive index, 79
Mount, immersion, viii, 28, 43–46, 49, 51, 52, 153, 211, 242, 252, 267, 290, 291, 309, 310, 318, 320
Munsell color, 67
Munsell color appearance system, 78
Muscovite, 94, 97, 182, 194, 209, 218, 320, 321, 329, 330, 194, 329

N.A., 34, 123, 186, 188, 189, 203, 204, 209, 224
N-symmetry plane figure, 126
Na_2SiF_6, 14
NaCl, 80, 83, 86, 331
Negative elongation, 152, 332
Nepheline, 154, 320, 331, 154, 331
Nesosilicates, 168
Newton, I., 1
  and dispersion, 25
Nickelsen, R.P., ix
Nicol, W., 36
Nicols, 36
Niter, 104, 105, 122, 132
Nitrate groups, 122
Normal incidence, 20, 100
nTl - nLi, 27, 216
Null position, 85
  in oblique illumination, 85
Numerical aperture (see also N.A.), 34, 123, 186, 188, 257
  defined, 34

O direction, 129, 137, 138, 148
O ray, 116, 117, 120, 122
Objective, 31, 33, 34, 43, 70, 72, 189, 282, 286
  centering, 33
  magnification, 34
  oil immersion, 33
  used for interference figures, 123
  working distance, 34
Oblique illumination, vii, 50, 60–63, 65, 67,

69–72, 75, 79, 147, 149, 211, 246, 311, 333–335
avoidance of, 246
shadow, 83
Oblique illumination test, 52, 57, 63, 76
practice, 57
principle, 59
rules, 58
Obtuse bisectrix (Bxo), 125, 163
defined, 163
Octahedral layers, 168
Ocular, 32, 33, 37, 38, 105, 112, 123, 209
description, 32
grid, 32
pinhole, 123
use and care, 32
Oil,
Cargille, 290
refractive index, 50, 290
Oil immersion, 33, 164, 310, 321
Oligoclase, 187, 330, 187, 330
Olivine, viii, xi, 79, 104, 203, 208, 209, 240, 300, 301, 304, 320, 321–323, 326, 330
Omega refractive index, 119, 145, 156, 314
ON, see optic normal, 163
One-symmetry plane figure, see 1-SP
1-SP figure, 126, 172, 174, 196, 197, 212, 213, 244, 267
CR and SR, 190
defined, 126
origin, 190–196
two-isogyre, 190, 196
Opaque material, 2, 48
Operations, 143, 144, 247, 248, 284, 285
of crystallography, 247
Optic angle, 2, 316 (*see also* 2V)
calculation from refractive indices, 165–167
defined, 163
determination, 207
estimation, 207
estimation from optic axis figure, 208
fortuitously zero, 200
from rotation angle, 209
in air, 164
in oil, 164
in the crystal, 164
measured directly, 285, 288
Optic axes, 8, 159
as propagation directions, 163
defined, 163
normals to circular sections, 160
Optic axial angle (*see also* optic angle, 163)
Optic axis, 102
of calcite, 12
defined, 117
Optic axis figure
biaxial, 194–196
centered, 126, 194–196
color curves in, 127, 195
estimation of 2V, 208
examples, 132, 196
isogyres in, 127, 194
uncentered, 131, 196
uniaxial, 126–132
Optic axis sections, 103, 104, 106, 141, 155, 212, 213
Optic normal (ON)
defined, 163
Optic normal figure (flash figure), 185
color curve spacing, 185
isogyres fuzzy, 185
vibration vectors, 185
Optic orientation, 30, 106, 316
biaxial, 228
defined, 168
expression of, 234
in feldspars, 265
in stereographic projection, 247
monoclinic crystals, 237, 257
orthorhombic examples, 260
related to symmetry, 168
triclinic crystals, 238, 250, 254
uniaxial, 118
Optic plane,
defined, 163
trace of, in Bxa figure, 176, 181, 183
trace of, in Bxo figure, 184
trace of, in flash figure, 185
Optic plane switch, 271, 273
Optic sign, 120, 316
and refractive indices, 167
defined, 119, 164, 167
determination, 142, 143, 179, 182, 197
determination illustrated, 145
direct determination, 204
from flash figure, 142
from optic axis figures, 143
from refractive indices, 206
rules, 142, 143, 205
structural basis (biaxial), 167
structural basis (uniaxial), 120
uniaxial crystals, 119, 141
Optical diffraction, 98, 99, 101
Optical Transform Kit, 99
Optics, 1–3, 12, 13, 20, 30, 36, 98, 99, 116, 159, 163, 266, 267, 268, 271, 274
crystal, 1
Order of interference color, 95, 106–108, 110–114, 128, 129, 133, 139, 140, 144, 145, 162, 183, 199, 203, 204, 288, 289
Ordering, 235, 268, 269, 271–275, 277–281
in K feldspar, 271

Ordinary ray, 116
Orientation diagrams, 169, 170
Orthoclase, 232, 239, 266, 267, 320, 322, 330, 331
Orthopyroxene, viii, 227, 263, 264, 300, 301, 304, 320, 321
  megacryst, 321
Orthorhombic crystals, 168, 201, 216, 218, 219, 231, 233, 236, 244
  extinction in, 231
  optic orientation, 168, 234
Orthorhombic system, 9, 231, 234, 239, 246
Orthoscopic image, 124, 131, 133–135, 143, 145, 173, 175, 176, 183, 184, 185, 187
Orthoscopic observation, 37, 123, 286
Out of phase, 89, 109
Outhouse, 75

Paragonite, 321, 329
Pargasite, 301
Partial refractive index, 117, 214
Path difference, 109
Path length, 90
Pericline twinning, 241
Peristerite, 101
Perpendicular incidence, 19, 21
Petrofabric study, 284
Phase, 87–89, 91, 109, 124, 261
Phase relationships, 87, 91
Physics, 1, 41, 98, 99
  crystal, 29
  quantum, 27
Pigeonite, 262–264, 328, 263, 328
Pinacoidal cleavages, 236, 242–244, 249
  frequency, 242
Plagioclase, vii, viii, xi, 55, 74, 106, 117, 214, 215, 230, 240, 241, 266, 280, 286, 290, 300, 319, 320, 321, 330, 331
  a-normal method, 234, 278
  determination in thin section, 234, 278
  composition, 278
  dispersion method, 304–307
  extinction angles, 278
  Michel-Lévy method, 278
  optic orientation, 278
  twinning, 240, 241
Plagioclase dispersion chart, 304
Plagioclase Dispersion Set, Cargille, 305
Planck, M., 1
Plane of vibration, 4
Plane polarized light, 5, 6, 78, 104, 116
Pleochroic colors, 137, 148, 201, 202, 257, 324
  related to composition, 202
Pleochroic formula, 148
Pleochroism, 29, 36, 135, 137–139, 148, 199, 201, 202, 321, 323, 327–329
  biaxial crystals, 201
  defined, 29
  determination, 138
  in mafic minerals, 323
  in refractive index determination, 148
  uniaxial, 137
Pleochroism and absorption, 138, 139, 199, 201, 202
  determination without interference figures, 138
Polarization, 5, 7, 9, 12, 113, 176
  by reflection, 7
  elliptic, 113
  rotary, 9
Polarized light, 1, 2, 5, 6, 8, 9, 29, 31, 77, 78, 90, 104, 112, 116, 138, 147, 313
Polarizer direction, 36, 91, 210, 288
Polarizers, 35, 36
Polarizers, transmission, 36
Polarizing devices, 6, 35, 36, 87
Polaroid, 7, 36, 94, 97, 141
Polars, 6
  crossed, 36
  described, 35
  orientation, 36
Polymorph, 2, 330
Positive elongation, 152
Posnjak, E., 73
Precision, ix, 42, 74, 85, 86, 207, 292, 293, 301, 304, 305, 308
  defined, 72
  in immersion method, 72
  of dispersion method, 74, 304, 308
Preferred orientation, 139, 151, 214
Prehnite, 209
Preparation of materials, 31, 43
Principal refractive indices, 49, 71, 117, 149, 159, 162, 167, 206, 214, 215, 288, 297, 317
Prism, 25, 26, 40–42, 145, 177, 235, 239, 249
  Bertrand-Amici, 40
Propagation direction, 4–6, 10–12, 17–19, 21–23, 29, 87–89, 91, 92, 97, 116, 117, 119, 159, 165, 166, 172, 176, 178
Pseudoisometric (pseudocubic) crystals, 78
Pseudozoisite, 281
Public health, 333
Pylon color, 67, 309, 311, 315–317, 309, 316, 334
Pylon color chips, 67, 72
Pyrite, 321
Pyroxene, vii, xi, 72, 168, 202, 214, 222, 240, 259–264, 283, 287, 321–323, 328
  optic orientation, 259

Quarter-wave plate, 107
Quartz, vii, xi, 47, 50, 53, 55, 58, 59, 67, 69, 72, 90, 105, 136, 140, 150, 152, 155, 200, 204, 215, 283, 319–321, 331
  grain orientation, 284
Quartz wedge, 95, 97, 110–114, 141, 145
  described, 107
  in optic sign determination, 144
  rule for grains, 111
  rule for optic sign, 144

Random search, 147, 148, 211–213
Range of refractive indices, 14, 51, 65
Ray,
  defined, 4
  E, extraordinary, 116
  O, ordinary, 116
Ray direction, 10, 11
Ray surface, 9–11, 17
Ray velocity, 9, 11, 12, 19, 116, 117
Reactions, 46, 82
Recipes for crystal identification, 309
Reciprocal dispersion, 290
Red light, 64, 106, 216, 217, 274
Reference temperature, 289
Reflection, 7, 12, 34, 49, 52, 55, 79–81, 98, 247, 248, 283
  defined, 22
  total, 22
Refraction, 16–19, 21, 22
Refraction correction, 207, 283, 285
Refractive index, 2
  alternate nomenclature, 280
  and density, 15
  and temperature, 16
  biaxial crystals, 159, 211
  defined, 14
  determination in thin section, 149
  direct measurement, 40
  effects of cleavage, 213
  epsilon and epsilon prime, 145
  in air, 14
  in immersion mounts, 145
  in thin section, 79, 215
  in vacuum, 14
  intermediate, 15
  labelling, 271
  mean, 16, 188
  omega, 145
  partial, 117, 161
  principal, 159
  range in solids and liquids, 14
  reference value, 14
  symbols used, 117, 159
  uniaxial crystals, 116, 145
  when mixed species present, 214
Refractive index data
  complete, 28
Refractive index liquids, 28, 39, 40, 46, 69, 73, 292, 333
  Cargille, suppliers, 290
Refractivity, specific, 15
Refractometer, 23, 40, 41, 72, 73
  Abbe, 40
  Jelley, 40
Relief, 49–51, 79–81, 136, 150, 214, 215, 319, 322, 324, 326, 327–332, 334
  defined, 49
  described, 49
  in thin section, 79
  in uniaxial crystals, 149
  negative, 50
  positive, 50
  twinkling, 150
Resin, mounting, 44–46
Resultant, 87–89, 91, 92
Retardation, 29, 38, 78, 87–94, 93–100, 107–109, 111–114, 124, 129, 130, 141, 182, 183, 203
  and birefringence, 89
  and thickness, 89
  between crossed polars, 91
  in anisotropic crystal plate, 89
  in Bxa figure, 182, 183
  in crystals, 29
  in interference figure, 129, 182
  in isotropic substance, 87
  in optic axis figure, 128–130
  measurement, 111, 112
  visual estimation, 111
Rhodonite, 234
Rhombic dispersion, 215–217, 219, 222, 273, 280
  and optic plane switch, 273
Rhombic section, 241
Rhombohedron, 148, 154
  negative, 154
  positive, 154
  steep, 154
Ribbe, P.H., x
Riebeckite, 333
Rising colors, 111, 143, 144, 284
Risk management, 333
Robinson, P., ix, x
Ross, M., x
Rotation angle, 36, 187–189, 209, 317
  in differentiating 2-SP figures, 187
Rules, 57–59, 63, 145, 167, 201, 205, 206, 218, 227, 234, 262

Rutherford, E., 41, 100
Rutile, 46, 136, 140

Salammoniac, 28
Same-rotating figure, see SR
Sample preparation, 43, 310
Sanidine, 266–269, 271, 272, 274, 280, 281, 330
Sapphirine, 257, 260
Scale, micrometer, 32
Scapolite, 331
Schairer, J.F., 75
Scheelite, 153
Schmidt sledge, 284
Schuster's Convention, 230
Section,
  crystal, 103
  normal to symmetry plane, 161
  one-SP, 161
  optic axis, 103
  two-SP, 162
  zero-SP, 161
Section, indicatrix,
  circular, 29, 117, 118, 127, 129, 160–164, 167, 180, 284, 285
  elliptical, 117
Sellmeier equation, 69, 292
Sénarmont compensator, 113
Sensitivity of match, 67
Separator, magnetic, isodynamic, 321
Septechlorite, 333
Sericite, 329
Serpentine, 330, 333
Shadow, 57–60, 72, 83, 85, 211, 224, 291, 311, 335
  crystal high, 58
  crystal low, 58
Shagreen surface, 79, 322, 323, 329
  illustrated, 322
Shape, 12, 45–47, 60, 72, 80, 81, 127, 150, 151, 155, 172, 216, 227, 278
Sheet silicates, 79, 122, 257
Shuttle, 69, 73, 85, 156, 291
Sight recognition, 3, 324
Sign of elongation, 334
Sign of extinction, 230
Sign of the birefringence, 120
Sign, optic; *see* Optic sign
Silhouette angle, 151
Silica tetrahedra, 122
Silicates, 76, 79, 122, 168, 257, 333
Sillimanite, 177, 320, 322, 327, 331
Size, crystal, 227
Slide preparation, 310
Slides, 2, 43, 45, 82, 151, 226
  gelatin coated, 45, 214
Slow ray, 103
Snell, W., 1, 17
Snell's Law, 16–19, 23, 25, 42, 54, 164, 209, 283
  and critical angle, 23
  defined, 17
Soda niter, 104, 105, 122, 132
Sodalite, xi, 76, 331
Sodium light, 70, 127, 134, 185
Sodium line, 28, 40
Solid solution, 73, 151, 200, 207, 210, 266, 289, 301
Solubility, 82
Sorby, H.C., 47, 110
Source, light, xi, 4, 7, 9, 10, 25, 26, 28, 31, 35, 37, 38, 40, 41, 57, 73, 87, 94, 98, 99, 291, 318
SP (*see* symmetry planes)
Spacing, 73, 98–101, 268
  interplanar, 100
Spectrograph, 25, 82
Spectrometer, 25, 41, 42, 82
  minimum deviation, 41
Spectroscope, 25, 26, 82
Spectrum, 25–27, 56, 62, 63, 69, 94, 95, 198, 216, 235
  absorption, 25
  continuous, 26
  defined, 25
  emission, 25
  Fraunhofer, 27
  line, 25
  solar, 27
Sphalerite, 80
Sphene, 216, 240, 258, 261, 319, 326, 240, 258, 326
Spin ellipsoid, 29
Spindle stage, viii, 34, 35, 75, 149, 207, 214, 253, 282, 287, 288
  description, 287
  manipulation, 288
SR figure, 174, 192, 193
SR1SP figure, 224, 257, 261, 267, 318, 257, 317
Stage, 31
  mechanical, 34
  spindle, 35
  universal, 35, 282
Standard glass powders, 72, 73, 290
Standard temperature, 28
Standard wavelength, 28, 56, 62, 64, 74
Statistical method, 147
Staurolite, xi, 263, 266, 319, 321, 326, 263, 326
Stellate twins, 240

Stereogram, 247–251, 254–257, 262, 271, 272, 278
  of crystal systems, 247
  rotation, 255
Stereographic projection, 239, 247, 283, 286, 287
  conventions, 250
  in optic orientation, 247
Stewart, D.B., x
Stibnite, 14
Strain shadows, 105
Strained crystals, 151
Strategy, xi, 50, 51, 202, 211, 311, 314, 317, 319
  biaxial crystals, 317
  isotropic crystals, 311
  uniaxial crystals, 314, 315
Structure, 8, 9, 14–16, 29, 98, 101, 120, 122, 167, 168, 199, 207, 253, 260, 268–271
Su, S.-C., ix, x
Su's method, 294, 295, 300
Sun, 9, 26, 27
  spectrum from, 27
Sylvite, 65, 66
Symmetry,
  crystal, 8, 119, 247
  optical, 247
Symmetry plane, 173
  coincident with crosshair, 172
Symmetry planes, 161
  and sections, 253
  biaxial indicatrix, 160, 161, 172
  uniaxial indicatrix, 118
System, crystal, viii, ix, 6–8, 13, 76, 78, 136, 154, 155, 159, 168, 199, 228, 231–234, 237–249, 268, 316

Tables, viii, xii, 76, 77, 85, 86, 136, 156, 199, 226, 314, 316, 318, 332, 336
  determinative, 50
  dispersion data, 290
  mineral, 313
  standard, 3
Talc, 257, 259, 330, 335, 336
Taylor, E.C., Sr., ix, 2
Temperature, 16, 28, 39, 42, 51, 73–75, 85, 156, 168, 263, 269, 275, 289, 290, 297, 304, 305, 318, 330
  reference, 289
Temperature coefficient, 16, 74, 75, 85, 289
  liquids and solids, 16
  of liquid, 39
Temperature correction, 51, 73, 289
Tetragonal crystals, 155
Tetragonal system, 9, 118
Textures, xii, 1, 48, 318
Thermal variation method, 75
Thickness, 90
  measurement, 113
Thin sections, viii, xi, xii, 2, 43, 46–48, 50, 79, 90, 105, 113, 127, 139, 141, 203, 204, 253, 283, 322
  crystal identification in, 322
  polished, 48
  preparation, 47
Thomson, J.J., 100
Titanite; *see* sphene
Topaz, 209, 224, 226, 263, 265
Total reflection, 12, 49, 55, 79–81
  at cleavage surfaces, 81
Tourmaline, 122, 136, 137, 139, 319, 321, 327
Tracy, R.J., x
Transmission efficiency, 6
Traps, refractive index, 319
Tremolite, 327, 333
Triclinic crystals, 106, 168, 201, 216, 219, 222, 233, 250, 252, 254, 286
  extinction in, 233
Triclinic system, 9, 199, 233, 238, 240, 243, 244, 247, 249
Tridymite, 332
Triphylite, 200
Tröger, W.E., viii, ix, xii
Trough, wave, 4
Tschermakitic substitution, 301
Twin law, 155, 239, 240
  in feldspar, 240
Twin planes, 103, 240, 286
Twinkling, 150
Twinned crystals, 151, 286
Twinning, 43, 78, 79, 81, 136, 154, 155, 187, 199, 230, 235, 239, 240–242, 286, 320, 322, 330, 332
  aragonite type, 235
  Carlsbad, 239
  hexagonal crystals, 154–155
  isometric crystals, 81
  monoclinic crystals, 239
  orthorhombic crystals, 239
  polysynthetic, 78, 154, 240
  pseudoisometric crystals, 78
  tetragonal crystals, 155
  triclinic crystals, 240
  uniaxial crystals, 154
Twins, 154, 169, 239–242, 245, 278, 326, 331
2E, 164, 165, 176, 183, 186, 207, 209, 223
2H, 164, 165
Two-symmetry plane figure, see 2-SP
2-SP figure, 126, 172, 173, 201, 213
  description, 175
  differentiating among, 185

2V, 163; *see also* optic angle
calculation, 167, 316
from rotation angle, 209
Mallard's method, 209
notations, 163

U-stage (*see* universal stage)
Undulatory extinction, 105, 151, 331
Uniaxial,
members of biaxial class, 199
rock-forming minerals, 136
Uniaxial character, 136, 137, 331
from interference figures, 137
Uniaxial crystals, 8, 11
axial character, 136
color, absorption, pleochroism, 137
identification, 136
optics of, 13, 116
relative abundance, 136
Uniaxial minerals,
abundance, 136, 142
Uniaxiality, 280, 281
fortuitous, 280
Unilateral screening, 69
Universal stage, 34, 35, 207, 209, 239, 240, 278, 282, 283, 286, 287
conoscopic observation, 286
manipulation, 283
uses of, 286
Urinary calculi, 2
USGS 1627, xii, 156, 224, 310

V, Abbe dispersion, 39
Vacuum, 14, 48
Vector, 7, 29, 87, 91, 92, 97, 117, 129, 159, 160, 162, 163, 178, 180, 181, 184, 268
electric, 29
magnetic, 29
vibration, 91, 92, 129, 179–185, 195
Velocity, viii, 5, 9, 11, 16–19, 21, 25, 29, 76, 90, 102, 103, 116, 117, 136, 159, 163
difference, 90
of light, 3, 8, 12, 14, 16, 17, 24
ray, 9, 11, 12, 19
wave, 5, 11, 14, 17
Vibration direction, 4, 6
E, 117, 133
E', 132, 143
high index, 29, 103
low index, 29, 103
O, 117, 132, 133, 143
Vibration directions,
as positions of extinction, 103
biaxial crystals, 159, 176, 178
constrained by crystal symmetry, 30
from Biot-Fresnel law, 165, 166, 176, 178, 179
high and low index, 102, 103
in Bxa figure, 176
in Bxo figure, 184
in CR and SR 1-SP figures, 190
in flash figure, 133, 185
in interference figures, 1123, 176
in optic axis figure, 130, 195
in the grain, 131
in uncentered optic axis figure, 132
in uniaxial interference figures, 134
names of, 103
origin of, 29
principal, 115
use of, 115
Vibration planes, 8, 21, 89, 91, 92
Vibration vectors, 91–93, 130, 179–185, 195
Video work station, 48

Water, 3, 7, 40–42, 44, 45, 310, 334
Wave front, 9–12, 17–19, 99, 100
defined, 9
Wave normal, 10, 11, 123, 129, 180
defined, 10
Wave propagation direction, 10–12, 17–19, 21–23, 29, 87–89, 97, 116, 119, 159, 165, 166, 176, 178
defined, 10
Wave velocity, 11, 14, 17
Wavelength, 4
and color, 24
dispersion method, 69, 74, 156, 292, 310, 318
of observation, 85
standard, D line, 85
Wavelength equations, 292
Wiebe, R.A., 177
White light, vii, 24, 25, 39, 40, 52, 63–65, 67, 72, 73, 83, 87, 94, 95, 97, 106, 155, 158, 201, 218, 219, 223, 224, 226, 273, 291, 292, 309, 310, 318, 323, 333
Wilcox, R.E., ix, 70
Willemite, 154
Wollastonite, 336
Work station, xii, 48, 292
monochromator, 292
video, 292
Worked example, 83, 155–157, 223, 225, 301, 303, 305, 307
biaxial crystal identification, 223
isotropic crystal identification, 83
uniaxial crystal identification, 155
Worksheet, 45, 83–86, 156, 157, 225, 226, 297, 299–303, 305–307, 318
biaxial crystals, 226

dispersion method, 297
isotropic crystals, 83
mafic minerals, 301
plagioclase, 305
refractive index, 297, 310
uniaxial crystals, 156
WPD (*see* wave propagation direction)
Wright (slotted) ocular, 112
Wright combination quartz wedge, 113
Wright, F.E., 73, 112

X axis, 180, 258
X vibration direction, 160
X′ vibration direction, 161, 182, 184, 185, 196, 197, 229, 230, 251, 254
X-ray diffraction, 3, 98, 100, 253, 321
Xylene, 45

Y axis, 161
Y vibration direction, 160
Y′ vibration direction, 197, 228

Z axis, 254
Z vibration direction, 160
Z′ vibration direction, 161, 182, 183, 184, 185, 196, 229, 251, 254
Zeolites, 153, 332
Zero-symmetry plane figure, see 0-SP
0-SP figure, 126, 197, 251, 254, 255, 258
Zies, E.G., 323
Zircon, 46, 136, 140, 200, 319, 326
Zoisite, 281, 328, 281, 328
Zone, crystallographic, 249
Zone axis, 249, 250, 286
Zone index, 249
Zunyite, 65, 66, 74